The Fax Modem Sourcebook

The Fax Modem Sourcebook

Andrew Margolis
Margolis & Co., UK

Published in association with EXE Magazine

JOHN WILEY & SONS
Chichester · New York · Brisbane · Toronto · Singapore

Other Wiley Editorial Offices

John Wiley & Sons, Inc., 605 Third Avenue,
New York, NY 10158-0012, USA

Jacaranda Wiley Ltd, 33 Park Road, Milton,
Queensland 4064, Australia

John Wiley & Sons (Canada) Ltd, 22 Worcester Road,
Rexdale, Ontario M9W 1L1, Canada

John Wiley & Sons (SEA) Pte Ltd, 27 Jalan Pemimpin #05-04,
Block B, Union Industrial Building, Singapore 2057

Library of Congress Cataloging-in-Publication Data

Margolis, Andrew.
       The fax modem sourcebook / Andrew Margolis.
          p.   cm.
       Includes bibliographical references and index.
       ISBN 0 471 95072 6 (alk. paper)
       1. Modems.   2. Facsimile transmission.   I. Title.
       TK7887.8.M63M35   1995
       621.382'35'0285—dc20   95-8378
                                                        CIP

British Library Cataloging in Publication Data

A catalogue record for this book is available from the British Library

ISBN 0 471 95072 6

Typeset in 10/12pt Palatino from author's disks
Printed and bound in Great Britain by Redwood Books, Trowbridge, Wilts
This book is printed on acid-free paper reponsibly manufactured from sustainable forestation,
for which at least two trees are planted for each one used for paper production

Contents

9 Programming Class 1 Fax Modems   199

10 Programming Class 2 Fax Modems   245

11 Programming Class 2.0 Fax Modems   297

Introduction

Why this book has been written

This is an unashamedly technical book about fax modems, and about how computer programmers and technical users can utilise them to transmit and receive faxes. Most of the latest generation of modems available now have fax as well as data capability, which is why computer-based faxing is currently undergoing something of an explosion. Anyone wanting to get the most out of their fax modem is going to need the sort of information that you can find in this book.

If you haven't any fax software

The most obvious reason for needing this book is if you want to write or implement your own fax software. Apart from the ever-present possibility that you simply may not like either the way existing software works or how it looks, there are a number of reasons that might give you no choice in the matter.

You may find there is no alternative to writing your own software if you need to use the modem under circumstances standard packages do not cater for. You may have a hardware platform different to those with which commercial software works. Alternatively, your operating system may be one that isn't supported. Even where the operating system and hardware platform are compatible with the software, it is always possible that some other aspect of the computer might present incompatibilities, thus preventing use of a standard software package.

If you have other software requiring fax functionality

With the almost universal adoption of fax in a number of commercial sectors, it is increasingly common to come across a request for fax functions to be incorporated in other software packages. For instance, an accounts package which automatically generated and sent faxes to customers with overdue accounts could quite reasonably expect bills to be settled more quickly than one which sent letters.

If you already have fax software that almost works

Detailed knowledge of how the technology works is an essential ingredient for proper support of an existing software/modem combination. If the fax modem or software is misbehaving under particular circumstances, the key to resolving the problem is providing the supplier details of what is going wrong. This book will help you put your finger on the nature of problems you may have. The software that is included with the book might well work where other software fails. Even if it doesn't, you will always be able to see exactly where in the process the problem arose.

If you have fax software that works but doesn't do what you want

It may be unfortunate if neither the fax software bundled with the modem nor any of the alternatives are suitable for your intended use, but it is hardly surprising. For good commercial reasons, packaged software is written to appeal to the widest number of users. If your needs are even a little unusual, you may well find you end up with a second-best software solution. While writing bespoke software is not cheap, in the long run it can often be cheaper than struggling with something that isn't doing the job properly or efficiently.

You may also need to use optional features found on standard fax machines but which are not supported in standard software packages, or are supported rather badly. These include more compact fax codings which reduce transmission times, support for error correction, where available, and the ability to dial a remote fax so as to receive a document rather than send one (polling).

If you already have fax software that works

Many modems come with perfectly satisfactory bundled software which sends and receives faxes. Also, a range of alternative fax packages is often available at reasonable cost for use in situations where bundled software is unsuitable.

If you are fortunate enough to have a fax modem with software that works, you will still need to read this book to find out how it works. Curiosity aside, one reason for needing a book on fax modems under these circumstances is because no amount of good software is ever a substitute for adequate technical information. No matter how good the software, you never know when access to the detailed information found here might be needed.

A background to any area of technology is never wasted. Existing or prospective users of fax modems will always find it useful to have a source with all the basic information on what the software driving these devices actually does.

You always need some technical information to install a package and select various options. If you don't have enough, you can end up with an installation which may have seemed to be trouble-free and simple, but has achieved this through the expedient of always opting for the easy answers. An optimal installation is always more difficult.

Why you can't find the information elsewhere

There has always been a reasonable amount of support available for anyone wanting to use data modems to exchange ordinary computer files. All the information that anyone could need is freely available from a number of different sources.

The structure of computer files has traditionally been included in technical reference manuals on a number of levels. ASCII charts giving basic character sets have been standard issue for many years. Details of file formats used by most popular software packages are available either in the manufacturer's documentation or textbooks. In cases of extreme difficulty, there have usually been lowest common denominator mechanisms, such as comma delimited data files and plain text files.

Manuals for data modems have always contained all the necessary command codes needed to control the process of dialling or answering, sending data and hanging up. The most commonly used protocols for data transfer between computers using data modems over normal telephone lines have long been in the public domain. Communication software and support has grown up along with the hardware technology, and there are a wide variety of toolkits and textbooks available for developers, a large number of commercial packages aimed at the corporate market, and an equally impressive number of shareware and public domain offerings for personal and home use.

However, the same sort of technical information has never been as widely available for fax communications, probably because fax machines are sold as plug-in-and-go solutions, while programmable fax modems are a much more recent development.

Information on the format of fax data and the structure of images has been buried in obscure standards reference books found only in the most technical of libraries. Detail on how faxes exchange information has met with a similar fate. Manuals for fax modems have never included the information that technical users need to control and use their fax functions. Consequently, the first hurdle for any technical fax modem user has been finding a comprehensive information source.

Until recently, there were no books available with any information on how to create faxes, transmit and receive them, and output them to screens and printers. While chapters on fax have begun to appear in some recent communications textbooks, the information has inevitably been less than comprehensive than that found in a book dedicated to the subject. It has also tended towards idealized descriptions of how fax modems ought to work, which are less than useful in the real world.

What you don't get in this book

- A general purpose communications textbook. There are a large number of these already available, and we don't attempt to compete with their comprehensive coverage of general communication issues.

- A general purpose programming tutorial which happens to take fax software as a case study. While working code with additional extensive documentation is provided on the accompanying disk, we neither list the code nor explain in detail how it works. Code fragments based on the software on disk are generally included in the text only where algorithms specific to fax are shown, such as operations involving extensive bit-manipulations.

- Page after page of program listings. There are some 8000 lines of code on the accompanying disk instead.

- A guide to specific items of fax software or models of fax modems. For that type of information, which becomes dated very quickly, you should look at magazine reviews and other more topical sources.

What you do get in this book

- Lots of standards information. Standards are what communications are about. They make communication possible. When you buy a modem or fax machine, one of the key items of information you look at is the standards your prospective purchase implements. A technically advanced machine that can send documents ten times faster than anything else is no good unless it also works with the type of fax that everyone else uses.

- Detailed information on the international standards that make fax machines work. As well as explaining those portions of the recommendations that everyone uses, we also provide information on the less frequently implemented parts and the latest extensions to the specifications. Where a feature is seldom implemented, we say so. Where a feature is (in our opinion) not worth implementing, we also tell you.

- Details of the command sets for programming the most common types of fax modem (EIA Classes 1, 2 and 2.0). We also give extensive accounts of the pitfalls awaiting anyone wanting to write software that will port across a range of modems. The coverage of commands is as comprehensive as possible.

- An account of the TIFF specification for storing fax images. Though this doesn't form part of the officially recognised corpus of fax or fax modem standards, it does offer what is probably the most widely accepted means of integrating fax with other applications.

- An introduction to basic information that fax modem engineers need to arm themselves with. Some of this may be obvious, but I've taken the view it isn't wise to take anything for granted. In particular, while much of this introduction can be found elsewhere, the emphasis on the fax side of the relevant issues cannot. I'd therefore recommend anyone without a fax background to read this chapter for orientation, even if they are experienced in other areas of communications.

- Finally, you get a disk of working fax software with full source code. Modules are provided for generating, sending, receiving and outputting faxes. To make the code as portable as possible, most of it is written in the highly portable and widely available ANSI C. Care has been taken to ensure that memory requirements and other overheads are as small as possible so as to maximize the utility of the code.

It should be emphasized that you don't get a complete working fax program suitable for unskilled users. You are getting a fax toolkit, complete with source code, which supports all the major fax modems and functions required for successful faxing. Though the executable programs included are only supposed to test various fax operations, they do make excellent diagnostic tools for testing problem installations, and can also be used in command-line form with operating system shells or scripts to add immediate fax functionality to any pre-existing program. Both of these applications were borne in mind when they were put together.

However, the user interface isn't up to the standards that most GUI operating system users have come to expect. That is probably why modern fax applications are so resource-hungry; it isn't the nature of the program that's is the problem, it's the nature of the environment in which it runs.

The structure of this book

The book follows a quite logical structure:

- It opens with an introduction to the technology behind modern fax and digital communications, and also to the main standards bodies responsible for providing the glue that binds different countries, manufacturers and systems together.

- The next section concentrates on the structure of fax images and sessions. The presentation follows the standards documents quite closely, for the simple reason that no deviation from them is permissible. Quite simply, a fax image is one that conforms to the relevant standard, and a fax modem is a device that has successfully implemented the standard fax session protocol.

- The final section of the book explains the differences between various classes of fax modem. For each class, we present typical modem dialogues for handling various fax operations, which should work on any modem conforming to a particular class. A complete listing of the command sets is also given, along with some detail on operational and manufacturing quirks.

Though this ordering is quite logical, it does not necessarily correspond with the reader's needs. Unless you have an academic rather than a practical interest, it would be a mistake to read this book right through; it has not been written with that in mind.

Each chapter begins with an introductory paragraph, explaining what it contains and under what circumstances it needs to be read. The basic problem with writing a book on fax modems is that the neat logical structure outlined above never corresponds to what any individual needs to know.

For example, while it makes perfect sense to discuss the fax session protocol before going into detail on programming fax modems that implement the standard, someone using a Class 2 fax modem may never to need to know the details of this protocol if they restrict themselves to the most basic operations. On the other hand, once a Class 2 programmer goes beyond a certain stage, he or she is plunged straight into some of the most complicated extensions to the standards. In contrast, programming a class 1 modem always requires an understanding of the fax protocol.

Someone with a Class 1 modem would probably move from Chapter 6 directly to Chapters 8 and 9, and would probably tackle Chapter 7 only if they had to. On the other hand, a Class 2 user need only read the first two pages of Chapter 6 before jumping straight to Chapter 8, and would then go on to Chapter 10. The rest of Chapter 6 and the whole of Chapter 7 would again be read as a last resort. While few people are going to read Chapter 7 right after Chapter 6, it is clearly sensible to group them both together.

So don't worry about hopping around chapters, reading parts out of sequence, and skipping whole chapters or sections that you find boring. Your own judgement about what you need to know is more likely to be accurate than my guesses.

The disk notes

Large chunks of potentially useful material can be found on the accompanying disk as text files. While I find it difficult to leave anything out, there is no doubt that too much detail over-complicates the issues. Yet ignoring minor details of specifications, standards and recommendations could lead to implementations that fail. Relegating these to the disk is a compromise designed to reconcile these two conflicting perspectives.

Your software license

Like all other published material, the software on the accompanying disk is under copyright. In this case, the copyright is held by the author, and all intellectual and other property rights are reserved worldwide. By purchasing this book you purchase a license to modify and copy the code only for your own personal use on only one machine at a time.

None of the code included with this book is to be reproduced, resold or distributed in any other form under any circumstances. It is not placed in the public domain, and is available for use only by the owner of this book. If you want to distribute programs incorporating this code or use it in any commercial product you must contact the author to negotiate a commercial license, which is possible on request at reasonable rates.

Figures

Listings

Tables

Acknowledgements

This book has taken much longer to write and has turned out rather differently to the way I originally envisaged it. The isolation and sheer drudgery of writing, revising, rewriting and rereading was something I hadn't expected, and I'm especially grateful to all those people who relieved it by freely sharing their resources, knowledge and expertise with me: in particular Joe Decuir, Lester Davis, Andrew Hurdle, Malcolm Jones, and Bill Pechey. Canon, Hayes, Kemamun, NEC, Racal, Sonix and US Robotics all kindly loaned me modems and other equipment which was invaluable in developing and testing the software on the disk, while Will Watts of EXE magazine must be credited with having the idea of my writing a communications book in the first place. I'd also like to thank everyone at John Wiley & Sons who helped in producing it, and to apologize to all those people whose help I've forgotten to acknowledge.

My family have had to put up with my irregular hours and moods, and I should like to say sorry publicly for all the times I've shouted at them, and to thank them for their support.

Finally, I'd like to dedicate this book to all those people who would be offended if I didn't dedicate it to them, especially Caroline.

Andrew Margolis

1

Fax Basics

Introduction

This introduction divides neatly into two: the first half begins with a potted history of fax, and moves on to a technical description of the development of modern modem technology. We concentrate on showing the relationship between fax and other types of modem, and briefly describe the key differences in the way that they transfer data.

The second half brings in the notion of industry standards, and gives an account of the main agencies responsible for setting and maintaining the specifications which govern how fax machines work and how fax modems are programmed.

A brief history of fax

The first fax was designed by Alexander Bain, an inventor from Caithness in Scotland, and patented in 1843. The telephone would not be invented for another 30 years, and while Bain's invention was planned around telegraph lines, the telegraph was still in its infancy, with Morse's first commercial service from Washington to Baltimore still a year away. Frederick Bakewell, an English physicist, patented a competing design in 1848. While models of both these devices were built for the 1851 Great Exhibition in London, neither of them went into commercial production. They both used electricity for the transmission and reproduction of images, but the fax machinery itself was driven by clockwork. The first commercial fax system was opened in France between Paris and Lyons in 1865, using a machine designed by Giovanni Caselli in 1861, but it lasted for only five years. Though Bell invented the telephone in 1876, the telegraph remained the only practical method of long-distance communication, until the invention of the telephone repeater in 1916 made it possible to build analogue fax machines which used phone lines. These were developed during the 1920s and 1930s primarily for the newspaper market, and their design specifications were the direct ancestors of the first fax machine standards. The best known of these standards were set by the *CCITT*, which was until recently the international body with responsibility for ensuring interoperability of communications equipment across national boundaries. It was part of the International Telecommunications Union (*ITU*). More details on these organisations can be found on pages 13–18.

Group 1 was the first internationally accepted fax standard, published by the ITU as specification *T.2* in 1968. It took about six minutes to send a page at the same vertical resolution as current faxes (3.85 lines/mm).

Group 2 fax, published by the ITU in 1976 as specification *T.3*, improved the speed to around three minutes per page. The signals used by groups 1 and 2 were analogue rather than digital. That is, the tones sent out by the machines didn't have only two possible values, used to indicate either all black or all white dots, but had an indeterminate value within a certain range, with the lowest frequency corresponding to all black portions and the highest corresponding to all white. Everything between indicated a shade of grey.

Group 3 fax was the first digital standard, published by the ITU in 1980 as specification *T.4*, which allows transmission of a page in only one minute. Group 3 fax was based on modern modem technology, and allows digital transmissions over *PSTN* (Public Switched Telephone Network) connections. Oddly enough, early machines based around the telegraph could be considered to have more in common with modern group 3 faxes, as they also used digital communications via the telegraph system. The group 3 specification went through the first of numerous revisions in 1984, when the specification for how two fax machines talk to each other (the fax session protocol) was split off to become recommendation *T.30*, while the specification for how a document is digitally encoded remained as T.4. Both standards (now in their downwardly compatible 1993 revisions) remain in place today, but group 1 and 2 faxes are now obsolete.

Group 4 fax was originally designed to work over *ISDN* (Integrated Services Digital Network) connections. Fax modems don't work or even interwork with ISDN. Many of the features of group 4 fax, such as denser encoding and higher resolutions, are being incorporated into group 3 machines. While we mention these when we come across them, group 4 fax is really beyond the scope of this book.

What modems are

Modem technology is what has made group 3 fax machines possible. This short summary of basic modem technology is optional reading, and concentrates on the sorts of modems used by faxes. Readers interested in a more accurate and detailed exposition should consult a communications textbook with more of a hardware bias than this book.

The word *modem* is a contraction of MOdulator-DEModulator. The telephone is designed to carry voice communications, which are basically characterised by alterations in the frequency of sound. It's an analogue device, in the sense that just as the human ear can discern a more or less infinite range of frequencies within its range, so the telephone can transmit and receive all those frequencies too. Because computers store information digitally, and every unit of information (each bit) can have only two values, modems have to be used both to modulate digital signals for transmission over voice telephone lines, and to reverse the process on reception.

The earliest modems used one specific frequency for a value of 0, and a second frequency for a value of 1. All a transmitting modem needed to send was to turn the data it received from its computer, already composed of a stream of binary 0s and 1s, into a sequence of the correct frequencies, and all a receiving modem needed to do was listen for the agreed frequencies and turn them back into 0s and 1s before sending the stream back to the receiving computer. Effectively, the sender whistles to the receiver, which worked out the message from the notes it hears.

The exact frequencies used, and the length of time each encoding lasts for, is an example of a *modulation scheme*. One of the first such schemes, known as *V.21*, worked at speeds of up to 300 bits/s, using a similar modulation scheme to that just described, the difference being that it specified four frequencies. Two of these frequencies, known as the *A channel*, were allocated to the modem that made the call, while the remaining two, known as the *B channel*, were used by the answering modem. This use of four frequencies enabled both modems to transmit and receive at the same time, without interfering with each other. This is known as *full duplex* communications.

In fact, the V.21 scheme is standard for use in all Group 3 faxes today. It isn't used to send the fax, but to enable the two fax machines to negotiate about how the fax should be sent. The faxes only use the B channel as they only need to communicate in one direction at a time. This is an example of *half duplex* communication. Using frequencies in this way is known as Frequency Shift Keying (*FSK*).

The number of changes of state in any one direction on the line each second is known as the *baud*. The V.21 modem therefore operates as a speed of 300 baud. This isn't necessarily the same as *bps* (bits per second); the baud rate is only the same as the bit rate when the modulation scheme uses only two states on the line, such as two frequencies. More line states would result in higher bit rates at the same baud rate. For example, if it were possible to use 256 frequencies on one phone line, it would be possible to encode an entire byte at a time instead of just one bit. Sending a character such as ASCII 'A', whose decimal value is 65, would simply be a matter of looking up frequency 65 out of all the 256 possible frequencies and sending that one. A hypothetical 300 baud channel that used 256 possible frequencies could achieve bit rates of 2400 bps.

Unfortunately, the phone line can't carry this number of frequencies. Faster modem speeds have to look for different methods of changing the signal on the telephone line, rather than cramming more frequencies onto a line that can't cope with them. While theoretical advances in modulation theory have continued to be made, they have gone hand-in-hand with more practical advances in modem and microchip technology. In particular, the advent of faster components such as *DSPs* (Digital Signal Processors) facilitated the development of techniques such as echo cancellation and adaptive and non-linear equalisation, which have made better signal-to-noise ratios possible. Without such advances, the more complex modulations discussed below would not have been feasible.

DPSK modulation schemes

The V.21 modulation scheme is too slow to permit the sending fax data, and FSK modulation is too unreliable at high speeds. Instead, the original T.4 fax specification relied on a different modulation scheme for sending data, *V.27 ter* (ter means third just as bis means second), which sends data at speeds of up to 4800 bps. It manages this using a system called Differential Phase-Shift Keying (*DPSK*). This method of sending bits changed the phase of the signal rather than the frequency. Bearing in mind that what is being transmitted over a telephone line when sound travels from one phone to another is a regúlar wave form, two otherwise identical wave forms can differ in phase if the crest of one happens to coincide with the trough of another (Figure 1.1).

This 180° phase difference is the simplest sort of phase shift, and is known as DPSK-2. When the transmitter changes the phase of a signal, it can be detected

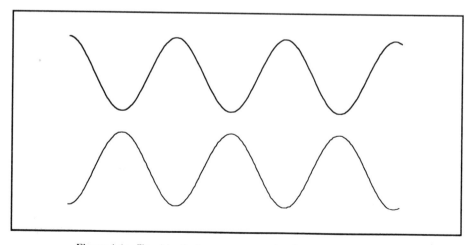

Figure 1.1: Two identical wave patterns showing a 180° phase shift

by the receiver just as easily as can a frequency change. Therefore a phase shift can be used to send data just like a frequency shift. However, a normal telephone line makes it possible for a receiver to detect more phase shifts than separate frequencies. In fact, the V.27 ter modulation scheme used by group 3 fax machines allows for eight different phase shifts of 0°, 45°, 90°, 135°, 180°, 225°, 270° and 315°, known as DPSK-8. Since there are eight possible phase values, the modulation scheme can actually encode three bits in just one phase shift.

Switching the phase of a signal is faster and more accurate than switching the frequency. V.27 ter can switch phases 1600 times per second. Since it uses DPSK-8 and each change of state on the line encodes 3 bits, the bit rate is 4800 bits/s, three times the baud. Recognising that this speed isn't always possible, V.27 ter includes a *fallback* modulation scheme using DPSK-4 at 1200 baud. As DPSK-4 can only encode two bits in each phase state, the V.27 ter fallback speed is only 2400 bps.

QAM modulation schemes

You may wonder why it isn't possible to combine phase shifts with frequency shifts to achieve even more states on a standard telephone line. Unfortunately, each time the phase of a signal changes, the frequency also changes momentarily, meaning that shifting the phase interferes with nearby frequencies. This makes combining the two schemes impractical.

However, it is possible to achieve more possible line states by shifting the amplitude of the signal (height of the wave) as well as changing the phase, known as Quadrature Amplification Modulation (*QAM*). While the modulation is actually more complex than just adding an amplitude change to a phase shift, QAM and similar schemes are usually shown diagrammatically as a *constellation*. Each star in the constellation shows one possible line state, with the distance from the origin representing the amplitude, and the angle from the horizontal axis representing the phase.

The constellation in Figure 1.2 shows the possibilities for the V.29 modulation scheme, used by group 3 fax machines to communicate at 9600 bps. The baud rate used is 2400, so each change state must be able to encode four bits at a time. This requires 16 possible states, made up of 8 possible phases, each of which has two possible amplitudes.

Note that while each phase permits two amplitudes, the actual amplitudes used aren't the same for any two adjacent phases. The phase has to shift a full 90° for the possible amplitudes to coincide. This makes the modulation scheme less error-prone. Where a fax machines finds a phone line too unreliable for 9600 bps communication, it falls back to V.29 at 7200 bps by dropping one of the amplitudes for each phase. However, as adjacent phases still have different amplitudes, the amplitude and phase are both required to identify a line state. Identification of possible states by phase alone is too unreliable.

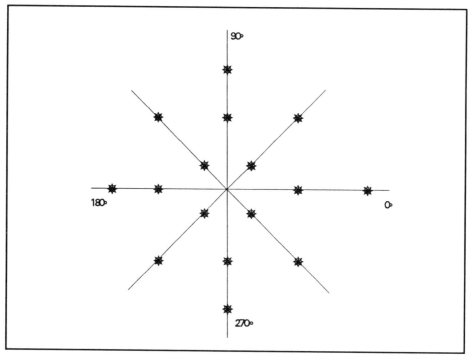

Figure 1.2: Symbolic diagram of constellation for 9600 bps V.29

TCM modulation schemes

Even higher speeds are achieved by fax machines with V.17 modulation schemes that use *trellis coding* modulation (*TCM*). The requirement that there should be a full 90° phase shift before an amplitude repeats means that only half the possible combinations in each constellation are used. The theory behind trellis coding is that extra combinations can be used not to encode more of the real data into each state change, but to encode an extra redundant bit which can be used to check the validity of the real data bits. The method used is forward error correction, where the previous data bits are one of the factors used in determining the precise modulation state to use next. Though all states are possible, all combinations of data bits and sequences of states are not. So while V.17 9600 is similar to the V.29 shown in Figure 1.2, and it uses the same baud rate, it makes use of the unused states that pure QAM modulation cannot.

The previously unused points in the constellation are effectively used to check the validity of the existing data, rather than to send more data at the same baud rate. This makes the V.17 implementation of 9600 bps far more reliable than the V.29 version.

To be announced

The latest generation of fax machines use the V.17 trellis coding modulation scheme not simply to achieve more reliability at existing speeds, but to achieve even higher speeds. V.17 can use up to 128 different points in a constellation. Allowing for the redundant bit used to check up on the real data, this still gives 64 points for encoding. This enables each state change to encode five data bits, rather the four possible with V.29. This gives a 25% increase in speed, giving good performance on normal phone lines at speeds of up to 14400 bps.

Even higher fax speeds, up to 28800 bps, are to be expected shortly, achieved by modifying the shape of the constellation used by trellis encoding. V.29 and V.17 used fixed gaps between modulations, of 3 and 5 for the four phases at 90° intervals starting from 0°, with $\sqrt{2}$ and $3\sqrt{2}$ for the other four phases starting at 45°. They specified the same number of points in the constellation at each amplitude. However, the ability of modems to discriminate between points near to the centre of the constellation is quite good, while more errors can creep in towards the outer limits. Two new techniques for improving the structure of a constellation take account of these phenomena. *Warping* is a technique that adds more points on the inside of the constellation to take advantage of the increased ability of the modem to distinguish phase shifts at these amplitudes, while *shaping* moves the outer amplitudes further away from the centre. Unlike the parallel V.34 data modem recommendation, international standards for these have yet to be approved.

Performance differences between full and half duplex modems

The fact that full duplex modems needed to be able to transmit and receive at the same time meant that for a long time they were not capable of the sorts of speed that half duplex modems could achieve over normal two-wire public telephone lines. For example, the V.22 bis full duplex standard enabled operation at speeds of up to 2400 bps using QAM-16 modulation at 600 baud. The same generation half duplex standard was V.29, able to achieve speeds of up to 9600 bps using the same QAM-16 modulation at 2400 baud. An exponential fourfold performance increase results from not needing to transmit in two directions.

Fax machines only need to transmit faxes in one direction at a time, so they have been able to use fast half duplex links, enabling them to achieve higher bit rates than data modems on identical telephone lines. However, the introduction of sophisticated *echo cancellation* technology in the 1980s meant that this fax modulation scheme advantage has been wiped out. Echo cancellation is a method whereby a modem can subtract the echo of data it is sending from the complete signal present on the line; it effectively enables a full duplex modem to isolate the received signal quite precisely. The adding of echo cancellation to the basic

V.29 QAM-16 2400 baud modulation enabled the development of the first generation of modems offering 9600 bps full duplex communications under the V.32 umbrella, which also included the first trellis coding modems, in 1984. The fast Digital Signal Processors (DSPs) needed to handle echo cancellation initially made these devices very expensive. It took a number of years for the price to drop to affordable levels, by which time the V.32 bis standard, which used trellis coding to achieve 14400 bps, had made V.32 QAM technology obsolete.

Modern full duplex modulation schemes all incorporate echo cancellation, so the half duplex V.17 modulation used for fax has no performance advantage over the same technology in its full duplex V.32 bis incarnation. It isn't impossible that in future a half duplex modulation scheme will be developed that once again enables the communications gap with full duplex to open up again, but to the best of my knowledge, nobody is working on it.

Synchronous and asynchronous communications

The fact the half and full duplex modems referred to above were historically associated with *synchronous* and *asynchronous* communications respectively, tends to blur the fact that the two distinctions are quite different. Synchronous and asynchronous have nothing to do with the direction of data flow.

A link used to send synchronous communication is always transferring data. Assuming the speed is 4800 bps, the receiver will sample the data stream 4800 times each second, and interpret the signal level as either a 0 or a 1 bit. The only way to stop sending data is to turn off the transmitter; as long as the transmitter is on, the receiver will be receiving data and trying to interpret it.

Notice that synchronous communication is inherently bit oriented – there is no structure implicit in the flow of data over a synchronous link. Any structure has to be imposed on the bit stream explicitly, as there is no way of distinguishing one bit as being more significant than any other.

In contrast, a link used for asynchronous communication does not transfer data all the time. The carrier is normally held in a high (*marking*) state, during which the line is idle and no data is transferred. When the transmitter wants to send data, it drops the signal for a period of time equivalent to 1 bit. At 4800 bps this would be for 1/4800th of one second. This dropping of the signal is known as the *start bit*. Following the start bit, all the data bits in a byte are transferred one by one, beginning with the least significant (rightmost) bit and ending with the most significant bit. Following the last bit, the line is held in the marking state for another period of time equivalent to at least one bit, known as the *stop bit*. As well as the usual single stop bit, some asynchronous data formats allow for two stop bits, or occasional 1.5 stop bits. Following this enforced idling period, the next bit could begin at any time.

Synchronous communication and half duplex links

The fact that many synchronous modems are half duplex, whereas many asynchronous modems are full duplex, is not simply coincidental. Synchronous data communication and half duplex communications go together for one very good reason—they are both highly efficient.

It is obvious that (given the common 8-bit word length) asynchronous communications immediately entail an overhead of a minimum 25% increase in communication time, simply because of the addition of start and stop bits. Where a synchronous modem sends only 8 bits, an asynchronous one will always send at least 10. We've already seen that making the full bandwidth available for communications in one direction where nothing is happening in the other direction results in an exponential gain in performance. Adding both these factors together, it is clear that the common ground between half duplex links and synchronous communications is that both maximise the efficiency of an information transfer.

The circumstance under which full duplex asynchronous communication makes more sense than half duplex synchronous communication is when efficiency is not the most important criterion. The classic use of a full duplex asynchronous modem is to link a terminal to a host for an interactive session. In this situation, data could flow in either direction in any quantity, and a large part of the session is often spent idling. There isn't much of a structure to the data, and the unpredictability of events makes it difficult to impose the sort of proper line turnaround protocol that bidirectional communication over a half duplex link requires.

In contrast, the bulk of the information transfer in a fax session is in one direction only; there is no idling to speak of, and the highly structured nature of the session protocol makes turning the line around at specific points quite easy.

Bits, bytes and words

There is one unfortunate side-effect of the difference between synchronous and asynchronous communications which has repercussions throughout the rest of this book. While synchronous communication transmits data one bit at a time, asynchronous communication transmits data one byte at a time, with the least significant bit of each byte sent first. The bits in a byte are sent in reverse order to the way they appear to a reader (Figure 1.3).

While the basic unit of information in a computer is undoubtedly the humble bit, the basic unit for manipulating data on most machines is the 8-bit byte. This generally corresponds to the size of the smallest register available to the CPU.

A fax is basically constructed out of a collection of bits, which has no natural structure serving to force it into groups of eight, and it thus has a natural affinity with synchronous communications. Sending a fax asynchronously requires all

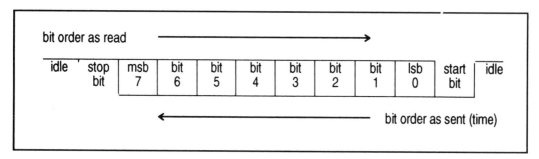

Figure 1.3:   Order of transmission of bits in an asynchronous byte

sorts of artificial contortions, such as reversing the order of the bits in each byte before transmission, or dividing a fax up into 8 byte chunks and storing them all backwards. This issue is visited time and time again, but for now it's worth remembering it as one area that can never be taken for granted.

Parallel and serial transmission

While discussing computer dichotomies, it is worth mentioning this one. *Serial* transmission sends a bit at a time over a single connection, while *parallel* transmission sends multiple bits at a time over multiple-way connectors. Typically, a parallel link can send all 8 bits of a byte simultaneously over eight separate wires, while a serial link has to chug away sending one bit at a time.

The reason for mentioning parallel and serial transmission in a book on fax is that recently, fax modems that connect to computers via bidirectional parallel ports have begun to appear. Their makers claim that as parallel ports are faster than serial ports, they get round serial port bottlenecks that can cause problems on other systems. This isn't really true. While in theory parallel transmission of data is faster than serial transmission, in practice the limiting factor on performance is never whether a link is serial or parallel. After all, networks such as Ethernet operate over single wire serial links, and their throughput is an order of magnitude greater than the fastest modems, being measured in Mbits/s. Further, if a computer is reading data a byte at a time, the time taken by the CPU to collect each byte from a port address will be identical no matter whether the port address references a serial or a parallel port.

Where there are problems on serial ports, they seldom affect fax communications. There is certainly a serial port speed limitation of 115200 bps on IBM PC systems. Some people claim this holds back communications in the same way that the 640K barrier held back other areas of PC computing, but it certainly doesn't affect fax.

Looking at how parallel modems claim to achieve their speed advantage, it seems clear that the key element is a software interface capable of buffering all received data and delivering multiple bytes per request, in much the same way as data is read from disk in blocks. However, there are enhanced serial ports available (such as the ESP from Hayes) which offer the same facilities over a serial connection. Any problems that exist are almost certainly due to specific serial port implementations rather than the fact that a port is serial and not parallel.

The black art of communications

That a question such as the utility of using a parallel port modem to avoid communications problems can arise at all is symptomatic of the fear, uncertainty and doubt that can so easily arise whenever communications and computers are mentioned in the same breath.

Communications has always had the reputation of something of a black art, requiring a particular set of skills not easily acquired. If your installation works you've been lucky, but if it doesn't then the sorts of problems you're going to have to solve are liable to begin just where the manuals stop.

Many first-time fax modem users are shocked to discover that while their simple fax machine sends and receives faxes day after day with no problems, their clever new fax modem in their fast computer fails to make simple transmissions and loses portions of received data. At first sight, the computer industry seems to have taken a perfectly good and highly reliable mode of communications and made a pig's ear out of it.

There are very good reasons why this seems to be so, and why things may well not change in the foreseeable future. Put simply, communications are unlike most other types of software in that the progress of the application is to some extent outside the user's control. Word processing, spreadsheets, programming, databases and most other applications are predictable. What happens depends on what the user inputs – what keys are pressed, what files are used, what installation options are selected, and so on. Hardware faults aside, the software user is in control of what is going on.

However, communications is a *real time* activity which depends upon more factors than simply what keys are pressed by an operator. There is a modem hanging off the serial port, a telephone link to another modem, and above all another computer, or a similar device, at the other end of the link. All these components have to work together, and unlike the other application types mentioned, this isn't simply a question of getting the software to work once and then making a note of how it was installed.

For instance, it is almost always the case that where a user dials different phone numbers to connect to different computers, they will also see different messages, with different prompts and different timings. Even where the same number is dialled every day, the response times might vary with the number of users on the system, or the line might be lost in the middle of the call.

This problem shouldn't occur with faxes, but a parallel situation can occur when a user has installed a multi-user or multi-tasking OS. While some applications may just take longer to complete certain functions, real-time applications like communications and fax can behave erratically when used in an environment where they're not guaranteed access to the processor.

Not only is each installation subtly different, but every communication session begins in the knowledge that there is the ever present possibility that the session may be different to the last one. It is this need for communications software to change in ways we cannot control that sets it apart from other applications.

The simple art of fax

There could not be a greater contrast between computer and fax communications. Over the last ten years, the growth in the number of fax machines has been phenomenal. The group 3 fax machine has become a commodity item in the business world, and is rapidly making inroads into the home. One of the main reasons for this fax revolution has been its ease of use and high reliability. Almost anyone who can use a fax can use a telephone: unpack the machine and plug it in; put a document in a fax; dial the number, and that's all. Faxes rarely to fail to arrive for any reason other than phone line problems. Receiving a fax is even easier – you don't even have to be in. Most fax users don't know that all group 3 fax machines contain the same sort of digital modem found in computer to computer communications.

It should be just as easy to write software that was just as simple to use. Plug in your modem, type or click on FAX, select your file and your telephone number, and that's all. The problem is that any software which was that simple would also fail to work. One reason is that computers and operating systems aren't that standardised. Little problems like needing to know the communications port to which the modem is connected, and the files and directories to be used, start multiplying. In addition, not all fax modems are alike, and the differences between them make a difference to how well computer faxing works and what facilities it offers.

Fax is easy to use mainly because all fax machines are standardised and behave in more or less the same way. The fact they are single-function machines undoubtedly helps, as does the fact they are built to do a particular job. As a general rule, the fewer tasks a piece of machinery has to perform, the easier it is to ensure it does its job properly. However, fax machines do have to adhere to standards in both manufacture and performance – there are even standard test charts to check fax reproduction.

By their very nature, computers are unlike fax machines, as they are inherently programmable, and can perform tasks in many different ways. This has led standards bodies to legislate for computer fax modems as well as fax machines. It is to the nature of all these standards, and the agencies that are responsible for them, that we now turn.

Introducing fax standards

There is no easy way of describing what standards are for and how they become established. There are a number of ways of looking at this fascinating area, where technology, commerce and politics come together, but the scope of this topic is unfortunately too broad to be dealt with here.

There are two main groups of standards and recommendations (we treat the two terms as equivalent) that anyone interested in fax will need to consider. The first group are the standards for the format of fax messages and for the fax session protocol, set by the ITU. The second group are standards for controlling fax modems, set by the Electrical Industry Association (*EIA*), a US body that provides standards for use in all areas of electronic manufacturing. More details on the EIA can be found on p. 17.

The mechanism for setting ITU fax standards is by now quite well oiled, and seems to work quite well. The existing group 3 fax standard is accepted worldwide, with only minor deviations from the specification ever arising. As we'll see when we come to look at the session protocol, it has a lot of inbuilt flexibility and room for almost indefinite expansion. Starting from a basic set of features that all faxes have to support, facilities and features have been added to keep pace with new technological developments, and to remedy errors, deficiencies and omissions in the original versions. Like all standards generated by a committee, it shows evidence of a few duff decisions, sometimes made for political reasons, but overall, the fax standard is a model for what an international standard ought to be.

EIA standards for controlling fax modems have a different history. Admittedly, the drive towards standardisation has been going for less than ten years, during which time the committee responsible has had to deal with a moving target in that computer technology has been changing rapidly. However, the history of the emerging standard has been marked by a split down the middle, resulting in two different and incompatible standards. The second of these standards was originally produced in a version that was widely adopted but subsequently withdrawn before ratification, with an incompatible approved version eventually being produced that hardly anyone is now using (we discuss this in more detail later).

A brief outline of these and other relevant standards bodies follows. A knowledge of the relationships between such bodies can help in understanding the limits to the recommendations each of them might make. A knowledge of their structure and decision making processes is also a useful aid in trying to guess what particular decisions they made were supposed to accomplish.

The ITU

The International Telecommunications Union may well be the international body with the longest continuous history in the world. Just as most people who see fax as a 1980s phenomenon are surprised to discover that the first fax patent was granted in 1843, so many people familiar with the ITU (today under the auspices of the United Nations) are equally surprised to discover that it was founded in 1865 in Paris as the International Telegraph Union. It predates not only the United Nations (by some 80 years), but even such venerable bodies as the Universal Postal Union.

The basic impetus behind the work of the ITU is the fact that unless internationally agreed technological standards are arrived at, it isn't possible to use equipment to transmit messages across national boundaries. While this was originally seen to apply to telegraph systems, it later took in telephone systems and all other telecommunication apparatus. As well as specifying standards for equipment and methods of use, the work has expanded to include co-ordination of international agreements on the allocation of radio frequencies.

The ITU is unlike any other standards organisation in the world, in that it is an intergovernmental treaty organisation. What this means in practice is that anyone who attends does so as a member of a national delegation rather than as a representative of a commercial or other non-governmental organisation. While governments may (and indeed do) nominate commercial or other organisations to represent them at ITU meetings, the attendees do so under the aegis of the government department responsible for such matters.

The ITU state that their recommendations are not binding, but "are generally complied with because they guarantee the interconnectivity of networks and technically enable services to be provided on a world-wide scale". In fact, the existence of the ITU and the standards it sets are intimately bound up with the nature of communications and connectivity. Communication without standards is almost impossible to imagine.

The CCITT

For a number of years, the ITU maintained two separate permanent bodies responsible for different areas of its standards work. The International Radio Consultative Committee (CCIR) handled all radio and television standardisation work, while the International Telegraph and Telephone Consultative Committee (CCITT – from the French 'Comité Consultatif International Telegraphique et Telephonique') was responsible for all non-broadcast telecommunications.

Historically, there was originally a distinct European bias to the work of the CCITT. This was primarily because the tradition in Europe was for telecommunications to be run as a government-licensed monopoly, while a combination of a less centralist political philosophy and the dominant role of AT&T in the United States prevented the US government from participating in the same way. There were competing fax and modem standards between the USA and the rest of the world until the 1980s. A hangover from those days can still be found in the command set of most fax modems, which generally include a command (commonly

ATB) enabling the modem to switch from supporting the Bell 212A standard to CCITT V.21 data modulation. Since the break-up of AT&T, such incompatibilities have vanished and the work of the CCITT has had universal co-operation and acceptance.

The CCITT used to meet in plenary session to ratify official standards every four years. Headings such as Geneva 1980, Malaga-Torremolinos 1984 and Melbourne 1988 can still be seen on many current recommendations. After each plenary, the recommendations were collected together and published as a series of coloured books. The 1980 plenary produced the Yellow Books, the 1984 plenary produced the Red Books and the 1988 plenary produced the Blue Books. The coloured books were multi-volume works, with each section being known as a fascicle.

Between plenaries, the work of the CCITT was divided into a number of different study groups. The most important of these for our purposes are Study Group VIII (SG8), responsible for fax standards, and Study Group XIV (SG14), responsible for modem standards.

Reorganisation of the ITU

The ITU was reorganised on 1 March 1993, with a new constitution taking effect from 1 July 1994. The main visible effect of this change is that the CCITT ceased to exist under that name, instead becoming known as the International Telecommunications Union Telecommunications Standardisation Sector. After a few hiccups, when it was sometimes known as ITU-TSS or ITU-TS, it is now officially the *ITU-T*. The structure and membership of the study groups remain intact.

The ITU has also developed a means of accelerating approvals procedures which can now be managed whenever a study group is ready. It was recognised that the four-yearly cycle of plenary approvals conferences was too elephantine to cope with the increasing pace of change in the world, and there was a danger that the standardisation enterprise itself could be put at risk if a much needed specification were delayed unnecessarily. The plenary sessions have been replaced by four-yearly meetings of a new ITU body, the World Telecommunications Standardisation Conference (*WTSC*), which establishes what topics should be considered by which study groups. Following this remit, the study groups themselves produce and publish the recommendations as and when they are ready. The first meeting of the WTSC was held in Helsinki in March 1993.

Among the ITU originated standards essential to fax are the V-series modem standards such as V.21, V.27 ter, V.29 and V.17. These are generally the concern of hardware rather than software engineers. However, a number of the T-series recommendations are used as primary references for all fax software, and are covered in some detail in this book. We have already mentioned the T.4 recommendation for the coding and compression of group 3 fax images, together with the T.30 recommendation defining the way in which the fax session protocol works. One other notable standard is the T.6 recommendation, originally for the coding of group 4 faxes, but now also supported by group 3. Other recommendations affecting fax include the T.434 standard for binary file transfer, whose implementation in turn depends upon an understanding of both the F-series recommendation F.551 for Telematic File Transfer (TFT), and Abstract Syntax Notation One (ASN.1), specified in the X-series recommendations X.208 and X.209.

National standard bodies

While communications standards such as those maintained by the ITU arise out of necessity, others arise out of a matter of public policy. Most countries maintain some sort of national standards setting body responsible for the development and certification of these standards. The most notable of these national bodies is the American National Standards Institute (*ANSI*), a voluntary non-governmental organisation responsible for standards promotion in the US. Similar bodies in other countries, such as the British Standards Institution (*BSI*), have neither the prestige nor scope of ANSI.

Among the numerous ANSI originated standards underpinning fax technology is ANSI X3.4, commonly known as *ASCII*, the American Standard Code for Information Interchange. This is an example of a standard that is technically aligned with an ITU standard, as ASCII is much the same as ITU's Alphabet No. 5. A subset of ASCII is used by fax machines to exchange ID codes and other textual information.

Another ANSI recommendation that plays a large part in this book, though not necessarily in fax technology, is ANSI X3.159, the specification for the dialect of the C language in which most of the code is written.

The ISO

The International Standards Organisation (*ISO*) dates from the end of the Second World War, and while it is run under the auspices of the United Nations, it isn't an intergovernmental treaty organisation. While countries can belong, non-governmental national standards organisations (such as ANSI) can also be nominated for membership by a particular country. However, the ISO is more than just an international umbrella body for standards organisations such as ANSI and BSI. It also does a lot of valuable work in its own right.

Surprisingly, the only pure ISO standard we refer to in this book is that for paper sizing. The A4 paper size is defined by the ISO, and is just one of a series of sizes used in designing the dimensions of fax paper sizes.

A number of standards are common to both the ITU and ISO, which work together in a variety of areas. For instance, the X.208 and X.209 ITU standards for ASN.1 are deliberately identical to ISO 8824 and ISO 8825, and both the ITU and the ISO co-operated in the work involved in producing them. Other standards that are replicated between different bodies include the ASCII character set, which as well as being ITU Alphabet No. 5 is also the ISO 646 7-bit code.

Trade Associations

Voluntary trade associations have historically played a large part in the development of many manufacturing standards, and facsimile is no exception to this. The British Facsimile Consultative Committee (BFICC) and the Communications Industry Association of Japan (CIAJ) have done most of their work via their participation in national delegations to ITU-T Study Group VIII, and have never been responsible for developing their own standards. However, possibly because of their early lack of participation in the then-CCITT, the situation was quite different in the US.

The most important trade association active in the fax standardisation area has always been the EIA. Its standards are voluntary, and are created by committees of specialist groups set up as needed by the EIA membership. In this respect, the structure of the EIA is similar to that of the ITU, with specialists in particular areas assuming responsibility for looking at the relevant issues and coming up with a standard. This professionalism is formalised by the fact that most EIA standards are accredited by ANSI, of which the EIA is a member.

Some of the EIA standards are now part of the everyday language of communications. For instance, RS-232 is actually one of the EIA RS-232 standards for 'Interface between Data Terminal Equipment and Data Communications Equipment Employing Serial Binary Data Interchange'. EIA standards are widely used throughout the world, and sometimes this adoption is formalised, as is in the case of RS-232, which was adopted by the ITU as recommendation V.24.

The telecommunications standards set by the EIA were originally the responsibility of their Information and Technology Group (ITG). In 1988 this body merged with another trade association, the United States Telecommunications Suppliers Association (USTSA), to create a single body, the Telecommunications Industry Association (TIA). This is now the main forum for US fax and modem standards activity. Though it refers to itself by both sets of initials, we'll simply call this body the EIA.

There are two main EIA committees relevant to fax. TR-30 has emerged as the main US body developing modem standards since the divestiture of AT&T; it did most of the preliminary work on the ITU V.34 standard, which ought eventually to result in fax machines that support 28800 bps transmission. The main EIA fax committee is the TR-29 Facsimile Equipment and Engineering Committee, set up in the early 1960s to bring some sort of order to the competing world of incompatible proprietary fax standards that had reigned until then. The TR-29 committee is similar in scope to ITU-T Study Group 8. They produced the RS-328 fax standard in 1966, but while this improved compatibility between faxes it didn't eliminate all inconsistencies. It was also incompatible with ITU Group 1 specifications. Meetings of TR-29 seemed to consist of different companies arguing that their proprietary standard was the best. Eventually, TR-29 was designated the US State Department's Technical Advisory Group for the ITU-T Study Group 8 on telephone facsimile, and future US fax standards were all channelled through the ITU. TR-29 played a large part in the development of both Group 2 and Group 3 fax.

The EIA TR-29 fax modem standards

The TR-29 committee also established a number of subcommittees with specific remits for computer fax applications. One of these, the TR-29.1 subcommittee on Binary File Transfer, prepared the BFT specification that was the basis of ITU T.434. Of more importance to fax modem users is the TR-29.2 subcommittee on Digital Facsimile Interfaces. It is this body that has been responsible for the development of the ANSI/TIA/EIA-578 and ANSI/TIA/EIA-592 fax modem standards, more generally known as Class 1 and Class 2/2.0, respectively. The basic difference between the two classes is that Class 1 modems leave more of the real-time work involved in a fax session to the computer controlling the modem, while Class 2/2.0 modems handle much of the real-time work themselves.

Unofficial public standards bodies

We have already commented on the way that private initiatives can evolve into *de facto* industry standards and sometimes metamorphose from there into official standards (the Hayes AT command set is one such example). Others don't get further than industry standard status, with IBM's PC architecture being an example. Other examples in the computer world include the Centronics parallel port and both Epson and Hewlett-Packard printer control codes. Despite the fact that all these examples were originally specified for private use, they have passed from private status into the public domain as *de facto* standards.

Another type of *de facto* standard is designed as being public from the beginning. Classic examples are file transfer protocols such as Xmodem and Kermit which, while they have always been public, have still been largely the work of one person (Ward Christiensen in the case of Xmodem and Frank da Cruz and Bill Catchings in the case of Kermit).

Other unofficial standards have emerged from unofficial standards bodies. One of these, discussed at some length in this book, is the TIFF file format developed by a group of companies coordinated by Aldus. The TIFF standard is the only one widely available that provides a method of storing native fax format images.

Conclusions

We've use this chapter to introduce a number of subjects. We've examined the history of fax and the development of basic modulations techniques used by modems integral to modern digital faxes. We've also discussed the concept of standardisation, and seen how vague, haphazard and arbitrary standards can sometimes be. Finally, we've been introduced to the main standards bodies whose work makes faxing and fax modems possible.

2

The Structure
of a Fax Document

Introduction

This chapter explains one of the essential standards of the fax world. In it, we describe how to turn a normal page of text into a facsimile image in accordance with the ITU T.4 standard.

This standard is so important because all fax modems expect to be provided with valid T.4 data when sending a fax, and in turn they provide valid T.4 data when receiving. In other words, if you can't create or decipher this type of data then you can't use a fax modem. The information in this chapter is therefore essential reading.

Our account is supplemented with a selection of code fragments at appropriate points in the text, providing useful techniques for implementing the more unusual portions of the specification.

Basic concepts

The T.4 standard provides a method of transmitting documents over ordinary telephone lines in black and white, with a standard page taking approximately one minute to transmit. It assumes that a page is broken down into horizontal lines, starting at the top left-hand corner and proceeding left to right, line by line, down the page to the bottom right-hand corner. Each line, known as a *scan line*, is then assumed to be broken into a series of black or white dots. This corresponds to the way that document scanners such as those built-in to stand-alone fax machines work.

The lines of dots are represented in digital computing terms as a two-dimensional array of bits with the values 0 or 1. We refer to this array as a *bit map*, and to the process of turning a page into a bitmap as *digitization*.

We use the terms *dots* and *bits* throughout, as most people are happy with them. Most computer users normally conceive of the resolution of scanners or printers in terms of dots, as in dots per inch (dpi). However, there is more than one way of talking about these things. For instance, the resolution of a screen is commonly given as a matrix, such as 1024×768, with each item in the matrix being termed a *pixel* rather than a dot.

Physical dimensions of a document

The ITU fax specifications use the international A4 paper size as a reference. The dimensions of the A series of paper sizes are themselves an ISO standard, and take as their basis the A0 sheet, a rectangle whose area is one square metre and whose sides are in the proportions of $1:\sqrt{2}$. This sheet has the highly useful property that the proportions of the sides remain identical if it is folded in half on the long side. The A1 sheet is thus an A0 sheet with the long side halved, an A2 sheet is half an A1 sheet, an A3 sheet half an A2 sheet, and the common A4 sheet is half an A3 sheet. It is thus a rectangle with an area one-sixteenth of a square meter with sides in the ratio $1:\sqrt{2}$. These dimensions never come out as exact integers, but are usually taken as being 210 mm wide and 297 mm long.

However, despite that fact that the ITU mentions A4 as being the minimum size that should be accepted, the basic dimension of the encoding scheme specifies a page as being 215 mm in width. Furthermore, while there is a requirement that a conforming group 3 fax can send an A4 page, there is no ban on longer or shorter pages being sent. In fact, no maximum or minimum length is specified at all.

There are benefits arising from this apparent peculiarity. The extra 5 mm width over A4 makes the standard width almost 8.5", which enables a number of common non-ISO standard paper sizes to be faxed, such as US legal (8.5" × 14"), half-foolscap (8.5" × 13.5") and US letter (8.5" × 11"). The fact that no length needs to be specified is simply a reflection of the assumption that documents are being scanned as part of a process generating a continual supply of scan lines. Conversely, we assume that at the receiving end, each line of a fax is printed out as it is received on a continuous roll of paper.

Resolution of a bitmap

The recommendation states that one scan line contains 1728 dots. As a scan line is 215 mm wide, the actual figure for the resolution is 8.04 dots/mm. The ITU also state that there is a standard vertical resolution of 3.85 scan lines/mm, with an optional doubling to a fine resolution of 7.7 scan lines/mm. This gives us either 1143 or 2287 lines to each page.

In terms of the standard units of resolution, which for some reason remain stubbornly based on Imperial rather than metric measures, a fax usually has a horizontal resolution of 204 dpi and a vertical resolution of either 98 or 196 dpi. So if you happen to be scanning a document into a computer with a scanner that can be set to scan an image at anything from 100–400 dpi, it's worth noting that the most sensible resolution to use would be 200 dpi. Going up to 300 dpi would possibly give a meagre 2% improvement in the horizontal quality of the image, while it is possible to go down to 100 dpi with no loss of vertical detail if the fax is to be sent in normal resolution.

Tolerances and margins for error

An accumulation of minor mechanical inaccuracies will always be present in machines that aren't built to military specifications. If they were to be built to such standards, many people who currently use faxes wouldn't be able to afford them.

In recognition of the fact that the scanners and printers used in fax machines are devices which have mechanical and analogue components prone to manufacturing error, and are never going to be as precise as digital technology, the resolutions to be used are specified as being ±1%. In other words, though the number of dots on one scan line is nominally 1728, it can vary by as much as 17 dots on either side of this figure and still remain within the recommendations. A similar 1% tolerance is allowed for variation in the vertical resolution.

The operational effects of the various tolerances built in to the recommendation can be seen in Figure 2.1. Horizontal margins of 6.7 mm on each side reduce the effective width of an A4 fax from 210 to 196.6 mm, while a top margin of 4 mm and a bottom margin of 11.54 mm mean that the length we can rely on is reduced from 297 mm to 281.46 mm.

However, adding up all the possible factors that cause losses makes this figure more explicable. I'll explain the horizontal losses first:

- The most obvious loss of width is caused by the fact that on a 210 mm wide page, a 1% horizontal tolerance could cause up to 2.1 mm to be lost.

- Another cause of such loss will be familiar to users of photocopiers. The paper being copied isn't always placed flush with the edge of the copier bed. Feeding a paper into a slot involves a similar inaccuracy in getting the paper aligned with the scanner edge, while when printing out a fax the print head

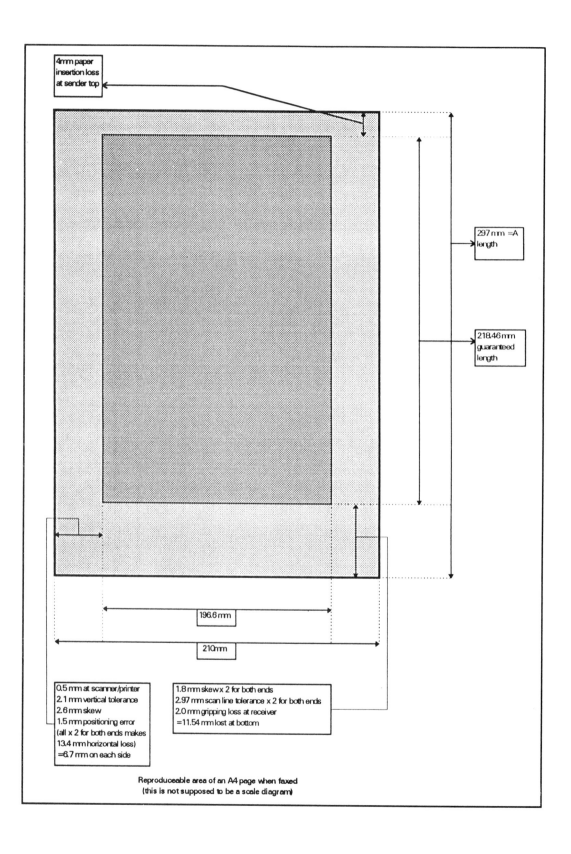

4mm paper insertion loss at sender top

297 mm = A length

218.46 mm guaranteed length

196.6 mm

210mm

0.5 mm at scanner/printer
2.1 mm vertical tolerance
2.6 mm skew
1.5 mm positioning error
(all x 2 for both ends makes
13.4 mm horizontal loss)
= 6.7 mm on each side

1.8 mm skew x 2 for both ends
2.97 mm scan line tolerance x 2 for both ends
2.0 mm gripping loss at receiver
= 11.54 mm lost at bottom

Reproduceable area of an A4 page when faxed
(this is not supposed to be a scale diagram)

might possibly move over the edge of the paper. The ITU allow ±1.5 mm for these sorts of positioning errors on a fax machine.

- Horizontal loss is also caused by skewing, when the paper being fed through the scanning or printing mechanisms doesn't always go through at a right angle. Up to ±2.6 mm should be allowed for this type of error.

- Finally, the scanner being used to read the document or the printer being used for output, may not be entirely accurate. We need to allow ±0.5 mm for this.

Adding all the possible causes of horizontal loss together, we find that a fax machine could lose up to 6.7 mm from the width of a page and still remain within tolerable limits. Since this sort of loss could, at worst, be encountered at both the transmitter and receiver, we should be prepared for a total horizontal loss as high as 13.4 mm.

So only the middle 196.6 mm of a 210 mm wide sheet can be guaranteed to reproduce after transmission. This doesn't prohibit attempts to send a full width fax, but it does mean that failure to receive it at the other end cannot be taken as an indication that either machine is at fault.

Similar problems are encountered with the vertical component of a fax image:

- Anything within 4 mm of the top of the page is not guaranteed to be transmitted. Some loss due to unreliability of the paper insertion mechanisms is always a possibility.

- Similarly, we cannot realistically expect friction-feed scanning mechanisms to always grip the paper perfectly. Where there is some slippage, a non flat-bed scanner might think it has scanned the full 297 mm when it has actually scanned less. Up to 2 mm gripping loss per page is permissible.

- As well as these losses, totalling 6 mm, which will only be encountered at the transmitting station, we have to allow for the sort of loss we found in the horizontal component of a fax occurring at both ends of the link.

    The vertical resolution nominally specified (3.85 scan lines/mm normal resolution, 7.7 scan lines/mm fine resolution) doesn't have to be exact, as manufacturing inaccuracies of up to 1% are tolerable. The bottom 2.97 mm of a 297 mm A4 page cannot be relied on. Also, the sort of paper skew that causes text to edge over the horizontal margins will also cause text to drop off the bottom of the page on one side, and ride up towards the top margin on the other. The recommendation allows ±1.8 mm at the end of the page for such loss. This gives a total possible loss of 4.77 mm, which, if found at both the transmitting and receiving ends, makes a combined total of 9.54 mm.

When the 6 mm loss at the transmitting end alone is added, we see why we can rely on only 281.46 mm from a 297 mm page.

Figure 2.1: Dimensions of a fax page

Code for digitizing a line of text

Putting all this into practice is not as difficult as it may seem, and the techniques needed for turning an ordinary ASCII text file into a fax image are really quite straightforward. The first step is to digitize the text, and we are already in a position to develop a code fragment that does this.

The process of translating a single character into a matrix of bits is very common, and is handled as routine by all font-based screens and printers. A font is actually nothing but the mapping of a complete character set onto a series of bit maps. As an example, Figure 2.2 shows one possible entry in an 8×8 font for the letter A.

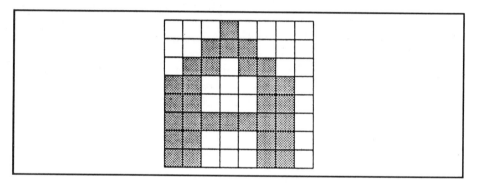

Figure 2.2: 8×8 bitmap for the letter A

Representing this particular map as a series of bytes is quite simple. Reading from left to right and top to bottom, the bits are represented as in Table 2.1.

The array {0x10,0x38,0x6c,0xc6,0xc6,0xfe,0xc6,0xc6} is therefore a digitized representation of the letter A on an 8×8 bitmap. Digitizing a

Table 2.1:   Representation of the font data from Figure 2.2

binary	hex
00010000	10
00111000	38
01101100	6c
11000110	c6
11000110	c6
11111110	fe
11000110	c6
11000110	c6

sequence of letters (a line) uses exactly the same principle, and a text file is simply a sequence of lines.

Font size

The main additional requirement for turning a text file into a faxable image is a font of the correct size. The dimensions of the fax page are the critical figures here. Were an 8 × 8 font to be used, a single 1728 × 1143 fax page would contain over 140 lines, each with over 200 characters. You would need a magnifying glass to read it.

To arrive at the correct font size, let us assume that the text file is designed to fit on an A4 page, and that it can be printed out using standard printer defaults of 10 characters per inch horizontally, and 6 lines per inch vertically. Since the paper size is 8.27" × 11.69", our text file can (in theory) be formatted as fax pages containing 70 lines of 82 characters each.

We use these figures as the basis of character digitization. The required font height is quite easy to work out. The vertical resolution (for a normal fax) is specified as 3.85 lines/mm. Since a page is 297 mm long, it consists of 1143 scan lines, and as we have 70 text lines, each line of text consists of 16 scan lines.

The width of the font is only slightly more difficult to arrive at. The specification for a fax is that it has 1728 dots on a 215 mm line. Since a page is 210 mm wide, it consists of 1688 dots, and as we have 82 characters per line, each character is, therefore, 20 dots wide. The size of a fax font should therefore be 20 × 16.

Generating a font

The skill needed to design a font is something I lack; I suspect this is true of most people. Good fonts are typically copyrighted, and property rights are jealously guarded. However, an easily accessible font with no restrictions on its use is the standard VGA font found on virtually all PCs, which is 8 × 16 dots, with no intercharacter gaps (that is, a couple of characters like MM side by side will run into each other). On PCs a ninth blank bit is usually added for the gap, making the font 9 × 16. For our purposes, we can conveniently start with the 8 × 16 font, and then add an extra dot on each side of a character. This makes the font 10 × 16. Finally, we can double this up to the 20 × 16 font that our fax image requires. Doing this font conversion 'on the fly' requires a couple of fairly trivial lines of code, which we deal with later.

The fragment of assembler code in Listing 2.1 shows how to download the VGA font on a standard PC. Alternatively, an 8 × 16 font data can be found in `font8x16.dat` on the accompanying disk. This font has been specially modified for fax. The shapes of many of the letters have been altered for better transmission and reproduction.

```
push        ds
push        es
push        si
mov         ax,1130h        ; function 11h subfunction 30h
mov         bh,6            ; bh=6 for 8x16 font pointer
int         10h             ; returned in es:bp
push        es
push        ds
pop         es
pop         ds
mov         si,bp
mov         di,offset font  ; store the array at the address font
mov         cx,256*16/2     ; size (in words)
rep         movsw           ; copy the font array
pop         si
pop         es
pop         ds
```

Listing 2.1: Locating and copying a VGA font table

Digitizing a complete line of text

One great advantage of an 8×16 font for digitizing is its convenient byte-wide size. Finding the dot pattern for any character is simply a question of using the value of a character as one index into a two-dimensional font array, with the font row being the other index.

The fragment in Listing 2.2 reads a line of text from a file to **text_line**, and stores a digitized version of the text at **bit_image**. It consists of two loops: the outer loop processes the entire line once for each row of the font, with the row being the control variable; the inner loop steps through each character in the line, with the character position being the control variable. The purpose of this loop is simply to store the value in the two-dimensional 256×16 font table indexed by the character and the row in **bit_image**. The end result of the two loops is that the original line of ASCII text is transformed into a series of scan lines.

```
fgets (text_line,79,infile) ;
for (row=0 ; row<16 ; row++)
{
     for (i=0 ; i<strlen(text_line) ; i++) if (text_line[i]>31)
     bit_image[image_offset++]=fonts[text_line[i]][row] ;
     bit_image[image_offset++]=EOLFLAG ;
}
```

Listing 2.2: Digitizing a line from a text file using an 8×16 font table

Each of the 16 scan lines we generate ends with a unique **EOLFLAG** sentinel, which can be any configuration of bits not occurring in the fax data (the code on disk uses 0x01). If we wanted to generate a complete bitmap, we would have needed to pad out each of our scan lines to the full 1728 bits. However, the type of image we are generating is an intermediate one which is going to be further processed before ending up as a fax image, so there is no reason to insist on fixed length scan lines at this stage. In fact, the use of a unique end of line code to end scan lines, coupled with the use of a byte-wide font, is what gives this encoder a great deal of speed.

This is the core of the digitization code that you'll find in **QAXIFY.C** on the accompanying disk, though the version there has a few extra tweaks to pad out tabs (in columns of 8) and to ignore extraneous control codes.

Encoding the bitmap

So far we have been developing a fairly straightforward bit-mapped representation of text aimed at a normal resolution fax page of .1143 lines of 1728 dots. However, the highest speed common to all group 3 fax apparatus is 4800 bits per second. It doesn't take a genius to see that this page consists of 1975104 separate dots. Assuming a representation of one dot by one bit, it would take a little over 411 seconds to send the page as a continual bit stream, which is seven times longer than the recommendation aims for. The bit map thus has to be compressed before sending. Before we look at any code to do this we should explain the theory behind the type of data compression used by the fax specification.

The method used to compress the bitmap derives from work originally done by D. A. Huffman in 1952, and is referred to as modified Huffman encoding. Accounts of Huffman's work can be found in many books, including most standard texts on algorithms, data compression or graphics. It isn't essential to understand it at all if all you want to know is how to generate fax images, so only a short technical description account is given here.

Huffman codes

Huffman discovered that if the components of a message are first sorted by the frequency with which they occur, and if a binary tree is then constructed with the most frequently occurring components at the root, then the optimal method of compressing the message is to replace each occurrence of a component within it by a bit string representation of the path leading to its occurrence in the tree.

The rules for encoding a bitmap constructed according to the dimensions given above are straightforward. The actual dots are not themselves transmitted. Instead, each line in the bitmap is broken down into consecutive runs of alternating white and black dots, referred to as run lengths. For both white and black run .

lengths, there is one unique code. The compression achieved is often quite significant. For instance, a blank line is sent as the 9 bit code 010011011, which indicates a white run length of 1728 (a completely white line) and gives a compression ration of 192:1. Blank lines are of course quite common in most documents, so this is not an artificial example.

The codes used for the runs are of variable length, ranging from two to twelve bits. They have been statistically selected (using Huffman techniques) so that the most common run lengths generate the shortest codes. The optimization was made on the basis of analysing a mix of documents, including business letters typed in English, pages of Japanese and French text, handwritten memos, black and white artwork, and graphs. It is quite easy to illustrate the usefulness of this optimization procedure.

Figure 2.3 shows a very small horizontal slice marked in the text. It consists of quite short black runs (where the slice passes through letters) interspersed with longer white runs. Slices taken through any text will exhibit the same property, which simply reflects the fact that most of the area of any document consists of white space. If you look at the code tables we use for fax, you'll see that this is reflected in the fact that the very shortest codes are for the small black runs of two and three, while the longest codes are all for the black runs over fourteen. The white runs come somewhere between, as while short white runs are found less often than short black runs, medium and long white runs are much more common than any but the shortest of black runs.

The ITU compression codes for both white and black run-lengths appear in Tables 2.2 and 2.3. There are two types of code: terminating codes, for run lengths in the range 0 to 63, and make-up codes, which are multiples of 64 used to generate codes for longer runs. The codes for run lengths below 64 can occur by themselves, while the make-up codes must be followed by a terminating code of the same colour, which makes up the difference between the run represented by the make-up code and the actual run-length being encoded. Thus a run length of 1000 would be represented by the make-up code for a run of 960 dots followed by a run of 40 dots. If a run happened to be an exact multiple of 64, and could therefore be represented by a solitary make-up code, it must still be followed by a terminating code. Hence the need for run-length codes of zero.

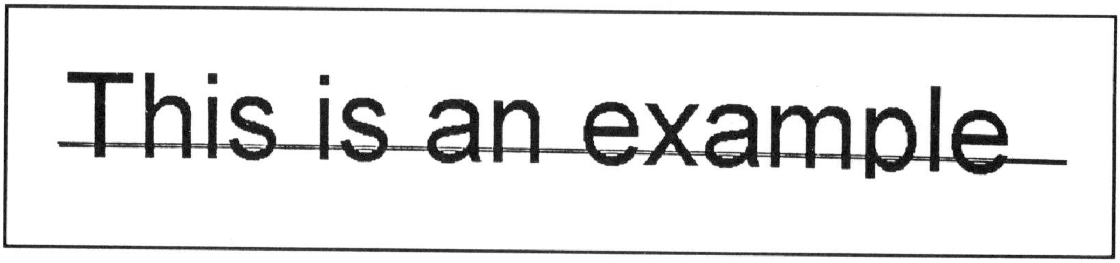

Figure 2.3: Relative lengths of black and white runs in text

Table 2.2: Terminating codes

run	white	black	run	white	black
0	00110101	0000110111	32	00011011	000001101010
1	000111	010	33	00010010	000001101011
2	0111	11	34	00010011	000011010010
3	1000	10	35	00010100	000011010011
4	1011	011	36	00010101	000011010100
5	1100	0011	37	00010110	000011010101
6	1110	0010	38	00010111	000011010110
7	1111	00011	39	00101000	000011010111
8	10011	000101	40	00101001	000001101100
9	10100	000100	41	00101010	000001101101
10	00111	0000100	42	00101011	000011011010
11	01000	0000101	43	00101100	000011011011
12	001000	0000111	44	00101101	000001010100
13	000011	00000100	45	00000100	000001010101
14	110100	00000111	46	00000101	000001010110
15	110101	000011000	47	00001010	000001010111
16	101010	0000010111	48	00001011	000001100100
17	101011	0000011000	49	01010010	000001100101
18	0100111	0000001000	50	01010011	000001010010
19	0001100	00001100111	51	01010100	000001010011
20	0001000	00001101000	52	01010101	000000100100
21	0010111	00001101100	53	00100100	000000110111
22	0000011	00000110111	54	00100101	000000111000
23	0000100	00000101000	55	01011000	000000100111
24	0101000	00000010111	56	01011001	000000101000
25	0101011	00000011000	57	01011010	000001011000
26	0010011	000011001010	58	01011011	000001011001
27	0100100	000011001011	59	01001010	000000101011
28	0011000	000011001100	60	01001011	000000101100
29	00000010	000011001101	61	00110010	000001011010
30	00000011	000001101000	62	00110011	000001100110
31	00011010	000001101001	63	00110100	000001100111

Table 2.3: Make-up codes

run	white	black	run	white	black
64	11011	0000001111	960	011010100	0000001110011
128	10010	000011001000	1024	011010101	0000001110100
192	010111	000011001001	1088	011010110	0000001110101
256	0110111	000001011011	1152	011010111	0000001110110
320	00110110	000000110011	1216	011011000	0000001110111
384	00110111	000000110100	1280	011011001	0000001010010
448	01100100	000000110101	1344	011011010	0000001010011
512	01100101	0000001101100	1408	011011011	0000001010100
576	01101000	0000001101101	1472	010011000	0000001010101
640	01100111	0000001001010	1536	010011001	0000001011010
704	011001100	0000001001011	1600	010011010	0000001011011
768	011001101	0000001001100	1664	011000	0000001100100
832	011010010	0000001001101	1728	010011011	0000001100101
896	011010011	0000001110010			

Table 2.4: End of Line code

EOL	code word	00000000001

Error limitation and the EOL code

As well as compressing a bitmap, the encoding process performs one other very useful function, which is that it enables error detection and limitation to take place. Since the recommendation is designed to allow faxes to be sent over normal telephone lines, transmission errors are quite likely to occur at random. Without such a mechanism, any corruption in the encoded data, even if it affected just one bit, might result in the loss of a complete document.

This dire circumstance is a consequence of the fact that Huffman encoded data consists of variable length sequences of bits and contains no redundant information at all. Altering just one bit can have very serious consequences. What is needed is not only a way of detecting when this might have occurred and recovering from it, but also a way of making sure that even undetected errors are limited in their effects. The fax recommendation does this by enforcing three additional rules:

1. Each scan line must begin with a white run length, and thereafter black and white runs must alternate. If the scan line really does start with a black run length then it has to be preceded by a white run length of zero.

2. The end of each scan line is always marked by a unique code that can't ever be interpreted as anything but an end of line. This code sequence is 000000000001, invariably referred to as an EOL. The EOL sequence of at least eleven zeros is unmistakable as no run length ever contains more than six consecutive zero bits.

3. No run length can ever be longer than one scan line.

The effect of these rules, when taken together, is that any corruption in the encoded bitmap introduced by the transmission process can both be easily detected, and its effects can be limited to just one line.

Error detection becomes a simple matter of adding up all the run lengths since the start of any line and watching out for the unique EOL code. If the total of all the run lengths since the line began are not exactly 1728, which is the magic number of dots in one scan line, then we know we have some data corruption. If at any point our accumulated run-lengths exceed 1728, all we have to do to recover is discard the current line and watch the rest of the data for an EOL code, when we can safely begin decoding the bits with the white run at the start of the line.

Encoding the bitmap

We can now get back to the job of developing code to turn a text file into a fax image file. We had reached the stage of having turned a line of characters from the text file into a bit map, and are now in a position to complete the task by encoding the bit map of the original line.

There are three steps in encoding a bit map:

- Counting the bits in each run length.

- Finding the Huffman code corresponding to that run.

- Outputting the code to the fax file.

We now deal with each of these in turn.

Identifying run lengths

The first step is to go through each of the scan lines octet by octet identifying the black and white run lengths. By convention, these are always composed of bits with the values 1 and 0, respectively, as if a screen-based font was used for the digitization step. Note that we refer to a grouping of 8 bits as *octets* rather than bytes when the sequence of bits is arbitrary.

We identify a run length by counting each bit out of the image individually. Undoubtedly, the quickest method of counting bits is to write a routine in assembler to shift each octet left 8 times; the carry flag will reflect the colour of each bit. Regrettably, no high level language in common use gives access to the carry flag, and so alternative methods have to be sought.

The traditional method is feasible using almost any language, and involves getting at the bits in each octet by means of a logical AND. While shifting the octet and ANDing with a constant is one option, the most elegant method is to define `char c=0x80`, and use that to examine each bit from an octet in turn. We do this by ANDing the octet with `c` and then shifting `c` right after each iteration. This means that the value we AND our octet with can double up as the control variable in a loop via `for (c=0x80 ; c>0 ; c>>=1)`. Coding this iteratively would of course execute slightly faster, but we use a loop for convenience.

Like most repetitive tasks, it is fairly easy to write a routine to do this if the underlying data structures have been simplified. In the earlier case of digitizing text, the use of a font which was 8 bits wide made life very simple. In the current case, where we need to identify runs in an image, the simplification comes through the way our control variable `c` is shifted right rather than decremented to zero, coupled with the fact that way the `color` variable is defined. In our code, we use the value `0x00` for white and `0xff` for black.

For each iteration of the loop, this enables us to perform a logical AND of our control variable with the octet we are looking at and compare the result with a logical AND of the control variable and the colour. If the two are the same, then the run length is increased by one; if the two are different, then we are at the start of a new run.

At the start of each new run, we check the current colour; if it was white then we are about to start a black run, but if it was black then we are about to start a white run. The two subroutines we have to deal with these cases are called **nextwhite** and **nextblack** but could just as easily have been called **lastblack** and **lastwhite**.

Before dissecting each octet, we check that it isn't an EOLFLAG sentinel. If it is, we end the scan line via an **endaline** function (see the following subsection).

We also begin and end each octet with an extra white dot. We digitized our text using an 8×16 font, which we now want to transform to a 20×16 font for fax. Adding an additional dot to the inter-character gap handles the first part of this transformation by turning our original 8×16 font into a 10×16 font.

Note that the fragment in Listing 2.3 makes use of the **C** **?:** ternary operator, mostly for cosmetic reasons. I find that replacing **?:** with the equivalent **if ...else** makes the code look more complicated.

```
for (i=0 ; i<image_offset ; i++)
{
        octet=bit_image[i] ;
        if (octet==EOLFLAG) endaline ;
        else
        {
          color ? nextwhite : run_length++ ;
          for (c=0x80 ; c ; c>>=1)
          {
              if ((color&c)==(octet&c)) run_length++ ;
              else color ? nextwhite : nextblack ;
          }
          color ? nextwhite : run_length++ ;
        }
}
```

Listing 2.3:   Code fragment to identify run lengths

Coding run lengths

The second step in encoding the bit map is to turn each run length into its Huffman code equivalent. Before we do this, each run length is doubled, completing the transformation of our 8×16 screen font to a 20×16 fax font.

We also keep track of the remaining number of dots on each line by subtracting each run length from a counter **dots_left** which we initialize to 1728 at the start of each line. As we shall see shortly, this is needed for the **endaline** function that handles the EOLFLAG case.

Listing 2.3 has two separate functions for coding black and white runs. These functions are complementary, so we look in detail at only one of them. (The full program these fragments are taken from can be found in **QAXIFY.C** on the accompanying disk.) The core of the **nextblack** function which ends each white run follows; all it does is use **run_length** as an index to a table of modified Huffman codes pointed at by **whiterun**, and pass the address of the correct entry to the function **shiftin**. As its name implies, this shifts the variable-length bit codes from code table into a coherent sequence.

Note that we check for runs greater than 64 and call **shiftin** with the address of the makeup code table pointed at by **whitemakeup** if needed.

The **nextwhite** function is fundamentally the same; it just looks up the code in the black run length tables and flips **color** to **WHITE**. Both functions return with **run_length** set to 1, as they are called once a dot of a different colour has been located in **bit_image**.

```
run_length *= 2 ;
dots_left -= run_length ;
if (run_length>63)
{
        shiftin (&whitemakeup[run_length/64)-1]) ;
        run_length%=64 ;
}
shiftin (&whiterun[run_length]) ;
color=BLACK ;
run_length=1 ;
```

Listing 2.4: Coding a white run length

Ending a scan line

The **endaline** function is called when we hit the EOLFLAG sentinel. It needs to perform three tasks:

1. Tidy up our incomplete digitization. While all scan lines in a fax have to be exactly 1728 dots wide, the bitmaps resulting from the lines of text read from the original file are of variable length. We have to ensure they conform to the specification, and that the proper white space is added at the end of the line, as shown in Figure 2.1. Therefore, if **color** is **BLACK** we first call **nextwhite** to end the black run. We are then guaranteed to be in the white run at the end of the line. We know the length of this run to the exact dot, as we've been keeping track of each run we have encoded by subtracting its length from **dots_left**. So all we need to do is to set **run_length** to half the number in

```
{
                if (color==BLACK) nextwhite ;
                run_length=(dots_left/2) ;
                nextblack ;
                shiftin (&eol) ;
                color = WHITE ;
                dots_left = 1728;
                run_length = (74/2) ;
}
```

Listing 2.5:   Ending a scan line

dots_left (which takes account of the doubling-up we have just seen) and call **nextblack** to code the last run in the line.

2. The second task is more straightforward. We have to place the end of line code in the output stream, which simply requires a call to **shiftin** with the address of the EOL code word.

3. The third task is to prepare for the next scan line. We do this by setting **color** to white and **dots_left** to 1728. We don't set **run_length** to 0, but we set it to reflect the fact that there is a left-hand margin on the page. A full line of 79 characters uses up only 1580 of the 1728 dots on a scan line, leaving space for (say) two margins of 74 dots on each side. The **endaline** function sets **run_length** to half this. The value of 37 will be doubled up to 74 by **nextblack** when it is encoded.

Constructing an array for the Huffman codes

Before we turn to the **shiftin** function, we need to look in more detail at the possible methods of including the modified Huffman data used in encoding fax images. The fragments shown so far use a run length as an index to an array of integers, and result in constructions such as **whiterun[run_length]**. While this is basically sound, there are complications arising from the fact that Huffman codes aren't really integers at all, but are variable-length bit codes. We know that no code is longer than 12 bits (the length of an EOL) or shorter than 2 bits, so there is no problem with representing the codes as 16-bit short integers. But it is not possible to construct an array whose elements consist simply of integer values based on the bits in Table 2.5.

If you look again at the black run lengths from 0 to 10 in Table 2.5, it is clear that variable-length bit codes can't just be padded out with leading zeros and treated as numbers. For instance, the codes for a black run of 2 (11), a run of 4 (011), a run of 5 (0011) and a run of 7 (00011) are quite different, but all would be represented as integers with the value 3 if all we did were to treat the code as a simple integer.

Table 2.5: Integer representations of the first ten black run-length codes

run	bit code	integer value
0	0000110111	55
1	010	02
2	11	03
3	10	02
4	011	03
5	0011	03
6	0010	02
7	00011	03
8	000101	05
9	000100	04
10	0000100	04

Since the problem arises from the fact that the codes are of variable length, some method of indicating the number of bits in each code is essential. There is more than one way of doing this. One method I've seen used is to insert a single 1 bit at the start of each code, and build an array which starts as in Listing 2.6.

```
short int blackrun [] =
{
        0x437, /* 10000110111 */
        0x00a, /* 1010 */
        0x007, /* 111 */
        0x006, /* 110 */
        0x00b, /* 1011 */
        0x013, /* 10011 */
        0x012, /* 10010 */
        0x023, /* 100011 */
        0x045, /* 1000101 */
        0x044, /* 1000100 */
        0x084, /* 10000100 */
          .    .
          .   ➤
```

Listing 2.6: Flagging the start of the significant bits of Huffman codes

The **shiftin** function would need to isolate each bit in turn and discard all bits up to and including the first 1; the remaining bits in each code are those which would count. This approach doesn't require any extra data storage, but it does require 16 bit shifts for all codes, irrespective of the number of bits they might contain, and the shortest codes will actually take the longest to identify.

Since the whole basis of the Huffman coding technique is that the shortest codes occur most frequently, this method is extremely inefficient in terms of speed.

The preferred solution trades off an increased memory overhead for quicker execution speed. It gets round the problem by including the length of each code in the array as a separate item from its integer representation. This means that each element of our array will consists of the integer representation of the bit code itself together with the number of bits that the code contains. This is handled quite simply using a structure such as that shown in Listing 2.7.

```
struct code
{
    char count ;                /* number of bits in the code  */
    unsigned short int bits ; /* bit code as a 16-bit integer */
} ;
```

Listing 2.7: Structure definition for holding Huffman code values

To avoid shifting the integer to discard the padding bits, it also makes sense to left-justify the bits in each code before representing it as an integer. The array of codes for the black runs constructed according to these conventions begins as in Listing 2.8.

```
struct code blackrun [] =
{
            {10,0x0dc0},     /* 0000110111000000 */
            {3,0x4000},      /* 0100000000000000 */
            {2,0xc000},      /* 1100000000000000 */
            {2,0x8000},      /* 1000000000000000 */
            {3,0x6000},      /* 0110000000000000 */
            {4,0x3000},      /* 0011000000000000 */
            {4,0x2000},      /* 0010000000000000 */
            {5,0x1800},      /* 0001100000000000 */
            {6,0x1400},      /* 0001010000000000 */
            {6,0x1000},      /* 0001000000000000 */
            {7,0x0800},      /* 0000100000000000 */
```

Listing 2.8: Storing the number of bits in the code followed by the value

Shifting in the Huffman codes

The last of the key techniques in our first encoder is the method for combining the various codes from the one-dimensional coding table. This is handled by the **shiftin** function used above, to which we pass the offset of the **code** structure we are interested in. The function makes a local copy both of the count of bits and the padded code word to be shifted. We want to store the coded run-lengths consecutively in a character array at **fax**. The variable **offset** is used to index the octet in this array which is currently being worked on.

The key to this function is to keep track of the number of unused bits in each octet being composed (at **fax[offset]**). We do this using a static variable called **spare** which is initialized to 8 at the start of each new octet. Each time **spare** is decremented to 0, we store the octet we have just finished, re-initialize to 8, and go on to the next octet.

We test successive bits in the code word by ANDing with 0x8000, and shifting the code word left between iterations. If the result was 0, we reset the least significant bit of our target to 0 by ANDing with 0xfe (11111110); otherwise we set the same bit to 1 by ORing with 0x01 (00000001). Finally, we also shift the target octet left between iterations; which means that the least significant bit in each octet is promoted to the most significant bit once spare is decremented to 0.

Listing 2.9 shows the important parts of the **shiftin** function. A full version can be found in **QSHIFTIN.C** on the accompanying disk.

```
count = code->count ;
codebits = code->bits ;
static int spare ;
for (;count;count--)
{
            if (codebits&0x8000) fax[offset]|=1 ;
            else fax[offset]&=0xfe ;
            codebits<<=1 ;
            if (!(--spare))
            {
               ++offset ;
               spare=8 ;
            }
            else fax[offset]<<=1 ;
}
```

Listing 2.9: A portion of the **shiftin** function

We can check to see if the address passed to **shiftin** is the same as the address of the EOL code word structure. The code is slightly different if we are shifting in an EOL code, as we can take the opportunity to end each EOL on a byte boundary. As well as being preferred by some image readers and class 2 fax modems, this also enables each scan line to be written to disk as it is generated (after each call to **endaline**). The memory overhead of the quick encoder can thus be kept extremely small as there is no need to hold an entire image in memory.

To byte-align an EOL, we pad out any spare bits in the current octet at **fax[offset]** simply by shifting left; then we reinitialize **spare** to 8, and place **0x0** followed by **0x01** at **fax[offset]** (Listing 2.10).

```
{
            if (spare!=8)
            {
               if (spare!=1) fax[offset]<<=(spare-1) ;
               offset++ ;
               spare=8 ;
            }
            fax[offset++]=0x0 ;
            fax[offset++]=0x1 ;
}
```

Listing 2.10:   Catering for byte-aligned EOL codes in `shiftin`

The RTC sequence and the end of a fax page

The EOL code has other functions apart from enabling error detection and limitation. An initial EOL is always used to marks the start of a page, and the end of a page is always indicated with six consecutive EOLs in a transmission stream. These six final EOL codes are collectively referred to as the return-to-control (RTC) sequence. Note that the EOL which ends the final line of each page is counted as one of the six in the RTC sequence. The rules that govern the use of the EOL are actually a little more comprehensible if we view the code as marking the start of a line rather than the end. The EOL at the start of a page then makes perfect sense, as it is needed before the first line of a page, while the question of whether the EOL after the final line is included in the RTC doesn't arise.

However, there is one further aspect of the EOL code which arises from a quite definite end of line requirement. The eleven 0 bits at the start of the EOL code at the end of each line can be padded out with more 0s. The T.4 recommendation refers to this as 'fill'. The reason for wanting to add fill bits is to ensure that each line takes at least as long as the minimum recommended transmission time to send. This is so that a receiving fax machine has time to perform various mechanical overheads associated with any line it receives, such as moving its print head and feeding the paper. Obviously, these tasks need to be handled after a line has been received and not before.

The standard minimum time that each line must take when being transmitted is 20 milliseconds (one-fiftieth of a second). This can be negotiated upwards or downwards by two fax machines before they exchange image data. A minimum scan line time of 0 ms is possible, and is used by fax modem software capable of storing data on disk as fast as it is received. However, most fax machines are unlike computer faxes and cannot print lines out as fast as data comes in. Fax communications are essentially half duplex, and data is streamed continuously from the transmitter to the receiver. Under these circumstances, no flow control is possible, and fax machines therefore require a minimum scan line time to work properly.

At a speed of 9600 bits per second, a 20 ms minimum transmission time means that there must be at least 192 bits in each line. A blank line would have to be sent as a white make-up code of 1728 (010011011), followed by a white make-up code of zero (00110101), followed by an EOL (000000000001). This only constitutes 29 bits and would take approximately 3 ms to transmit. Sending this line as it stands may well cause a receiving fax to start falling behind with printing out data. The solution is to pad out the EOL code with at least an extra 163 zero bits before the final 1. This brings the total number of bits taken to send the line up to 192, and the 20 ms minimum line transmission time is thus satisfied.

Makeup Cod	Terminating Code	Fill	RTC					
			EOL 1	EOL 2	EOL 3	EOL 4	EOL 5	EOL 6
White run of 1728	White run of 0							
010011011	00110101	0000000000000000000000000000 0000000000000	00000000 0001	00000000 0001	00000000 0001	00000000 0001	00000000 0001	00000000 0001

Figure 2.4: Last line of data on a page

Notice that fill bits ought not be included between the six EOL codes that comprise the RTC sequence, though they may precede the RTC if it is necessary to pad out the final line. Again, this is quite logical, since the last line is like any other as regards mechanical overheads.

Clearly, a fax image will need to have an RTC added at the end when it is sent, but this is usually done at the time the fax image is sent rather then as part of the coding process. In any event, you should note that a loop such as

```
for (i=6 ; i ; i--) endaline ;
```

would not do the job properly, as the EOL codes in an RTC sequence should not contain any fill and must therefore not be byte-aligned. The proper byte sequence is

```
0x00,0x10,0x01,0x00,0x10,0x01,0x00,0x10,0x01
```

and should always be inserted as such.

Don't be confused by the endpage function in the code on the accompanying disk, as it has nothing to do with the RTC code. Its main purpose is to hold a simple routine which adds a bottom margin at the end of a fax image.

One last point that must be mentioned is that, as well as a minimum transmission time, there is also a maximum transmission time of 5 s. If the transmission time of a line exceeds 5 s then the receiver must disconnect the line and hang up. (We'll see later on that smart fax modems and fax software can use this line stretching feature to avoid transmitter underrun by padding out the end of a line up to the 5 s limit if they haven't another full line to send.)

3

Fax Documents: Advanced Topics

Introduction

This chapter deals with the various additions and extensions to the fax image specification which have been made over the years. Since the basic fax image described in Chapter 2 is the lowest common denominator of the fax world, and constitutes something that any fax machine can understand, none of this chapter is essential reading. If all you want is to write basic fax software, then you may skip this chapter.

But anyone who wants to know how to enhance the capabilities and performance of their fax software will find this chapter very worthwhile. In it, we outline the parameters for generating images of different size and resolution, and explain how different coding methods can be used to create more highly compressed images which will take less time to send.

Optional extensions to the basic T.4 recommendation

We have already mentioned three of these (which are negotiable at transmission time) in Chapter 2: transmission speed and modulation (2400 bps using V.27 ter upwards); image resolution (normal or fine); and minimum line transmission time (which defaults to 20 milliseconds). There are now many more fax options, most of which are also negotiated at the time of transmission through the T.30 session protocol. Those outlined here consist of modifications to the way the fax data is generated and decoded.

The options and extensions fall into two groups:

- The first is fairly straightforward, and consists of additions to the specification designed to enhance the physical capabilities of fax machines. Options such as the ability to transmit documents that aren't 215 mm wide, and the enhancement of resolution to allow more detail to be sent, are not difficult to implement. Some of the techniques have already been seen in various listings, while others appear later in the chapter. Adaptation of the code is left as an exercise for the reader.

● The second group alters the way in which an image is encoded. This changes the underlying logic involved in generating faxes, and has a correspondingly greater effect on any software implementation. We therefore need to develop more detailed code for these options later in the chapter.

Physical extensions

Superfine and inch-based resolutions

The most recent 1993 version of the T.4 recommendation officially extends the possible resolution of fax transmissions to include so-called 'superfine' resolutions.

The dimensions of the new official superfine resolution are double that of fine resolution. The horizontal resolution is 3456 dots across a 215 mm line, with a vertical resolution of 15.4 lines/mm.

The ITU have also recognised inch-based resolutions. These are now rather quaintly supported in terms of dots per 25.4 mm (1 inch = 25.4 mm), which is a convention we shan't follow here.

Normal resolution is given as 200 dpi horizontally × 100 dpi vertically, with fine resolution specified as 200 × 200 dpi and superfine resolution as 400 × 400 dpi. All these resolutions are considered to be equivalent, even though the differences between the inch-based 200 × 100 dpi and the metric 204 × 196 dpi are both outside the normal 1% tolerance, and are also something that will lead to distortion of received images. An additional 300 × 300 dpi resolution specifically for inch-based machines is also added, which has no metric equivalent.

The official definitions of these inch-based resolutions are given in Table 3.1.

Table 3.1:   Official ITU definitions of inch-based resolutions

200 dpi	**1728 dots per 219.46 mm**
300 dpi	**2592 dots per 219.46 mm**
400 dpi	**3456 dots per 219.46 mm**

Don't be misled into thinking that inch-based machines can have scan lines longer than 215 mm. The reason for the odd figure of 219.46 mm occurring in the definition is simply that with a resolution of 200 dpi, 1728 dots will take up

219.46 mm. This is a matter of arithmetic. The standard could equally well have stated that 200 dpi was equivalent to 1693 dots per 215 mm.

The recommendation also states that machines that want to implement the 200×100 dpi normal resolution must also be capable of higher resolutions. Metric-based machines without fine resolution are permissible, while inch-based ones are not.

Larger paper sizes

The 215 mm line is designed to enable both overscanning of A4 pages (which are 210 mm wide) and the sending of common 8.5" wide paper sizes. There are a number of optional extensions to these specifications which allow for larger or smaller paper sizes while keeping the same horizontal resolution. These have to be carefully negotiated, as the width of a fax image is one of the key parameters which must be known for the image to be properly decoded.

There are two larger sizes that have been specifically catered for for many years. The ITU allow a 303 mm line to be encoded as 2432 dots (or 4864 dots in superfine), which enables an A3 page to be faxed. A further option is to allow a 255 mm line to be encoded as 2048 dots (or 4096 dots in superfine), which enables a 250 mm wide B4 page to be sent.

As the original make-up coding table only goes up to 1728 dots, the ITU extended the range to allow larger widths to be encoded. However, there is only one extended set of make-up codes, which has to be used for both black and white run-lengths (Table 3.2). The exact colour has to be inferred from either the terminating code which follows, or from the colour of the preceding run-length.

Table 3.2: Code words for long run-lengths

run	code
1792	00000001000
1856	00000001100
1920	00000001101
1984	000000010010
2048	000000010011
2112	000000010100
2176	000000010101
2240	000000010110
2304	000000010111
2368	000000011100
2432	000000011101
2496	000000011110
2560	000000011111

The reason for the shorter run-length codes being different for each colour wasn't to minimise errors, but to reflect the different frequencies with which they occurred.

Dealing with very long run-lengths

Because there are no make-up codes for run-lengths longer than 2560 dots, it would seem that the longest possible run length would be one coded with a 2560 make-up code followed by a 64 dot terminating coded, which totals 2624 dots. This is clearly insufficient to deal with a superfine A3 page with scan lines containing 4864 dots. The method for dealing with longer runs is to issue multiple make-up codes for 2560 dots until the remaining run-length can be encoded in the normal way. Only the make-up code for 2560 can be used in this way.

Smaller paper sizes

The recommendation was later extended to enable sizes smaller than A4 to be sent. The width of an A5 page is half the length of A4 at 148 mm, and is catered for by allowing a 151 mm line to be encoded as 1216 dots, while the width of an A6 page is half the width of an A4 page at 105 mm, and can be sent by encoding a 107 mm line as 864 dots.

According to some sources, the development of A5 and A6 fax machines was part of a Japanese effort to sell a fax into every home. Since the smaller size fax machines cost about the same as normal sized devices, the miniature versions never really caught on. However, the restricted sizes are still catered for in the extended fax specification.

Logical extensions

Two dimensional coding

The coding scheme outlined so far is generally referred to as one-dimensional Modified Huffman coding (MH). The term 'one-dimensional' (1-D) is used because the dots in the bit image are related only to adjacent dots in the same line, not to the dots in the lines above or below their own.

Two-dimensional (2-D) Modified Read coding (MR) is an optional alternative method of compressing fax data than typically results in images that are between 20% and 40% smaller. The term 'two-dimensional' is used because, as well as the dots being related in the horizontal dimension, the lines are also coded in the vertical dimension, with every run on each line being compared to the runs on the same portion of the previous line. This type of coding relies on the fact that where many of the lines in a particular bitmap are to be related in some way to the previous lines, it is often going to be quicker to simply describe a line in terms of any differences that there might be between it and the previous one.

During the period that ITU Study Group VIII were debating the form the fax specification should take in the late 1970s, the American and European delegations were pushing for the 1-D modified Huffman coding scheme to be adopted. This was suggested by Plessey in 1976. The Japanese delegation was pushing 2-D encoding. A compromise was eventually reached, making the one-dimensional coding the standard, with a modified version of the two-dimensional scheme included as an optional extension.

Apart from the politics involved in this decision, there is an excellent technical reason for 2-D coding being implemented as an extension. A genuine 2-D coding method would run into the same problems with transmission errors and line corruption that led to the need for error detection and limitation methods being introduced via the EOL code. Errors in one line would propagate throughout the rest of the image as each new line is coded in relation to the previous one. The following rules govern the interworking of the two coding schemes:

- The first line of the bitmap must always be coded one-dimensionally, and all lines, no matter how they are coded, must end with an EOL.

- At least one line which has been coded one-dimensionally is transmitted every K lines, where K is a variable parameter set to 2 for normal resolution and 4 for fine resolution. This means that no more that one line in every two can be coded two-dimensionally for normal resolution, and no more than three lines in four for fine resolution, enabling the effects of transmission error to be limited to a few lines.

- When 2-D coding is used, each EOL code is always followed by a tag bit, set to 1 if the next line is to be coded one-dimensionally, but reset to 0 if the line is to be 2-D. This applies to the EOL before the first line, which is thus sent as EOL + 1. The six EOL codes in the RTC sequence are also all sent with the extra tag bit.

Implementing two-dimensional coding

Two-dimensional (2-D) coding is not as difficult to understand as it is either to explain or to read about. Conceptually, you need to visualise the 1728 dots of the previous line of the bitmap, known as the *reference line*, as being placed directly above the 1728 dots of the current line, known as the *coding line*. An imaginary cursor is placed at the start of the line being coded.

Beginning at the cursor position, the next pair of run-lengths starting are examined. It is usual to refer to these by their offsets in the coding line as a^0a^1 and a^1a^2. The convention is that the first offset is the position of the first dot in the run-length, while the second is the position of the first dot of the succeeding run-length.

The reference line is then searched from the corresponding cursor position, and the location of the next run-length in the reference line which is of the opposite colour to a^0a^1 in the coding line is found. If we happen to begin the search in the middle of a run of the opposite colour, then we ignore the rest of that run and look for the next one of the opposite colour once we reach the end of it. It is usual to refer to the correct run in the reference line by its offsets as b^1b^2.

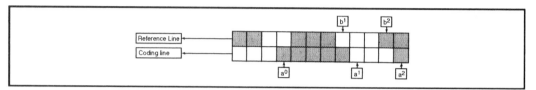

Figure 3.1:   Example of how reference points are set for 2-D Read encoding

The positions of b^1b^2 on the reference line in relation to the positions of a^0a^1 and a^1a^2 on the coding line determine the way in which we proceed with encoding. There are three possibilities to be considered in turn:

1. *Pass mode* is used whenever b^2 falls completely to the left of a^1 ($b^2 < a^1$). In this case, the difference between the coding and the previous line can be ignored. We don't need to bother with a run b^1b^2 on the reference line if it does not overlap the start of a run length a^1a^2 of the same colour on the coding line. The most obvious instance of this is when we are receiving lines of text and we come to the end of one line of text. We then begin encoding the blank lines (perhaps with the odd descender) that separate the lines on the printed page, and the little black runs on the reference line that were part of the letters can be ignored. Figure 3.2 shows an example of pass mode.

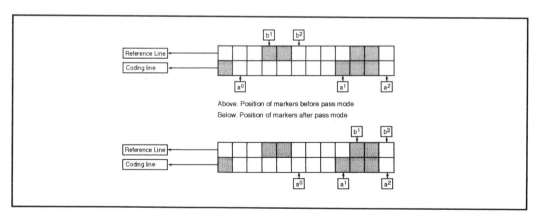

Figure 3.2:   Example of pass mode

Wherever a corresponding run length on the reference line can be ignored, the four bit code 0001 is inserted in the coding line. The imaginary cursor is placed on the dot in the coding line immediately beneath b^2 and the encoding continues.

2. *Vertical mode* is used when there is an overlap between a run b^1b^2 on the reference line and the start of a run length a^1a^2 of the same colour on the coding line, provided that the distance a^1b^1 is less than or equal to three dots in either direction. An obvious instance of this is when a sloping line is being encoded, as the start of the slope on each successive line in the bitmap will be only a few dots away from the start of the slope on the preceding line. This is likely to be the most frequent type of code used in a line encoded by the 2-D method. Figure 3.3 shows an example of vertical mode

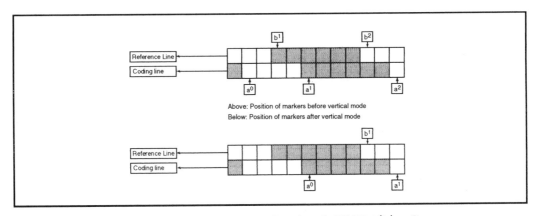

Figure 3.3:   Example of vertical mode code 000011 ($a^1b^1 = -2$)

To use vertical mode, a single variable length code taken from Table 3.3 is inserted in the coding line. The imaginary cursor is placed on what was the a^1 position, which is the start of the overlapping run-length of the same colour as b^1b^2 in the reference line, and the encoding continues.

Table 3.3:   Vertical mode code words

a^1b^1	code
0	1
−1	011
−2	000011
−3	0000011
+1	010
+2	000010
+3	0000010

3. *Horizontal mode* is used where there is an overlap between a run b^1b^2 on the reference line and the start of a run length a^1a^2 of the same colour on the coding line, but the distance a^1b^1 is more than 3. As the name implies, horizontal mode is essentially a temporary return to the 1-D encoding method.

Horizontal mode is flagged by the three bit prefix 001 inserted in the coding line, to be followed by the 1-D codes for the run-lengths a^0a^1 and a^1a^2.

Predictably, the imaginary cursor is placed on what was the a^2 position, and the encoding continues. An example of horizontal code is shown in Figure 3.4.

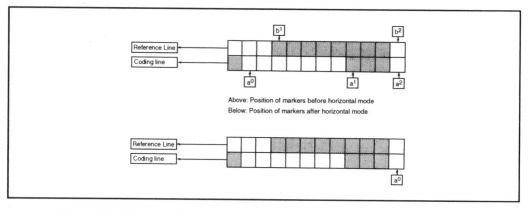

Figure 3.4: Example of horizontal mode (code 001 + run lengths 7 white and 3 black)

The rules for the starting and ending dots of each line in the bitmap are quite simple. The line starts with the imaginary cursor placed on an imaginary white dot just before the start of the line, and the first a^0a^1 run-length of the coding line is therefore always reduced by one. Each line (both coding and reference) is deemed to end with an imaginary dot of an opposite colour to the last real dot in the line.

Code fragments

The series of fragments presented in the next section are taken from files in the ENCODE directory on the accompanying disk, and show how to digitize an ordinary text file and turn it into a 2-D fax image. While there are some similarities with the techniques used in generating 1-D images, there are substantial differences.

Font width

The need for a 2-D coder to have access to a fully digitised version of the previous scan line also means that using an 8×16 font, adding a extra element on each side to get 10×16 and then doubling all the run lengths to get 20×16, will not work this time. That technique makes generating a full scan line impossible.

It is useful to maintain the convenience of a font table exactly 8 bits wide. While it is possible to use the 8×16 font and stretch it 'on the fly', doubling individual bits isn't a trivial operation. Therefore having a ready-made 16×16 font makes more sense, and this is what we do in the code fragments.

Handling variable length lines of text

If you refer back to the previous listings, you'll see that as soon as the digitiser we used previously hit the end of a line, it stuffed an end-of-line sentinel in the bit map; and as soon as the encoder came across an EOL sentinel, it worked out the remaining number of dots in the line and encoded that as a white run before adding an EOL code word. The advantage of this method is that for normal text, it is extremely quick and very economical.

Unfortunately, this technique cannot be used to generate two-dimensionally coded images, because 2-D coding requires each scan line to be compared with the previous one. Therefore, every line must be coded in its entirety, and stopping at the last printable character is not a viable option. All the white space at the end of a line (as well as all the margins) must be filled in the digitised bitmap.

Digitising the text

The fragment in Listing 3.1 shows how this has to be done if a full bitmap is to be generated. Note that we start by initialising the entire bitmap for one text line to white space (0 bits). There are some magic numbers here, notably 1728 for the bits in each scan line and 16 for the font height (giving 16 scan lines per text line). We use the formula 1728/8 rather than the integer 216 as our count of 8-bit bytes to make it easier to see how the width of a fax is incorporated in the code.

Instead of using an EOL sentinel, we keep track of the pointer **image_offset** to the start of the bitmap for each scan line, and when we reach the end of the line we can simply encode the next scan line beginning 216 bytes further on. In this way, we ensure each scan line is padded out with white space at the end. Similarly, we start the line with a suitable left-hand margin by adjusting our copy of the pointer (**temp_offset**) as needed. The next fragment uses **MARGIN** to achieve this, which can be defined to any suitable value:

This fragment contains the code needed to expand the original 16×16 font to 20×16. It is perhaps unfortunate that the best size for our font isn't an exact number of bits wide, but that can't be helped. The expansion to 20 bits wide is handled in the fragment above not by adding bits to each side of every character as it is digitised, but by stretching out every alternate character, as tested by **if (i&1)**, to occupy three octets rather than two.

If you don't think well in Boolean terms, this is the most difficult part of the code, so it is probably worth explaining it in more detail. As with our method for handling the margins at the start and the end of the line, the method depends upon the fact that our storage at **bit_image** is pre-initialized to white space.

```
for (i=0 ; i<((1728/8)*16) ; i++) bit_image[i]=0 ;
image_offset=0 ;
fgets (text_line,79,infile) ;
for (row=0 ; row<16 ; row++)
{
    temp_offset=image_offset+(MARGIN/8) ;
    for (i=0 ; i<strlen(text_line) ; i++) if (text_line[i]>31)
    {
            if (i&1)
            {
            bit_image[temp_offset++]=fonts[text_line[i]][row*2] ;
            bit_image[temp_offset++]=fonts[text_line[i]][(row*2)+1] ;
            }
            else
            {
              c=fonts[text_line[i]][row*2] ;
              bit_image[temp_offset++]|=(c>>4) ;
              bit_image[temp_offset]|=(c<<4) ;
              c=fonts[text_line[i]][(row*2)+1] ;
              bit_image[temp_offset++]|=(c>>4) ;
              bit_image[temp_offset++]|=(c<<4) ;
            }
    }
image_offset+=(1728/8)
}
```

Listing 3.1: Digitising a line from a text file using an 16 × 16 font table to generate a full bitmap

We work with a copy of the font data for both speed and readability. We shift the top four bits of the first half of the font into the bottom four bits of the first of our three data bytes. Then we shift the bottom four bits of the first half of the font data into the second of our three data bytes. Next we pick up the next half of the font data. The top four bits of this go into the bottom four bits of the second data byte, with the bottom four bits going into the top half of the third data byte.

The bottom half of the third data byte, like the top half of the first, is left as white space. Since both the previous character and the next one are digitised as two bytes without any additional padding, we are coding two character widths into five bytes; which gives us our 20 bits for one character.

Picking the right encoder
Once we have a complete bitmap of a line of text, we loop through it once for each scan line. As well as the usual start-of-line housekeeping tasks to be performed

at this stage (including re-initialising necessary variables and ensuring that every line begins with an EOL code), if we want to be able to handle 2-D encoding we must also include code that copies the previous scan line so that its contents will still be available as a reference line.

Picking the right encoder for each of the scan lines is simply a matter of following the specification. We maintain a variable **k** which we initialise to the value specified for the maximum number of two-dimensionally coded lines. A modification is made to the **shiftin** function shown in Listing 3.2. At the point where we make a check to byte-align an EOL code, **shiftin** decrements **k** and resets it to the default value once it reaches 0. All we have to do in this fragment is inspect **k**. If it is at the default, we code the line one-dimensionally with **mhcoder**; for all other values of k we code the line two-dimensionally using **mrcoder**.

It's also worth noting that, while the quick encoder handled blank scan lines at the top and bottom by inserting extra EOL sentinels in the digitised bitmap, the full encoder needs a special routine for this. When using 2-D coding, the way in which a blank line is handled depends upon the contents of the previous line.

```
for (i=0 ; i<((1728/8)*16) ; i+=(1728/8) )
{
            for (i=0 ; i<(1728/8) ;i++) lastline[i]=thisline[i] ;
            shiftin (&eol) ;
            dots_left = 1728;
            run_length = 0 ;
            thisline=&(bit_image[i]) ;
            if (k!=default_k) mrcoder ;
            else mhcoder ;
```

Listing 3.2: Looping through a bitmap for each line and picking the right encoder

Coding 1-D lines

The code for 1-D lines is simpler than the corresponding version used in our quick encoder, as we don't have any partial lines to consider. We don't need either to worry about EOLFLAG sentinels or keep track of the bits left in each line. The use of a 16×16 font also means that we don't need to double up our run-lengths. As before, we use the value **0x00** for white and **0xff** for black. The **nextwhite** and **nextblack** routines, with their associated tables of code words, are almost identical to the quick encoder versions, so we shan't duplicate them here.

The line after the body of the loop ensures we include the final run length of each scan line.

```
color = WHITE ;
for (i=0 ; i<1728/8 ; i++)
{
            octet=thisline[i] ;
            for (c=0x80 ; c ; c>>=1)
            {
                if ((color&c)==(octet&c)) run_length++ ;
                else color ? nextwhite (T4) : nextblack (T4) ;
            }
}
if (run_length) color ? nextwhite (T4) : nextblack (T4) ;
```

Listing 3.3: Coding the 1-D lines in a 2-D coding

Coding 2-D lines

Two-dimensional coding requires extra entries in our data tables for the various 2-D code words. The **HUFF.DAT** file in the ENCODE directory includes these additions after the same 1-D code words as the quick encoder used. The same technique of including the number of bits in the code word followed by the code itself as a left-justified short integer is applied.

While the 2-D coding method certainly looks more complicated than 1-D coding, it is actually a fairly mechanical process. The key is the accurate identification of the markers a^0, a^1, a^2, b^1 and b^2.

As implied in Listing 3.2, the 2-D coder has access to both the current and previous scan line in the two arrays pointed at by **thisline** and **lastline**. These are all arrays of bits, while the markers a^0, a^1, a^2, b^1 and b^2 are all stored as integers. Unsurprisingly, the Boolean operations required to access individual dots in each line are likely to cause the most trouble with this listing, so we look in detail at how this is done in the following fragment. Its purpose is to identify the colour of a particular bit in the current line identified by the integer a^0.

```
i=(a0/8) ;
this_bit=(0x80>>(a0%8)) ;
if (thisline[i]&this_bit) color=BLACK ; else color=WHITE ;
```

Listing 3.4: Isolating the colour of one dot in a line

The integer a^0 is first divided by 8. The quotient from this operation is used as an index to the byte which contains the bit we want, and the remainder is used to shift the mask 10000000 to the right before storing it in **thisbit**. Finally, the Boolean expression **thisline[i]&this_bit** isolates the required bit. It is true if the bit is 1 (black) or false if the bit is 0 (white).

Listing 3.5 contains virtually a complete encoder for a 2-D line. Before the coding starts, the marker a^0 is set to the imaginary position just before the start of the

Listing 3.5: Coding a 2-D line

```
a0=-1 ;
color=ref_color=WHITE ;
while (a0<1728)                           /* main loop */
{
          if (a0!=-1)
          {
            i=(a0/8) ;                    /* first find out colour */
            this_bit=(0x80>>(a0%8)) ;
            if (thisline[i]&this_bit) color=BLACK; else color=WHITE;
            if (lastline[i]&this_bit) ref_color=BLACK;
            else ref_color=WHITE;
          }
          for (a1=a0+1;a1<1728;a1++)          /* next find out a1 */
          {
            i=(a1/8) ;
            this_bit=(0x80>>(a1%8)) ;
            if ((thisline[i]&this_bit)!=(color&this_bit)) break ;
          }
          for (b1=a0+1;b1<1728;b1++)          /* next find out b1 */
          {
            i=(b1/8) ;
            this_bit=(0x80>>(b1%8)) ;
            if ((lastline[i]&this_bit)!=(ref_color&this_bit)) break ;
          }
          if (ref_color!=color) for (b1++;b1<1728;b1++)
          {
            i=(b1/8) ;                    /* b1 must be opposite
                                              colour */
            this_bit=(0x80>>(b1%8)) ;
            if ((lastline[i]&this_bit)!=(color&this_bit)) break ;
          }
          for (b2=b1+1;b2<1728;b2++)          /* next find out b2 */
          {
            i=(b2/8) ;
            this_bit=(0x80>>(b2%8)) ;
            if ((lastline[i]&this_bit)==(color&this_bit)) break ;
          }
          if (b2<a1)                      /* check for pass mode */
          {
            shiftin (&passmode,T4) ;
            a0=b2 ;
            continue ;
          }
          if (((b1-a1)<=3)&&((a1-b1)<=3))    /* check for vertical mode */
          {
            shiftin (&vertical_mode[b1-a1+3],T4) ;
            a0=a1 ;
            continue ;
          }
          for (a2=a1+1;a2<1728;a2++)          /* must be in horizontal
                                                 mode */
          {
            i=(a2/8) ;                    /* so find out a2 */
            this_bit=(0x80>>(a2%8)) ;
            if ((thisline[i]&this_bit)==(color&this_bit)) break ;
          }
          shiftin (&horizontal_mode,T4) ;
          run_length=(a1-a0) ;
          if (a0==(-1)) run_length-- ;
          color ? nextwhite (T4) : nextblack (T4) ;
          run_length=(a2-a1) ;
          color ? nextwhite (T4) : nextblack (T4) ;
          a0=a2 ;
}
```

line. Both the colour of this imaginary dot stored in `color` and that of the same bit on the reference line stored in `ref_color` are set to be WHITE. The normal coding definitions of WHITE as 0 and BLACK as 0xff apply.

The encoder is basically a loop which relies for its functioning on the correct setting for the marker a^0; once this reaches 1728 the loop terminates. The first task of the encoder is the identification of both the colour of the a^0 bit on the current line, and the colour of the corresponding bit on the reference line. Next, the value of the first element of the opposite colour on the same line is saved as a^1, and the first changing element of the opposite colour on the reference line is saved as b^1. This last step can require two attempts, as the first changing element on the reference line isn't always of the opposite colour to a^0. Finally, the next changing element after b^1 on the reference line is saved as b^2.

As pointed out earlier, once the key values have been identified, 2-D coding is quite simple:

- If b^2 is less than a^0, we shift in the code for pass mode, set a^0 to the value of b^2 and start the loop again.

- If the difference between b^1 and a^1 is no greater than 3, we shift in the correct code for vertical mode. This is taken from an array of the possible codes, indexed by the expression b^1-a^1+3. We then set a0 to the value of a^1 and start the loop again.

- If neither pass mode nor vertical mode applies, we shift in the code for horizontal mode, then identify the next changing element after a^1 on the current line as a^2, and (using the colour of a^0) shift in the normal T4 code word for the run a^0a^1, followed by the code word for the run a^1a^2 of the opposite colour. After setting a^0 to the value of a^2 we start the loop again.

Optional extensions to 2-D coding

The 2-D coding scheme has been designed to allow for up to eight further extensions to the coding modes available if neither pass, horizontal nor vertical modes are adequate. On a 2-D coded line these optional extensions are flagged with the bit sequence 0000001xxx, where xxx is a three-bit code identifying the mode used. Unfortunately, this sequence can't be used on a line which happens to be coded one-dimensionally, as it already occurs no less than 65 times in the existing 1-D code table. If a line is being coded one-dimensionally, the flag sequence 000000001xxx is used instead.

However, this doesn't overcome all potential problems, as if the flag sequence leading in to one of the extensions were to be used immediately after a code such as a white run-length of 3 (1000), this would lead to the sequence 000000000001 being introduced into the bitstream, which is the same as the EOL code. Because the EOL code is used for error detection, it has to be unique. Even if we knew there were no errors, we would be able to tell from the context that a particular occurrence of the sequence 000000000001 was not an EOL, but the whole point of using a unique EOL code for error recovery is that it must be usable where the context doesn't make sense. The recommendation thus states that it is not permissible to switch into any extension immediately following any of the 12 code words ending in 000.

In fact, only one extension to 2-D coding is described, which uses the bit pattern 111 for the xxx bits listed above, and which is for uncompressed mode. All other bit assignments are unspecified, and the ITU say that they are reserved for further study.

Uncompressed mode

Uncompressed mode is useful because the Huffman encoding method used for compression in 2-D horizontal mode and all 1-D coded lines can sometimes be worse than no compression at all. This is because a number of the standard 1-D run-length codes are actually longer than the runs with which they are associated. Though the clearest examples of this are the zero run-lengths (00110101 for white and 0000110111 for black), these particular instances should be regarded as coding overheads as they only ever occur at the start of a line or after a very long run that happens to be an exact multiple of 64.

The situation is different when we consider an example such as the code for a white run-length of one (000111), which is six times longer than the run it encodes. If a sequence alternates this white run with a black run of one (010) the worst possible encoding scenarios could result.

A line consisting entirely of alternating black and white dots would actually take 7776 bits to encode the original 1728 dots if this were done in 1-D mode. If the line followed an all-white one and was being coded two-dimensionally, the situation is worsened, as each pair of dots would have to be coded with a horizontal mode prefix of 001. In this case, the original 1728 dots would expand to 10368 bits. If this line were followed by a blank all-white line, this may well be coded entirely as a series of pass mode codes and would take up another 6912 bits.

Though this seems to be a disaster, we must remember that it is true of every compression scheme that some data set can be found which will expand when we attempt to compress it. A pre-defined code table just happens to be a particularly easy scheme in which to find a loophole. While the example just given is rather extreme, it does illustrate the point quite dramatically. A glance at the 1-D encoding table shows that as well as black and white runs of 0 and 1, white runs of 2 and 3 are also affected. In terms of pairs of run-lengths, any pair starting with a white run of 5 or less followed by a black run of 1 is going to be expanded rather than compressed.

Using uncompressed mode

Uncompressed mode is assigned the extension pattern 111. This means that entry to uncompressed mode on a line being coded two-dimensionally is flagged with the sequence 000000111, and on a line being coded one-dimensionally the flag sequence 000000001111 is used instead. Following entry to uncompressed mode, Table 3.4 defines how dot patterns are be encoded.

Table 3.4: Uncompressed mode code words

Dot pattern	Uncompressed Code
1	1
01	01
001	001
0001	0001
00001	00001
00000	000001

The first line in the table shows that black dots (1 bits) are used directly in uncompressed mode. All the remaining dot patterns (except the last) are short white run-lengths followed by a black run-length of 1. The longest of these pairs is the pattern 00001, which would be coded in compressed mode as 1011010. Using uncompressed mode has saved us 2 bits, but the final pattern seems a little odd as it apparently allows us to code a white run of 5 in 6 bits. To modify a bit of jargon from Chapter 6, uncompressed mode involves 1 bit stuffing after every sequence of 5 consecutive 0 bits. Decoding uncompressed mode involves removing any 1 bits that occur after any sequence of five consecutive 0 bits.

The reason for this is quite easy to work out. We've already seen there's a minimum 9 bit overhead involved in entering compressed mode, and we'll see shortly that there's a 7 bit overhead in leaving it. Clearly, we wouldn't want to enter uncompressed mode unless there was a substantial penalty being incurred by continuing with encoding what is presumably a highly complex bit map. However, it is also true that once we've entered uncompressed mode, we ought to be equally reluctant to leave it just because we hit a couple of short patterns that it couldn't handle.

If we only had access to the first five codes in the uncompressed mode table, black runs would present no problems, but as soon as we hit a white run of more than 4 dots, we'd have to incur a 16 bit penalty arising from the overhead of leaving or entering uncompressed mode. The point of the last entry in the table is that, even though we are apparently 1 bit worse off by having to code a run of 5 dots as 6 bits, it could enable us to avoid the larger cost of having to exit uncompressed mode and then re-enter it.

Leaving uncompressed mode
Uncompressed mode is ended with the code sequence 0000001 which must always be followed by a tag bit indicating the colour of the next run-length. Sensibly enough, a tag bit of 0 means that the next run is white, while a tag bit of 1 means it is black.

If you want to leave uncompressed mode following a white run length of 4 or less, you may insert the uncompressed run at the start of the code word. This gives the table of codes which cover all the methods of leaving uncompressed mode, with x indicating a tag bit for the colour of the next run (Table 3.5).

Table 3.5: Code words for leaving uncompressed mode

Dot pattern	Exit Code
any	0000001x
0	00000001x
00	000000001x
000	0000000001x
0000	00000000001x

Problems with uncompressed mode

Uncompressed mode has something of a reputation as being problematic, and the specification is less precise than it could be. Clearly, one important difference between a bit map coded using the option for uncompressed mode and one coded without it is that the former could result in many possibilities for equally efficient bitmaps while the latter can only result in one most efficient case. This is because entry into uncompressed mode and exit from it is up to the person writing the software that implements the scheme.

Some bit patterns, such as black run lengths of 2 and white run lengths of 4, come out the same whether uncompressed mode is used or not. Where these occur immediately before or after a sequence which could definitely be shortened by using uncompressed mode, it makes no difference in which mode they are coded. A sequence such as 11000010101010101010101010101000011 is a good example of this indeterminacy. It can be argued that an algorithm which has multiple possible outcomes is one that has not been specified properly in the first place.

Another more serious ambiguity in the recommendation is that it is silent as to what should happen if the last dot on a line is one that is part of an uncompressed mode sequence. Obviously, the exit pattern 0000001x should precede the EOL code, but the value of x is problematic. Since the next row is going to begin with a white run, one could argue that x should take the value 0. On the other hand, it could also be argued that since a run-length has just ended, x should set to match the colour opposite to that of the last dot on the row, whatever that might be. Since it makes no difference, as the bit should be ignored anyway, it is possible to write software that works either way. However, if it really doesn't matter, then the specification ought to have said so. The ambiguity means it is possible for an otherwise valid decoder to fail on certain bitmaps simply because the coder was working on an equally valid but incompatible interpretation of the same specification.

T.6 coding

The ITU T.6 recommendation defines the method used to transmit facsimile data using group 4 fax machines. It can be described fairly simply once we have understood the basics of how T.4 works, as the specification is derived from the group 3 standard.

T.6 images are assembled using the 2-D coding method specified as an option in T.4. However, none of the characteristics listed on page 47 apply. There are no 1-D coded lines, and lines are not separated by EOL codes and tag bits – they simply run into each other. No padding or fill bits within or between lines are allowed either. As all lines have to be coded two-dimensionally, the first line is coded by reference to an imaginary all-white line. The T.4 coding tables are still used, but only for those portions of lines that need to be coded in horizontal mode. The T.6 coding scheme is sometimes known as 2-D Modified Modified Read (MMR) as opposed to the 2-D Modified Read used in T.4.

The uncompressed mode extension to 2-D coding is permissible under T.6, and is implemented in almost exactly the same way as in T.4 . The sole difference is that, while it is obviously not possible on a 2-D scheme for a run length to cross a line boundary, it is permissible for those dots at the end of one line and the start of the next to be taken as a continuous sequence when uncompressed mode is being used.

Each block of compressed data is terminated with a special EOFB (End Of Facsimile Block) code, and consists of the 24- bit pattern 000000000001000000000001. This may be followed (not preceded) by padding bits to ensure that it ends on byte or frame boundaries.

As observed earlier, a purely 2-D coding method is easily disrupted by telephone line noise, and coding without any EOL synchronisation points makes this method even more susceptible to transmission errors. That this isn't considered to be a problem is due the fact that group 4 fax was not originally designed for use with normal telephone lines, but was intended for use with ISDN channels.

ISDN (Integrated Services Digital Network) connections are made using special adapters, but the humble fax modem can only use the *PSTN* (Public Switched Telephone Network). By comparison with ISDN lines, which are fully digital and capable of carrying error-free data at 64K bits/sec, the analogue PSTN used to appear slow, unreliable and noisy. However, telephone lines are becoming more reliable, and error-correcting fax machines capable of guaranteeing perfect transmission are now widely available. Given this combination of circumstances, the use of T.6 coding on normal telephone lines is a practical proposition.

Current versions of the ITU recommendations explicitly provide for the use of the T.6 coding method, and it is now supported as a recognised option for group 3 machines, with its own place in the T.30 negotiating protocol. However, the T.4 recommendation states that it can only be used together with the optional facsimile error correction mode. In view of the fact that the naked use of T.6 compression is so easily disrupted if a telephone line is only slightly noisy, the enforced use of error correcting mode (*ECM*) seems to be an eminently sensible precaution.

Though a portion of the ECM specification can be found in the T.4 recommendation, it can't be properly understood without some knowledge of the T.30 fax session protocol. Since whether ECM is being used makes no difference to the coding of a file, details of its workings are deferred to Chapter 7.

4

Storing Fax Images

Introduction

This chapter is about the TIFF standard for storing fax images. It is different from all the other standards discussed in this book. While a fax that ignores ITU recommendations won't communicate with another fax, and an application that disregards the TR-29.2 standards won't be able to talk to a fax modem, there is no obvious penalty for a fax application that chooses not to follow the TIFF specification.

Why have a standard fax file format ?

We have seen that the ITU specify in exact detail the ways in which a fax image can be generated, but there is no equivalent recommendation governing how an image should be stored as a file..This is not really surprising, as the whole fax enterprise, and the T.4 recommendation in particular, can be seen as being a transport mechanism, performing the same function the old telegraph system. It is designed to transport the image of a message from one sheet of paper and reproduce it on another. Arguably, the way the image is stored should be as uninteresting to a user of a fax machine today as the exact Morse code representation of a message was to anyone sending a telegram.

Part of the reason why fax modems are so useful is because they represent the convergence of two new technologies, fax and PCs, that both came to sudden maturity in the 1980s. A fax sent directly from a computer through a fax modem may never previously have appeared on paper before being printed out by a receiving fax machine. Furthermore, a fax received by a computer might never have to be printed out. It can be displayed on a screen, or incorporated into a word processor (either as an image or via an OCR as text), or retransmitted to another fax machine or computer instead. It can also be filed digitally on disk with other computer files and documents.

A standard method for storing fax images as computer files would therefore seem to be a good idea. It would enable any fax received by any modem or software, on any computer, to be retrieved and used by any other computer or any other software, in the same way as ASCII represents an almost universal medium of information exchange for English text.

Problems with saving fax images directly

The simplest storage method is dumping a raw fax image on disk. Unfortunately, this isn't really adequate for the following reasons:

- An application reading the file would have no way of knowing what resolution was used, whether the fax was of standard width, which of the T.4 coding methods was used, and so on. If you only permit normal resolution faxes using 1-D T.4 coding you might be able to get away with a simple image dump; but once you add support for fine resolution and 2-D coding you have four possible formats instead of just one. If you add T.6 coding and the recent T.30 additions for superfine and inch-based resolutions, you have at least 32 formats. Each new option doubles the number of possible formats a fax image could take.

- Most faxes consist of more that one page. Until the fax is decoded, the file is basically an unstructured bit stream which hasn't any clear page breaks. Dumping images to disk without any means of separating pages wouldn't be a good idea. An end-of-page marker (such as an RTC code) inserted between pages isn't by itself sufficient either, as the resolution of a fax could change from one page to the next at the request of the transmitter.

It might seem that a solution to these problems could be to save each page of a fax in a different file. A portion of the filename could be used to sequence the pages, with either a separate portion of the filename or possibly the location somehow specifying the image format. This creates at least three further problems:

- The one-file-per-page method is unwieldy and complicates computer fax management in much the same way as storing separate pages of a document in different files would complicate word processing.

- The method of using a file name or location to identify the order of the pages and the type of image is highly unsatisfactory. Simple file operations such as moving a file or renaming it are liable to make an image impossible to decode. Transporting images between different systems also becomes highly problematic.

- The overhead on closing and opening multiple files is much greater than the overhead involved in simply reading and writing within a single file, which is quite an important consideration when a series of images is being read in real time for transmission or written on reception.

Encapsulating images in TIFF files

The obvious solution to all these problems is to add a header to each fax image, conveying both the details of the options used to encode the image and a pointer to where the next image in the fax can be found. This main standard for such storage of fax images is the Class F subset of TIFF.

TIFF (Tagged Image File Format) can be used to store a wide variety of image types, including colour and grey-scale. It is also an extensible format, which lends itself very easily to customisation and adaptation to new technologies.

The reasons why I recommend that all fax applications software support TIFF as a storage format, are as follows:

- TIFF recognises T.4 compression algorithms and thus permits the storage of fax images without the need for further processing. No other publicly specified file format can do this. The benefit of storing a fax image in native form is that it removes the onus for interpreting and assembling the fax data from the communications portion of a fax application, where it would have to be handled in real-time.

- As we'll see shortly, the addition of a small 114-byte file header and image description is all that is required to turn a single page raw fax file into one that any TIFF reader would recognise – this is a very small storage overhead.

- TIFF includes a method for keeping more than one image in a single file, which simplifies the storage of faxes consisting of multiple pages. It is even possible to store the same image in a number of different ways, so that both 1-D Huffman and 2-D Read versions of the same image can be stored in the same file. The decision as to which will be used can then be deferred until transmission, when it is known what options the receiver supports. This makes it possible to conduct an optimised fax session without overburdening a computer with the substantial (and sometimes impossible) overheads of handling complex image compression 'on the fly'.

- TIFF is machine independent; it makes very few assumptions about the type of hardware on which an image will be displayed or printed. This can be contrasted with other formats such as PCX (and its multi-image extension to DCX) which were primarily intended for use with IBM screen displays.

- Options already exist in the TIFF specification which cater for existing fax extensions such as different coding algorithms, resolutions and alternative document widths. This flexibility means that, as and when fax hardware, software and specifications change, the TIFF format is ready to cope.

- TIFF is designed for use by applications that only implement those portions of the overall format they need. It is thus quite feasible, and wholly within the tenor of the TIFF specification, to use a restricted version of TIFF purely as an internal storage method without having to commit to full TIFF reader/writer status.

TIFF is a well-defined format with a specification that is widely available in the public domain. While it hasn't attained the status of a formal international standard, it is explicitly recognised by the ITU as being the only universally available transfer format for bidirectional fax.

Adding TIFF support to fax image handlers

Code enabling any fax bit-image handling software to support the TIFF format is straightforward, and can be found in the TIFF directory on the accompanying disk. This contains two TIFF-specific modules. Their aim is to provide all the necessary general-purpose TIFF-related services that might be needed by the other portions of our fax code. We therefore isolate as much as possible of the TIFF-specific code from the fax code proper. The TIFF file is regarded as a wrapper whose sole purpose is to contain one or more fax images. One of the modules does the unwrapping. It contains a `tiffread` function, which returns the next fax image inside a file passed to it. The other module contains a couple of complementary `tiffwrite` functions which do the wrapping, and inserts a fax image inside a file.

Fragments of code based on these modules are used in the remainder of this chapter to show how various portions of the TIFF specification can be implemented.

The structure of TIFF file headers

While most TIFF files have the file extension. TIF their distinctiveness lies not in the name but in the structure. All TIFF files start with an 8-byte header, and contain one additional header, called an Image File Directory (or *IFD*), for each image in the file.

0	1	2	3	4	5	6	7
byte order		42 = TIFF ID		offset of first IFD			

Figure 4.1: TIFF file header

We begin by looking at the 8-byte file header shown in Figure 4.1. Bytes 0–3 are used for identification purposes, while bytes 4–7 contain a pointer to the IFD for the first image in the file. The first four identification bytes in the header are divided into two 16-bit words:

- The first word, contained in bytes 0–1, indicates the byte order used to contain integer and other values in the rest of the file. Two possibilities for ordering bytes are Intel ordering, where the least significant byte comes first, and Motorola ordering, where the most significant byte comes first. The only two possible values for the first word in the header are II (4949 hex or 0100100101001001 binary) and MM (4D4D hex or 0100110101001101 binary). As you might expect, II files have Intel's little-endian format and MM files have Motorola's big-endian format. This applies to both long 32-bit and short 16-bit integers. The byte sequence 01020304 may be read as either a bigendian 01020304 or little endian 04030201, but a hybrid format with the interpretation 02010403 or 03040102 is not allowed.

- The second 16-bit word, contained in bytes 2–3, always has the short 16-bit unsigned integer value 42 decimal, or 2A hex. If bytes 0–1 were "II" then byte 2 will be 2A hex and byte 3 will be 0, while if bits 0-1 were "MM" then byte 2 will be 0 and byte 3 will be 2A hex. The number 42 does not indicate a TIFF version number – there are no TIFF version numbers, merely an ever-increasing number of possible optional fields and tags. Hunting in the TIFF version 5.0 specification reveals the whimsical statement that 'the number 42 was chosen for its deep philosophical significance', while version 6.0 states that it is 'an arbitrary but carefully chosen number.'

Whatever the reason for the choice of ID bytes, it enables any software to identify a TIFF file with the minimum of bother. Any file suspected of being a TIFF needs to have only the first four bytes examined. If they don't contain either the hex values 49492A00 or 4D4D002A, then the file is not in proper TIFF format and no attempt to decipher it should be made.

However, once the first four bytes in the header have been verified, the last four bytes can be read. They comprise a long 32-bit unsigned integer (with bytes ordered according to the values contained in header's first two bytes) consisting of the offset in the file where the IFD for the first image in the TIFF file is located. The offset is calculated from the start of the file so the header is located at offset 0 and the lowest possible image offset is 8. A TIFF file must contain at least one IFD, and each IFD must be located on a word boundary (the pointer to its location must not be an odd number). The fact the IFD offsets are expressed as unsigned 32-bit integers means that the maximum size of any single TIFF file should be considered as 2^{32} bytes, or 4 Gbytes. This can't be considered a restriction given the technology that exists today as a fax totalling 4 Gbytes would take around a month to transmit at normal speeds.

Code for processing TIFF headers

Writing a TIFF header is straightforward. The following fragment writes the correct header for any II or MM system (assuming 16-bit short integers and 32-bit long integers) – see Listing 4.1.

```
struct
{
            short int order_id ;
            short int tiff_id ;
            long int first_ifd ;
} header ;

header.tiff_id=42 ;
if ((char)header.tiff_id==42) header.order_id=0x4949 ;
else header.order_id=0x4D4D ;

/* insert an appropriate header.first_ifd=&ifd here */

fwrite (&header,1,sizeof(header),outfile) ;
```

Listing 4.1: Writing a TIFF header

Reading a TIFF header is more complex. The fact that systems such as IBM PC clones and Apple Macintoshes can be networked together means that it is possible for a fax server to use a different byte ordering convention to a workstation, so an image handler must to be able to process images written using either II or MM format. There is generally no problem reading II files on an Intel processor, or MM files on a Motorola processor, but special provision has to be made for reading a TIFF file where there is a mismatch between the byte ordering of the reader and writer.

The first four bytes in a TIFF file need to be carefully examined to check for such mismatches. Our next fragment implements this in a processor-independent way by establishing two variables: **file_order**, which is set for Intel byte ordering if the first two bytes in the file are **II**, but is set for Motorola byte ordering if the byte is **MM**. Whatever the ordering, we provisionally set our second variable, **cpu_order**, to the other possibility, and then read the integer contained in the next two bytes. If this is 42, **cpu_order** is reset to match **file_order** and the rest of the file can be read with no problem. But if **file_order** and **cpu_order** don't match, the ordering of the bytes in all the values in the IFD will have to be reversed.

As well as detecting such mismatches, our code also checks the first four bytes to ensure that we have a valid TIFF file. The fragment returns with either the off-set of the first IFD if the file is acceptable, or with a 0 error code if a problem is encountered – Listing 4.2.

```
fread (&tiff_id,1,2,infile) ;
if (tiff_id==0x4949)
{
          file_order='I' ;
          cpu_order='M' ;
}
else if (tiff_id==0x4D4D)
{
          file_order='M' ;
          cpu_order='I' ;
}
else return (0) ;
fread (&tiff_id,1,2,infile) ;
if ((short int)tiff_id==0x2A) cpu_order=file_order ;
else
{
          reverse(&tiff_id,2) ;
          if (tiff_id!=0x2A) return (0) ;
}
fread (&ifd,1,4,infile) ;
if (file_order!=cpu_order) reverse(&ifd,4) ;
return (ifd) ;
```

Listing 4.2: Checking the validity of a TIFF header and returning the address of the first IFD

The **reverse** function in Listing 4.2 will be needed throughout the TIFF reader code to reverse the order of the bytes in any value read from the file **if(file_order!=cpu_order)**. It is called with the address of the value to be reversed, and the number of bytes used to hold the value. While in our code this number is always going to be either 2 for a short or 4 for a long integer, the **reverse** function in Listing 4.3 is capable of reversing the order of any number of bytes.

```
void reverse (void *pointer,int countdown)
{
int countup=0 ;
unsigned char store ;
unsigned char *address = (unsigned char *)pointer ;
if (pointer==NULL) return ;
if (countdown%2) return ;
while (countup<countdown)
    {
    countdown-- ;
    store=address[countdown] ;
    address[countdown]=address[countup] ;
    address[countup]=store ;
    countup++ ;
    }
}
```

Listing 4.3: Reversing values read from a TIFF where the file and CPU ordering don't match

The structure of TIFF IFDs

Just as the TIFF header is not too complicated, neither is the basic structure of an IFD. The overall format is simple, starting with two bytes containing an integer count of the total number of 12-byte directory entries, or TIFF fields, each of which starts with a unique tag value. This is followed by each of the fields in turn, in ascending tag order. Finally, after all the fields, the last four bytes contain a 32-bit integer offset pointing to the next IFD in the file. If there are no more IFDs, this pointer is set to zero.

Many different images can be found in a TIFF file, each with their own differing IFDs. Since each IFD contains a pointer to the IFD for the next image in the file, a TIFF file is a good example of what are known as *linked lists*, with the header containing the pointer to the first item in the list. The fact that the chain of IFD consists of pointers means that an IFD isn't tied to any specific location in the file, and could precede or follow an image. The actual location of an IFD in a file is entirely up to the person writing the software that is responsible for generating it. The reader must take responsibility for following all pointers wherever they lead, as with the exception of the 8 byte header at offset 0 in the TIFF file, there is no fixed location for any data structure relative to anything else.

It also follows from this that a fax application reading a TIFF file should, if it discovers that a particular image isn't a fax file, skip to the IFD for the next image and look at that instead. There may some faxable image later on in the file.

Constructing an IFD entry for a fax image

The sample IFD constructed here is for a file containing just one faxable image. The IFD contains the minimum of 7 directory entries, beginning with a 2 byte short integer 7 (in either big- or little-endian format) followed by the entries (totalling 84 bytes). Bytes 86–90 of the IFD would contain a long integer pointer to the next IFD if there were one, but in this case they contain 0 as there is only one image in the file (Table 4.1).

Things become a little complicated when we come to the 12-byte directory entries in the IFD. These entries, together with the values they refer to, collectively comprise the TIFF fields, and all the fields, taken together with any remaining defaults in the TIFF specification, describe an image.

The best way of understanding how TIFFs, tags and IFDs work is to start in the middle. Let's stipulate we want to write a complete A4 page of fax data, generated using standard 1-D T.4 coding and normal resolution, as a TIFF file. The default for TIFF files is that an image is bilevel (consists of black and white dots), so at least this doesn't have to be spelled out explicitly. We can leave out that information, not because it isn't necessary, but because the default happens to suit us. However, there are a fair number of things about our image that a TIFF reader does need to have spelled out, as either there is no default or the default doesn't suit our needs.

The most obvious things that have no default are the dimensions and resolution. Not only is there no such thing as a default size for a graphics image, but even faxes have options for the width, length and vertical resolution. Since we're talking about a normal resolution fax, the page width is 215 mm, rather than the 210 mm of a normal A4 page. An A4 page has a fixed length of 297 mm. We also know that all lines consist of 1728 dots, and that there are 3.85 lines/mm.

This is rather a mixed bag of information. While we know the dimensions of the original physical page, a TIFF file is a digitised image and needs to contain

Table 4.1: A minimal 7-entry IFD

Byte	Contents
0-1	number of entries
2-13	entry 1
14-25	entry 2
26-37	entry 3
38-49	entry 4
50-61	entry 5
62-73	entry 6
74-85	entry 7
86-90	pointer to next IFD

the dimensions of the bitmap. We know that the width is 1728 dots, and this happens to be the first of the directory entries in the IFD. So let's begin by constructing an IFD entry for the width.

Sample IFD entry for ImageWidth entry

The first item in each entry is an identifying tag that says exactly what information it refers to. The TIFF specification defines the tag for the width of a bitmap as 256. This tag is stored as an integer in the first two bytes.

The next item contains the storage type of the information – an integer, fraction or an ASCII string. If stored as an integer, we also need to know if it is short or long. Since the width is known to be 1728, it can be stored as a 16-bit short integer. This type is identified with the value 3, which is stored in the next two bytes of the entry.

The third item in the entry is a count of the number of items that make up the information. Obviously, there is only one item that makes up this particular entry as 1728 is only one integer. The count for most entries in an IFD is in fact 1. The most common instance of a count being more than one is if an entry refers to an ASCII string, in which case the count is of the number of characters in the string. This item is always stored as a long 32-bit integer. Even if (as in this case) we only need to store a 1, the count still takes up four bytes of each entry.

The last item in the entry is the location of the information itself in the file. This is a file offset, and like all other TIFF file offsets, it is stored as a long 32-bit integer. However, if the information to be stored (in this case one short 16-bit integer) itself fits into four bytes, then the TIFF specification states that the last item should contain the data itself rather than a pointer to where it can be found. If the data (as in this case) occupies less than four bytes then it is left-justified, i.e. the trailing bytes are ignored. This provision is made to save time and space. A TIFF reader is never in any doubt as to what this item consists of, as it knows from the count and the type entries how many bytes a particular item of information will take up.

Our two 16-bit and two 32-bit integers total 96 bits, which is why each directory entry takes up 12 bytes. Assuming "II" little-endian format, the entry for the width of our fax image looks as shown in Table 4.2.

This seems straightforward, and indeed it is. The complexity of TIFF arises from the number of different cases it has to cater for, and the number of different

Table 4.2: IFD entry for width of an image

12 hex bytes	Description
00 01	16-bit integer containing the tag 256=width
03 00	16-bit integer containing the type 3=short integer
01 00 00 00	32-bit integer containing 1=the count of items
C0 06 00 00	32-bit integer containing 1728=width fits into 4 bytes

tags fields needed. Version 6.0 of the TIFF specification contains over 70 different tags. Luckily, black and white images are among the simplest, and can require as few as seven entries in an IFD. The entries required are shown in Table 4.3.

Table 4.3: Minimal IFD for a black and white image

Tag Name	Number	Type
ImageWidth	256	SHORT or LONG
ImageLength	257	SHORT or LONG
Compression	259	SHORT
PhotometricInterpretation	262	SHORT
StripOffsets	273	SHORT or LONG
Xresolution	282	RATIONAL
Yresolution	283	RATIONAL

The entries in an IFD must appear in ascending tag order, so the table lists them in the order in which they really would appear in a seven-entry IFD. To complete our introduction to TIFF, we also need to know the value and size details for the five most common TIFF information types (Table 4.4).

Table 4.4: TIFF field types

Type	Value	Size in bytes	Description
BYTE	1	1	8 bit unsigned integer
ASCII	2	1	ASCIIZ string (see below)
SHORT	3	2	16-bit unsigned integer
LONG	4	4	32-bit unsigned integer
RATIONAL	5	8	Fraction

An ASCII string is stored as one or more (depending on the count) strings of characters terminated by a binary 0. Final 0 bytes are included in the count portion of the IFD. Fractions are stored as RATIONAL types, which consist of two LONGs comprising a fraction in numerator/denominator order. Three-quarters would be stored as a 32-bit long integer 3 followed by a 32-bit long integer 4. Vulgar fractions (where the numerator is larger than the denominator) are likely to be very common.

Completing the IFD entries for a fax image

Let's work through each of our IFD entries in turn. We've seen how to store the width of a fax image as an **ImageWidth** entry. For **ImageLength** we have to work out the length of the image in dots; in the case of the single A4 page we are assuming, the length is 1143 and is stored as an entry with the tag 257 in the same way as the width.

The **Compression** entry tells a TIFF reader what method (if any) has been used to compress the image. As our fax file is compressed using the group 3 T.4 fax encoding method, we indicate this using the value 3 in the Compression field.

PhotometricInterpretation is an impressive name for a simple item of data. We know that our image is a bilevel one containing only black and white dots, as this is the default for TIFF images. What we don't know is whether a decoded 1 in a bitmap indicates a white or black dot. A value of 0 for this entry indicates that White is zero, and a value of 1 shows that Black is zero. Fax images, like most print-based graphics, assume that white is zero. However, most screen-based graphics programs assume that black is zero, as a memory-mapped graphics screen containing nothing but zero bytes is normally completely black. As there is no way of telling whether an image is destined for screen or printer, there is no default for this item.

StripOffsets contains a pointer to where the fax image is stored in the file. This isn't called ImageOffset because it is possible to divide a large graphics image, which won't all fit into memory, into a number of smaller strips, and to store their locations as an array of pointers. The default for TIFF files happens to be one strip, so we only need one pointer; the count for this field is set to 1 and the value points directly at the location of the raw fax image.

Table 4.5: Basic IFD for a fax image

offset	hexadecimal data	meaning
0	4D 4D	MM big endian byte order
2	00 2A	TIFF ID byte
4	00 00 00 08	Offset to first IFD (must be even)
8	0007	IFD has seven entries
10	0100 0003 00000001 06C00000	ImageWidth 1728
22	0101 0003 00000001 04770000	ImageLength 1143
34	0103 0003 00000001 00030000	Compression T.4 1-D fax
46	0106 0003 00000001 00000000	PhotometricInterpretation Whiteiszero
58	0111 0004 00000001 00000072	StripOffsets -> offset 114
70	011A 0005 00000001 00000062	Xresolution -> offset 98
82	011B 0005 00000001 0000006A	Yresolution -> offset 106
94	00000000	Next IFD in file is set to zero
98	000139F4 0000018A	Xresolution 80.372 dots/cm=80372/394 dpi
106	00009664 0000018A	Yresolution 38.5 dots/cm=38500/394 dpi
114		start of fax data

The last two entries in our IFD contain the image resolution. The **Xresolution** entry contains horizontal and the **Yresolution** vertical resolutions. We know this latter figure is specified as 3.85 lines/mm, so we could set Yresolution to the 385/10 if we were defining this in dots/cm. However, the TIFF default for units used for both these entries is in dots/inch. As 1 cm is 0.394 inches, we can use the fraction 38500/394 instead. We have to infer the horizontal resolution from the fact that we have 1728 dots across a 215 mm page. This is 8.0372 dots/mm, so using the same conversion factor, we set the Yresolution fraction to 80372/394.

Since all fractions are stored as RATIONAL types consisting of 8 bytes, there is no way they can be stored in the value part of the IFD, which thus contains a pointer to where they are to be found. Many TIFF writer programs find it easiest to store values such as 8-byte rational types immediately after the IFD entries themselves.

We can put together the TIFF header, the IFD and the image data for a single page A4 fax in Table 4.5, which could be used as a basic header for a fax image that would begin immediately after the header at offset 114. In other words, if the following 114 bytes were written before any fax data in a file, it should be readable by any TIFF reader. We'll assume a big-endian "MM" format for this example.

Other tags needed for a TIFF fax reader

The TIFF header we is about as basic as they come. Most TIFF readers should be able to recognise such a file and either read it or apologise for not being able to do so. However, though the information on TIFF fields and tags given so far is sufficient to enable us to store our own files in a valid format, it relies heavily on the default values for certain tags that we haven't listed. It cannot be emphasised too strongly that although our sample IFD is a perfectly valid one, and quite suitable for a TIFF writer, understanding how it works is not in itself sufficient to enable construction of a TIFF reader. This is because our seven entry IFD relies on the default values for a number of fields which we haven't used. A reader coming across a previously unknown tag cannot assume that the contents of that field don't refer to something crucial.

Some fields with unknown tags in a TIFF file can often be ignored. But other tags, including some not yet listed, cannot be ignored. For example, tag 258 indicates a field that gives the number of bits used to represent each pixel in the image. The default happens to be 1, which corresponds to a bilevel black and white image, but colour images are examples of TIFF files that would have multiple bits for each pixel. If we read a file and came across a tag of 258, we would need to know what it meant and what values the field could contain before knowing whether the file was faxable or not.

The converse applies to one of the tags we have listed. While our seven-entry IFD is almost the smallest possible, it isn't quite the smallest. One of the fields we used needed to be specified as it has a default that didn't suit us. The field is the one with the tag 259 (Compression), which we set to 3 to indicate that our file uses T.4 1-D modified Huffman fax encoding. This field has a default of 1, indicating an uncompressed file. This is an important piece of information; we need to know that a TIFF image whose IFD has no entry with a tag of 259 is one we can't fax directly.

Apart from the compression field, none of the other seven fields listed so far have any default values. They should be regarded as required, and their omission is a fatal error. They will be present in all TIFF files. To be able to distinguish files containing fax images from other TIFF files, we need to have a more comprehensive listing of all the relevant tags and a list of what the defaults in the field they denote might be.

Because of the space it would take up, this listing is included on the accompanying disk in the NOTES directory where all those tags and TIFF field definitions that a fax application should understand are explained. We include tags which extend the TIFF specification in areas where the TIFF Class F committee thought it was deficient, plus tags which need to be understood so that non-faxable images can be rejected by applications software. Most are quite straightforward, and some have already been explained, but a few are interesting or important enough to warrant more detailed explanation. These TIFF fields are described in numeric tag order, which corresponds to the order in which they ought to appear in an IFD.

In addition, Table 4.6 lists the tags in alphabetic order for easy reference. The table is self-explanatory, with the exception of the last entry, headed count. This refers to whether there is a default for the count portion of the IFD for any field. Most TIFF fields have a compulsory count of 1, which means they contain just one item of information. Unless the type is RATIONAL, this will fit into the 32-bit value field.

There's a rough sort of hierarchy of importance for these fields, stemming largely from the attempt to categorise the various types of TIFF files into classes (known as 'baselines' in version 6.0). All TIFF files are expected to have explicit values for the following fields (though where there is a default that happens to be the desired value the field need not be written out):

- ImageWidth
- ResolutionUnit
- StripOffsets
- XResolution

- ImageLength
- RowsPerStrip
- StripByteCounts
- Yresolution

The following fields and values are additional requirements for baseline TIFF Bilevel images (formerly known as TIFF Class B):

- SamplesPerPixel = 1
- Compression = 1 or 2 or 32773

- BitsPerSample = 1
- PhotometricInterpretation = 0 or 1

Table 4.6: Alphabetic listing of TIFF tags relevant to fax images

Name	Tag	Type	Count
BadFaxLines	326	LONG	1
BitsPerSample	258	SHORT	SamplesPerPixel
CleanFaxData	327	BYTE	1
Compression	259	SHORT	1
ConsecutiveBadFaxLines	328	LONG or SHORT	1
DateTime	306	ASCII	20
DocumentName	269	ASCII	
FillOrder	266	SHORT	1
ImageDescription	270	ASCII	
ImageLength	257	SHORT or LONG	1
ImageWidth	256	SHORT or LONG	1
NewSubfileType	254	LONG	1
Orientation	274	SHORT	1
PageNumber	297	SHORT	2
PhotometricInterpretation	262	SHORT	1
ResolutionUnit	296	SHORT	1
RowsPerStrip	278	SHORT or LONG	1
SamplesPerPixel	277	SHORT	1
Software	305	ASCII	
StripByteCounts	279	SHORT or LONG	StripsPerImage
StripOffsets	273	SHORT or LONG	StripsPerImage
T4Options	292	LONG	1
T6Options	293	LONG	1
XResolution	282	RATIONAL	1
YResolution	283	RATIONAL	1

The TIFF Class F specification needs the following fields in addition to those listed above:

- FillOrder
- PageNumber
- T4Options
- NewSubfileType

It further restricts the values that fields may have in line with the ITU T.4 fax recommendation as follows:

- ImageWidth, XResolution and YResolution must all conform to valid T.4 resolutions and dimensions.

- Compression must have the value 3 or 4 (indicating a T.4 or T.6 fax image).

- All EOL codes in the file must be byte aligned (this is specified via the T4Option field).

Please refer to the disk for detailed coverage of each field. Those fields not listed above can be regarded as optional.

Writing fax images inside TIFF files

As explained on page 63, the thinking behind the design of our TIFF support is that it should be as transparent as possible. Only two functions, which share a little static storage, need to be added to a raw fax image writer to turn it into a TIFF writer. The first is called with a file pointer after opening the file. It writes the TIFF header at the start (a suitable technique is shown in Listing 4.1) with a dummy pointer of zero at the location reserved for the first IFD in bytes 4–7. A static structure is initialised to hold the IFD, and the address of this structure is returned to the caller. The final task of this function is to initialise a static pointer to 4, which is the offset maintained by a TIFF header to the offset of the IFD for the first image in the file.

The IFD is initialised as shown in Listing 4.4. The defaults are broadly those for TIFF Class F, and assume a single-strip image 1728 dots wide, with 1-D coding and a horizontal resolution of 8.03 dots/mm with a vertical resolution of 3.85 dots/mm. The use of various defaults for the other fields such as **FillOrder** and **StripByteCounts** is explained more fully in the NOTES directory on the accompanying disk. Some aspects of the initialisation, such as the resolutions being given in inches using a conversion factor of 3.94 mm/inch, are set as a matter of personal preference.

The **StripOffsets** value is filled in with 8, which is the offset of the next byte in the file after the header is written; it is at this offset that the first image will be written to the file by the caller).

The fact that the address of the IFD is returned to the caller makes the function very efficient, as an IFD is easy to modify as needed. All the caller needs is to have the structure definition included in a suitable header file; for example, if the fax written is 2-D rather than 1-D, then **T4Options** can easily be altered to fit. If it is a fine resolution fax, then the value of the **Yres** fraction can be changed to 77000/394.

If the fax being written does actually conform to the lowest common-denominator assumptions on which the IFD was initialised, then the only modification required to the IFD is to insert the appropriate value for the number of scan lines in the fax at **ImageLength**. It can be argued that even that modification is unnecessary if the intended TIFF reader doesn't use this value.

After the first function has been called, the fax is written sequentially to the file. Once each image is completed, the second of our TIFF writer support functions is called. The only parameter it takes is a file pointer, as it already has access to the IFD which was initialised by the header writer (and may have been modified by the caller). Its purpose is to write this IFD to the file:

```
#define BYTE 1
#define SHORT 3
#define LONG 4
#define RATIONAL 5
static struct IFD ifd =
{
    16,                      /* number of entries in the IFD */
    {254,LONG,1,2},          /* NewSubfileType is 010 binary */
    {256,LONG,1,1728},       /* ImageWidth 1728 */
    {257,SHORT,1,0},         /* ImageLength filled in by caller */
    {258,SHORT,1,1},         /* BitsPerSample is a default */
    {259,SHORT,1,3},         /* Compression is 3 */
    {262,SHORT,1,0},         /* PhotometricInterpretation is 0 */
    {266,SHORT,1,1},         /* FillOrder_tag is a default */
    {273,LONG,1,0},          /* StripOffsets set to next free file
                                offset */
    {277,SHORT,1,1},         /* SamplesPerPixel is a default */
    {278,SHORT,1,0},         /* RowsPerStrip = ImageLength */
    {279,LONG,1,0},          /* StripByteCounts (image size)
                                filled in later */
    {282,RATIONAL,1,0},      /* Xresolution (pointer filled in
                                later) */
    {283,RATIONAL,1,0},      /* Yresolution (pointer filled in
                                later) */
    {292,LONG,1,0},          /* T4Options is a default */
    {296,SHORT,1,2},         /* ResolutionUnit is dpi */
    {297,SHORT,2,0},         /* PageNumber */
    0,                       /* nextifd (filled in later if
                                needed) */
    {80300L,394L},           /* Xres fraction */
    {38500L,394L}            /* Yres fraction */
} ;
```

Listing 4.4: An IFD structure definition

- The first step is to find out how many bytes are in the image, using the `ftell` function to obtain the current file position and subtracting the image start as stored in `StripOffsets`. The image length in bytes is stored in `StripByteCounts` in the IFD structure.

- The second step begins by writing a dummy byte if the current file position (or `StripByteCounts` for that matter) is an odd number, as all TIFF IFDs must be written on even byte boundaries. The resulting file position is where the IFD is going to be written. This value is written to the file at the offset stored at `lastifd`, which was initialised by the header writer to 4 (the place in the header where the offset to the IFD for the first image is held). We then return to the end of the file to prepare to write the IFD.

```
{
    char zero=0 ;
    long int thisifd ;

/* step 1 */

    ifd.StripByteCounts.value=(ftell(outfile)-ifd.StripOffsets.value);

/* step 2 */

    if ((ifd.StripByteCounts.value&1)!=0
    {
        fwrite (&zero,1,1,outfile)
        ifd.StripByteCounts.value++ ;
    }
    thisifd=ifd.StripByteCounts.value+ifd.StripOffsets.value ;
    fseek (outfile,lastifd,0) ;
    fwrite (&thisifd,4,1,outfile) ;
    fseek (outfile,0L,2) ;

/*step three */

    ifd.nextifd=0 ;
    ifd.RowsPerStrip.value=ifd.ImageLength.value ;
    ifd.Xresolution.value=(thisifd+2+192+4) ;
    ifd.Yresolution.value=(thisifd+2+192+4+8) ;

/* step four */

    fwrite (&ifd,1,sizeof(ifd),outfile);

/* step five */

ifd.StripOffsets.value=(thisifd+2+192+4+8+8) ;
ifd.ImageLength.value=0 ;
ifd.StripByteCounts.value=0 ;
(short int)(ifd.PageNumber.value++) ;
lastifd=(thisifd+192) ;
return ;
}
```

Listing 4.5:   Writing an IFD

- The third step is to update the rest of the IFD structure. The **nextifd** member is always reset to 0 in case there is no subsequent image. Since the writer only handles single strip images, **RowsPerStrip** is set to the same value as **ImageLength**. The file position is also used as the basis for updating those portions of the IFD structure that require file offsets, notably the pointers held in **XResolution** and **YResolution**. The fragment uses a series of apparently magic numbers for this updating to save space, but the formula is really quite obvious. For instance, the **Xres** offset is calculated from the beginning of the IFD as stored in **thisifd**. We add 2 for the short integer used to hold the number of entries, and add 12 (the length) for each IFD entry – this gives us 192 for our 16 entry IFD. Finally, we add an extra 4 for the pointer used for the offset to the next IFD. 2 + (16 × 12) + 4 = 198. The offset for **Yres** is 8 bytes further on as the **Xres** RATIONAL is stored as two LONGs.

- The fourth step is to write that IFD to the file.

- The final step is to re-initialise. Two values are crucial here. The first, **StripOffsets**, must be set to the current offset immediately after the IFD in case another image is to be written. We use our magic number for this as it is quicker than finding out from an **ftell**. The second is the static pointer which holds the address of the IFD for the next image in the file. This must be set to the offset where the **nextifd** member of the IFD structure was written in the file. Any subsequent image must update this link if it is to be successfully retrieved by a TIFF reader. Other IFD housekeeping, such as resetting **ImageLength** and incrementing the **PageNumber**, can also be handled at this point if the caller doesn't do this.

Reading a fax image stored in a TIFF file

Finding out what sort of images are contained in a TIFF file is not difficult. First, we check that the file really is in valid TIFF format and locate the address of the first IFD (Listing 4.2). We then **fseek** to the offset of the IFD and read the short integer at the start. As shown in Table 4.1, this integer holds the number of the entries in the IFD, and can be used to control a loop which reads through the IFD entries one-by-one. Each entry is processed by means of a large switch statement, shown in Listing 4.6.

We've left the cases in the switch statement empty because it's worth thinking in some detail about exactly what we are doing here. The obvious course of action for each IFD entry is for the reader to fill in an appropriate member of an IFD structure whose address has been passed by the caller. But while this approach complements that used in our TIFF writer, where we aren't permitted to deviate from the TIFF specification in any way, it is less efficient for a function designed to offer TIFF support for reading images.

TIFF was designed with portability in mind, and like most highly flexible formats, it is much easier to write information than to read it. This is most apparent when dealing with something like the resolution of an image, which takes up one bit of a fax negotiating frame, but occupies 52 bytes of an IFD.

```
struct
    {
    unsigned short int tag ;
    unsigned short int type ;
    unsigned long int count ;
    unsigned long int value ;
    } ifdentry ;

fseek (infile,ifd,0) ;
fread (&ifdcount,1,2,infile) ;
for ( ; ifdcount ; ifdcount--)
    {
    fread (&ifdentry,1,12,infile) ;
    switch (ifdentry.tag)
        {
        case ImageWidth:                        /* insert code */ break ;
        case ImageLength:                       /* insert code */ break ;
        case BitsPerSample:                     /* insert code */ break ;
        case Compression:                       /* insert code */ break ;
        case PhotometricInterpretation:         /* insert code */ break ;
        case FillOrder:                         /* insert code */ break ;
        case StripOffsets:                      /* insert code */ break ;
        case SamplesPerPixel:                   /* insert code */ break ;
        case RowsPerStrip:                      /* insert code */ break ;
        case StripByteCounts:                   /* insert code */ break ;
        case Xresolution:                       /* insert code */ break ;
        case Yresolution:                       /* insert code */ break ;
        case T4Options:                         /* insert code */ break ;
        case PageNumber:                        /* insert code */ break ;
        default:                                                  break ;
        }
    }
fread (&ifd,1,4,infile) ;
```

Listing 4.6: Reading through an IFD

The thinking behind the design of a TIFF support function for reading fax images should be that, when called with a file pointer, we get only the information we need to send an image from the file. If a file isn't a TIFF file, or if an image isn't in a valid fax format, the function should return with an error code. We don't want to bother with checking **BitsPerSample** or **Photo-metricInterpretation** in our application, as all that should be offloaded onto the reader function.

The information we actually need falls into four categories:

- Items that *physically* describe the image data. This is used in phase C of a fax session, and consists of only two items: the offset of the start of the image in the file, and its length in bytes.

- Items that *logically* describe the image data. This is used in phase B of a fax session, and includes variables such as the length of image, its resolution, and the type of coding used in its construction. If non-standard widths are to be supported, then the width of the fax must also be included.

- Items that *sequentially* locate the image. There is only one datum here; whether an image is the final one in sequence. This is needed for phase D of a fax session, and is necessary for sending the correct post-message command.

- Finally, we have a number of TIFF-specific items, the most important of which is the ordering of the bits within a byte.

The approach taken by the TIFF reader is to check the IFD entries for validity and range and only return successfully if an image is faxable. For acceptable images, we bundle up the information we need in a customised **IMAGE** structure, the address of which is received from the caller together with the file pointer. The structure definition is as shown in Listing 4.7.

```
struct IMAGE
{
unsigned long int offset ;
unsigned long int bytecount ;
unsigned short int width ;
unsigned short int length ;
unsigned char options ;
unsigned long int ifd ;
unsigned short int page ;
unsigned char bitorder ;
} ;
```

Listing 4.7: Information requested from a TIFF reader

The *physical* location of the image data is returned in the first two members. Reading an image is simply a matter of doing an **fseek** to the correct place in the file as passed in **IMAGE.offset** and reading the number of characters passed in **IMAGE.bytecount**.

The *logical* description of the image is contained in the next three members. The width and length of the fax are really provided more for completeness than out of necessity. If we were happy with default values, none of these would be needed, as a normal resolution 1-D A4 page is guaranteed to be acceptable to all fax machines. While few faxes correspond exactly to A4 length (as they start from an A4 original and add a header) this doesn't usually matter, since virtually all faxes accept pages of unlimited length. We also don't need to know the width unless we want to make use of non-standard fax dimensions; 1728 dots can otherwise be assumed, and the TIFF reader can check for this value before returning successfully.

In contrast, **IMAGE.options** is a much more targeted item of information. It is bit-mapped as shown in Figure 4.2.

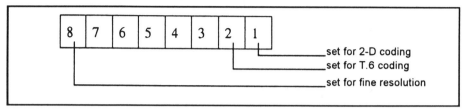

Figure 4.2: Bit fields in **IMAGE.options**

The reason why this member of the structure is bit-mapped is that the resolution and coding are independent parameters which translate directly to T.30 bit fields. While bit 1 of this field reflects the contents of **T4Options** and bit 2 reflects the value of **Compression**, bit 8 has no direct TIFF equivalent. It is certainly possible to insist (as does the Class F specification) that the resolution of an image is either 204 × 98 dpi for normal resolution or 204 × 196 dpi for fine resolution, but this seems to be unduly harsh. If there are 1728 dots per scan line, a fax will transmit no matter what the actual resolution. It makes more sense to set the resolution to indicate the aspect ratio. A ratio of 2:1 means that a fax can be sent in normal resolution, while a ratio of 1:1 means fine resolution must be used. Assuming that **case Xresolution:** and **case Yresolution:** in Listing 4.6 stored the offsets of the resolutions at **Hres** and **Vres**, respectively, the method of setting a correct resolution is shown in Listing 4.8.

```
long int res [2] ;
short int oures ;
long int Hres,Vres;

fseek (infile,Hres,0) ;
fread (&res,4,2,infile) ;
Hres=(res[0]/res[1]) ;
fseek (infile,Vres,0) ;
fread (&res,4,2,infile) ;
Vres=(res[0]/res[1]) ;
oures=(((float)Hres/(float)Vres)+0.5) ;
if (oures<2) image->options|=0x80 ;
```

Listing 4.8: Setting fine or normal resolution from an IFD

IMAGE.ifd is used to hold the offset of the next IFD in the TIFF file. As the last image in the file has a null IFD offset, this enables us to work out the type of post-page message to send if an image is being read for transmission.

The remaining TIFF-specific members of the structure have no real parallel in any other image format, but are quite straightforward. **IMAGE.page** would enables us to detect TIFF files that include identical pages in different formats. The ability of TIFF to allow this is one solution to the problems caused by the fact that making the optimum fax connection is impossible without real-time encoding of a fax image. Finally, **IMAGE.bitorder** is included in the structure so that we can read TIFF class F files no matter what convention they used for ordering the bits in each byte created by other applications. It simply returns the **TiffOrder** IFD entry, which enables the application reading the image to invert the data read as needed.

5

Decoding Fax Images

Introduction

This chapter should be of interest to anyone receiving faxes. Many computer fax installations are only ever used for sending faxes, and if you have no interest in receiving them this chapter can be omitted.

The decoder needs a chapter devoted to it because it has to cater for the complexities of various output device types which are controlled in different ways, fed with different types of data and have different resolutions.

A decoder is best seen as alternating between two distinct processes:

- The first process is concerned with the decoding proper. It takes fax data as input and produces scan lines composed of run-lengths totalling 1728 dots as output.

- The second process is concerned with reproducing the image; it takes the scan lines and run lengths as input and outputs a stream of instructions and graphics data which enable a specific output device to print or display an appropriate image.

The TIFF wrapper

We start by considering the different types of fax data a decoder needs to deal with. It has to be able to cater for images in either normal or fine resolution, composed of entirely of 1-D coded lines separated by EOL codes. In the case of a T.6 image, it might be composed entirely of 2-D coded lines with no EOL codes at all.

The final case is the T.4 2-D image, consisting of a mixture of 1-D and 2-D coded lines separated by EOL codes, each of which starts with a tag bit indicating the type of coding to be used on each line.

It is not possible to decode an image unless we know in advance which options were used in the encoding. This can't be deduced from the data itself, which is where the TIFF file format comes into its own. When we standardize images using TIFF, an image doesn't come to us naked but is enclosed in a wrapper. Examining the wrapper tells us what we need to know about the file. We saw in the last chapter that the `tiffread` routine does this for us. All our fax file decodings begin with a call to `tiffread`; this tells us the encoding options used to create the image, together with its size in bytes.

Since a TIFF file can contain multiple fax images, we also end each image decoding with a call to `tiffread`; this tells us whether we have to repeat the whole process for the next image in the file.

Structure of an image decoder

The basic structure of an image decoder is a loop through every bit in a file to isolate the code words used to put it together. Since we are always dealing with computer architectures oriented towards dealing with bytes rather than bits, this has to be handled using two loops. The outer loop uses the image size in bytes as the control variable and reads octets from the file, while the inner loop uses the number of bits in a byte as the control variable and isolates bits from each octet (Listing 5.1).

```
for (; image_size ; image_size--)
{
          octet= getc(faxfile) ;
          for (i= 0 ; i<8 ; i++, octet<<= 1)
          {
                .
                .
(/* here we decode, using the most significant bit of octet.)
note that we are assuming that the order of bits in an octet runs
from MSB to LSB */
                .
                .
          }
}
```

Listing 5.1: Basic decoding double loop

Setting up a decoding tree

There is more than one way of decoding the bits that go to make up an image. Suppose we know that the code word we are looking for is a black run-length. The start of the relevant table of codes for these run-lengths is shown in Table 5.1. For convenience, we ignore the code for a run-length of 0 and begin with 1.

One method of identifying these codes in a bitstream found in some textbooks, is to start by constructing an array containing all the possible codes stored as integers, together with the number of bits they contain and the run length to which they correspond. As an example, Listing 5.2 shows the start of such an array for the black run lengths in Table 5.1.

The search for a code word begins by initializing both a test integer and the number of bits in the code word to zero.

As each bit is extracted from the image, it is shifted into the test integer from the right, and the count of bits in the code is incremented. The array is then searched for a match of both the number of bits in the code and the test integer, and if no match is found, the procedure is repeated.

Though intuitively obvious, this method is actually very inefficient. Each successful search for a run-length must always be preceded by unsuccessful searches for all shorter run-lengths; in the case of the longer code words in particular, the time taken to find a particular run-length is arbitrarily determined by how the codes happen to be ordered.

A better way of decoding an image can be used if the code words are arranged in a binary tree. This method eliminates the need for unsuccessful searches through a list and requires just one test for each bit in the code. It also guarantees that the time taken to find the meaning of a particular code word depends only on its length, and not on any arbitrary ordering of the code words in a list. The fact that a tree is one of the best ways of decoding T.4 1-D images is not surprising. You may remember from Chapter 3 that the principle on which the T.4

Table 5.1: First twelve code black code words

run length	code word
1	010
2	11
3	10
4	011
5	0011
6	0010
7	00011
8	000101
9	000100
10	0000100
11	0000101
12	0000111

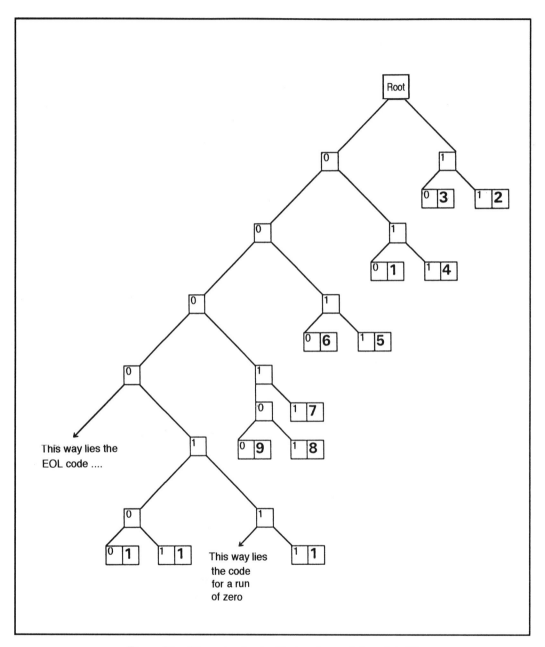

Figure 5.1:   Binary tree for the black code words from 1 to 12

```
struct codewords
{
            short int count_bits ;
            short int code_word ;
            short int run_length ;
} ;
struct codewords blackruns [] =
{
            {2,0x03,2},
            {2,0x02,3},
            {3,0x02,1},
            {3,0x03,4},
            {4,0x03,5},
            {4,0x02,6},
            {5,0x03,7},
            {6,0x05,8},
            {6,0x04,9},
            {7,0x04,10},
            {7,0x05,11},
            {7,0x07,12},
                  .
                  .
                  .
```

Listing 5.2: First twelve black codes stored in an array

Huffman codes are constructed is that the possible runs are first sorted according to frequency, and the frequencies are used to build a tree. This decoding method just reverses the process.

Our tree consists of nodes, each of which has two branches. A branch can lead either to another node (which has its own branches), or to a leaf which holds a run-length. The first node in the tree is called the root. A representation of a binary tree for the black codes for runs of 1 to 12 is shown in Figure 5.1.

The method for traversing the tree is quite straightforward. Beginning at the root, we take successive bits from the bit-stream. If the bit is a 0 we take the left-hand branch of the tree, and if the bit is a 1 we take the right-hand branch. If another node is reached, we replicate the procedure with the next bit, continuing until a leaf with a valid run-length is reached instead of another branching node. The tree illustrated has some apparently rotten branches that don't lead anywhere, but this is only because it is a partial tree that decodes only the black runs from 1 to 12. You can take any of the codes in Table 5.1 and trace them through the tree to see how the run length is extracted.

This sort of data structure doesn't contain any of the code words for the run lengths as data items; instead, they are implicit in the tree's structure. The Huffman codes for the run-lengths revert to their original nature as descriptions of the location of a run in the tree, with the most frequently occurring runs being at the top of the tree, and therefore being decoded in the shortest time.

Turning the tree into a data structure isn't terribly difficult. We can easily construct a 2-D array that reflects the tree structure. Listing 5.3 shows how this can be done; the array corresponds exactly to the tree shown above. Even the rotten branches (caused because the tree only contains black runs from 1 to 12) are present and marked as **xxx**.

```
int btree [][2] =
{
                {2, 1},                 /* 0 */
                {-3, -2},               /* 1 */
                {4, 3},                 /* 2 */
                {-1, -4},               /* 3 */
                {6, 5},                 /* 4 */
                {-6, -5},               /* 5 */
                {9, 7},                 /* 6 */
                {8, -7},                /* 7 */
                {-9, -8},               /* 8 */
                {xxx, 10},              /* 9 */
                {12, 11},               /* 10 */
                {xxx, -12},             /* 11 */
                {-10, -11},             /* 12 */
                   .
                   .
                   .
```

Listing 5.3: Binary tree coded as a data structure

The rows of the array consist of pairs of numbers, corresponding to the two branches from each node of the tree. We begin decoding by taking row 0 of the array, which is effectively the root of the tree. If the first bit is 0, we take the left-hand branch and examine the number at [0] [0], while if it is a 1 we take the right-hand branch and examine the number at [0] [1]. A positive number means we have found the row corresponding to the next node in the tree; a negative number indicates we have arrived at a run-length which is simply the negation of the number we have found. If we reach another node, we take the next bit and go through the same procedure, but if we reach a run-length we can output it, and start again at the root with the next bit.

Using a decoding tree

The data used with the decoding program on the accompanying disk is structured as three trees. One contains white run lengths (**wtree**), a second contains black run lengths (**btree**), and a third contains all the 2-D codes for pass, vertical and horizontal modes (**twotree**). A common pointer is set to whichever tree happens to be one that needs to be used.

The simplest decoding case to consider is a series of run-lengths coded one-dimensionally (we assume there are no errors). We begin (as with all fax lines) by setting the colour to **WHITE**, the **tree** pointer to the address of the white codes and the initial code to 0. Using the same basic double loop specified earlier, the decoding proceeds as in Listing 5.4.

The recursive nature of the redefinition of **code** using itself as one of the indices into the array is typical of the way in which binary trees are traversed. The rather awkward expression (**octet&0x80)>>7** effectively casts the left-most bit of an octet as an integer, and enables it to act as the other index to the array.

The line **code= (*tree)[code][(octet&0x80)>>7]** is repeated for each bit until **code** returns with a negative number. We know that we have hit a run length in this case, and then negate the value to get the proper run length, which is then output.

If the run was less than 64, the code was a terminating one. Both the colour and tree being used have to be changed. We use the same definitions of **WHITE** as **0x00** and **BLACK** as **0xff** as when we were encoding; and changing the colour is thus handled using ~color, just as in ENCODE and QENCODE. If the

```
color= WHITE ;
tree= &wtree ;
code=0 ;
for (; image_size ; image_size--)
{
    octet= getc(faxfile) ;
    for (i= 0 ; i<8 ; i++, octet<<= 1)
    {
      code= (*tree)[code][(octet&0x80)>>7]   ;
      if (code<1)
      {
        code= (-code) ; output_run_length(code) ;
        if (code < 64)
        {
            color= (~color) ;
            if (color) tree=(&btree) ; else tree=(&btree)
        }
        code= 0 ;
      }
    }
}
```

Listing 5.4: Walking the binary tree

run was 64 or greater, then the code was a makeup one and the colour remains the same.

In both cases, setting `code=0` ensures that the decoding process begins again at the root of whatever tree happens to be used next.

Messy but important details

The decoding routine in UNFAXIFY.C has the tree-based decoding method shown above at its core. However, as it stands, the code fragment can only deal with infinitely long error-free 1-D T.4 data.

To make the code function as an effective decoder, additional code must be added to handle the EOL code and any padding that precedes it. The start of the decode must always be a search for an initial EOL. Along with EOL handling, routines must also be added to handle the most common error conditions, including lines that are too long and data that doesn't decode properly. In both cases, a `resync` flag can be set and used to ignore all subsequent data until the next EOL code is detected.

The action to be taken when a decoding error is discovered is up to a programmer to decide. Fax machines often replace suspect lines with a duplicate of the previous good line; alternatively, blank lines can be inserted, or the line can simply be ignored. There is no way of telling (short of a visual inspection) whether a partial line contains good or flawed data. Generally speaking, the human eye is likely to make a better job of working out what the missing bits in any image ought to be, and leaving bad lines blank is often the best course; there is a lot of redundancy even in normal resolution faxes.

Enhancing a 1-D decoder to handle 2-D data isn't trivial, but it is fairly straightforward in programming terms. The key is to maintain and inspect the various flags needed to let the decoder know what coding scheme is being used and what state it has reached. These flags can get pretty complex:

- One flag holds the encoding scheme used to create the image. The EOL handling routine will need to be sensitive to this flag; in the case of 2-D T.4 data, the bit after an EOL is the tag bit which identifies the type of coding to be used on the following line.

- The tag bit itself will be used to set a second flag, which must be inspected if the first flag is set for 2-D data, and which holds the coding scheme for any particular line.

- A third flag is needed for 2-D coded lines which indicates the mode being used, and in the case of horizontal mode, which of the two possible runs is being assembled.

The point of all these flags is that while the application of our basic T.4 decoding routine to 1-D images was quite straightforward, extending it to handle the needs of 2-D data means that the fragment has to be able to handle four distinct cases rather than one. Apart from normal 1-D lines, there are also 1-D coded lines in 2-D data; and within 2-D coded lines, horizontal mode involves two distinct runs of T.4 data. All these cases use the same T.4 decoding fragment, but the program flow in in each case is different.

Two-dimensional decoding

A 2-D decoding tree can easily be constructed from the codes for pass, vertical and horizontal modes. Instead of requiring a special routine, all that is needed to decode 2-D data is to set a suitable flag and ensure that the root of this tree is selected using the statement **tree=&twotree** at the start of all 2-D lines.

When the 2-D code for horizontal mode is found, the decoder simply drops into the 1-D decoding routine for two run-lengths, while pointing **tree** at the correct root for the colour of the next run; it is just as easy to keep track of the colour when decoding 2-D coding as it is when decoding one-dimensionally.

Whenever a 2-D image is being decoded, a copy of the last scan line always needs to be kept, which is initialized to an all-white line at the start of each page. If two arrays are set aside for this, each of which is 1728 bits long, a pair of pointers to the current line and the reference line must be kept, and swapped over each time an EOL is reached. The reference line needs to be used to determine what to do when the codes for pass mode and vertical mode are encountered, while the array for the current line needs to have each run-length mapped into it as it is decoded.

Listing 5.5 is a code fragment showing one way of writing run-lengths to a bit array. A 1728 bit (216 character) array at **current_line** has its bytes indexed by **byte_inx**, and each byte has its bits indexed by **bit_inx**. Both of these indices have to be declared as static variables, as they need to retain their value between calls. At the start of each row, **current_line** is initialized to be all-white, while the byte index **byte_inx** is initialized to 0 and the bit index **bit_inx** is initialized to 8.

The fact that the scan line defaults to all white means that for a white run, all we need to do is adjust the indices for the value **run_length** which is passed to the routine. The byte index is increased by the value of the run length divided by 8, while the remainder is subtracted from the bit index (which is wrapped around to 8 again if it becomes negative). Try this out with a few numbers to see how it works.

The black runs are slightly more complex. We may have to OR in some individual bits, but for long black runs we can simply place bytes with the value **11111111 (0xff)** in the array (Listing 5.5).

The process of working out the runs embedded in pass mode and vertical mode uses the same code for identifying the positions of b^1 and b^2. The two key preliminaries are to have access to the current position on the scan line being decoded (which has already been done with the **byte_inx** variable) and use

```
if (color==WHITE)
{
    bit_inx-=(run_length%8) ;
    byte_inx+=(run_length/8) ;
    if (bit_inx<0)
    {
        bit_inx+=8 ;
        byte_inx++ ;
    }
}
else for (;;)
{
    while (bit_inx>0)
    {
        current_line[byte_inx]|=(1<<(bit_inx-1)) ; bit_inx-- ;
        if (!(--run_length)) break ;
    }
    if (run_length==0) break ;
    byte_inx++ ;
    for (; run_length>8 ; run_length-=8) current_line[byte_inx++]=0xff ;
    if (run_length==0) break ;
    bit_inx=8 ;
}
```

Listing 5.5: Updating pointers to the current position in the current line

that as the value of a^0, and to keep a copy of the last line that was decoded as the reference line **ref_line**, so that b^1 and b^2 can be identified. The code in Listing 5.6 works out the position of b^1 starting from a^0 under these circumstances. Code for computing the position of b^2 starting from b^1 is similar.

```
for (b1= a0+1 ; b1<1728 ; b1++)
{
    bit_inx= (0x80>>(b1%8)) ;
    if ((ref_line[b1/8]&bit_inx)!= (color&bit_inx)) break ;
}
```

Listing 5.6: Identifying the next changing element on the current line

Two-dimensional pass mode does not affect the colour of the next run, but codes a run length computed using b^2-a^0 and carries on with the same colour. In contrast, vertical mode codes a run length computed using $b^1-a^0\pm3$, and changes to the opposite colour. Horizontal mode, which codes the next two run lengths one-dimensionally, ends with the decoder using the same colour with which it began, but only because of the two intermediate colour changes.

As with encoding, decoding 2-D images is significantly slower than decoding 1-D images. Apart from the extra chore of constructing a reference line for each scan line, the decoding of 2-D images doesn't directly result in series of run-lengths. The run-lengths always need to be worked out at one level of indirection; the 2-D codes simply tell us how to find the next run, not what it is. Though 2-D data offers significant savings on storage space and transmission time, it does require more computation and is thus always going to take a little longer than simpler coding methods.

Outputting scan lines

The second of the two processes we divided the program into has the task of reproducing the image on an output device. This is generally quite complicated. Not only does more than one type of output device (screen and printer) normally have to be catered for, but a number of different printer models have to be supported.

Accurate reproduction of a fax image on an output device isn't a question of simply sending out a scan line. The normal resolution of an A4 fax page, at 1728 × 1143 dots (approx. 200 × 100 dpi), isn't matched by any output device in common use. Some devices are capable of greater resolution, while others can't hope to match the clarity of fax. In addition, while some devices can output a single scan line at a time, others either use or are descended from dot-matrix technology, and will need to output a series of scan lines as image slices, which are composed of multiple lines.

Table 5.2 shows the devices catered for in the code on the accompanying disk. While the same basic routines are used for all output devices, the number of scan lines (**n**) output at a time, together with the **X** and **Y** resolutions (dots/page), are quite different.

Table 5.2:   Resolution of various output devices with the number of scan lines output at a time

Output Device	n	X=width	Y=height
Uncompressed image	1	1728	1143
CGA graphics screen	1	640	200
EGA graphics screen	1	640	350
VGA graphics screen	1	640	480
Hewlett-Packard (HPGL) 100 x 100	1	802	1069
Hewlett-Packard (HPGL) 300 x 300	1	2406	3207
Epson 9-pin dot matrix 120 x 72	8	960	792
Epson 24-pin dot matrix 360 x 180	24	2880	1980
Canon Bubblejet 360 x 360	48	2880	3960
Canon CaPSL printer 300 x 300	1	2340	3390

Once the selection of a specific output device is made, we simply use the correct routines for each of these possible devices.

The initialization routine has to take care of any steps necessary to place a device in graphics mode, and also includes the setting of three program variables for lines, width and height as shown in Table 5.2. The width and height variables are used for scaling the image, with `width/1728` being the horizontal scaling ratio and `height/1143` being the vertical scaling ratio for normal resolution images, while the number of lines **n** is used to tell how many scan lines are output at one time.

Scaling

Unusually, we show the two scaling routines almost exactly as they appear in the code; this is because they are short and compact, as well as important.

Horizontal scaling is performed as each run-length is generated. The fundamentals of this are straightforward arithmetic, in that each run is first multiplied by the X-resolution of the device stored in `width` and is then divided by 1728, which is the width of a fax line. If floating-point arithmetic were used, the computation (which is basically administering a scaling ratio) could be performed the other way round; but using long integer variables and doing the multiplication first is quicker.

There are two complications, both of which are only really important when scaling down. The first is that, to avoid inaccuracies accumulating, we keep any remainder from each operation and add it back into the next one. If this isn't done, discarding all the remainders means each scan line will be shortened by an apparently random amount, and vertical alignment of features on adjacent lines will be distorted by the preceding runs.

The second complication is a particular fax requirement. We assume that, as fax is primarily designed for printed text, preserving the alternations of black and white runs is more important than accuracy in reproducing any individual run-length. To take one device as an example, when scaling the 1728 dots of an image for reproduction on an IBM PC VGA screen, which is only 640 dots wide, any runs less than three dots long are not guaranteed to be reproduced. If no remainder is available for adding in, the scaling ratio of 640/1728 as applied to a run of 2 invariably results in a run of 0.

We remedy this by ensuring that every run-length in the fax will be reproduced by at least a run of one in the reproduced image. This is done by borrowing a dot in advance from the next run-length if we have scaled a run down to 0, and setting a flag to indicate what has happened; the borrowed dot is repaid the next time we have a run greater than 1. We never borrow more than one dot at a time. Both the remainder and the borrow flag can be initialized at the start of each row by calling the scaler with a run-length of 0.

While the code for vertical scaling looks similar to that for horizontal scaling, it is quite different in concept. The simplicity and accuracy of calling a function

```
int hscale (long int width, long int run)
{
    static long int remainder ;
    static char borrow ;
    if (!run)
    {
      remainder=0 ;
      borrow=0 ;
      return(0) ;
    }
    run=(run*(long int)width);
    run=run+remainder ;
    remainder=run%1728 ;
    run=run/1728 ;
    if ((!run)&&(!borrow))
    {
      run++ ;
      borrow++ ;
    }
    if ((run>1)&&(borrow!=0))
    {
      run-- ;
      borrow-- ;
    }
return (int)(run) ;
}
```

Listing 5.7: Scaling horizontal runs with remainder adjustment and short-run borrow

with one run-length and getting back the correct run-length to use instead cannot be replicated in the vertical dimension.

Instead, we call a function each time we have completed decoding a scan line, but before the line is actually output. We get back a value corresponding to the number of times the scan line should be output. If the output device's resolution is less than that of a fax, we have to miss out some lines to reproduce the page accurately; if the resolution is greater than that of a fax, some line will have to be output more than once. Inevitably, inaccuracies of vertical reproduction will be more noticeable than those of horizontal reproduction; if you think about the problems involved if we had to decide whether to output each individual dot in a run, rather than how many dots to output, you can get an idea of why this is so.

The function outputs the first line of each image once for each time the **height** variable (indicating the number of lines the device used to output A4) is divisible by the height of the fax image. For the second and subsequent lines, we add the remainder to the height of the fax and do the same again. For instance, assume we are outputting a fine resolution A4 fax (with 2286 scan lines) to a 300 dpi laser. If we assume an 11" page, this gives a device with a nominal `height` of 3300. The sequence for scaling up is:

- For the first line, we simply divide 3300 by 2286. The result is 1 with a remainder of 1014, so the line is output once.

- For the second line, we add the remainder to `height` before starting, and so divide 4314 by 2286. The result is again 1, but with a remainder of 2028; the line is again output once but the remainder has doubled.

- For the third line, adding this new remainder to `height` means we divide 5328 by 2286. This time the result is 2 with a remainder of 756, so the line is output twice.

- The sequence carries on with the fourth line being output once, the fifth line twice, and so on. Each time, addition of the remainder makes any errors in the vertical scaling self-correcting.

This works whether we are scaling up or down. Let's assume we are outputting a normal resolution A4 fax (with 1143 scan lines) to a 9 pin dot matrix printer, whose 72 dpi gives us a `height` of only 792 lines on an A4 page. The sequence for scaling down is:

- For the first line, dividing 792 by 1143 gives us 0, so the first line isn't output at all, but we have a remainder of 792.

- For the second line, we add the remainder to `height` and then divide the result of 1584 by 1143. The result is 1, so the second line is output once.

- The new remainder of 441, when added to the `height` of 792 for the third line, gives us 1233; so the third line is also output once.

- Unfortunately, the new remainder of 90, when added to `height`, gives us only 882, so the fourth line isn't output at all. A remainder of 882 is carried forward.

- At the fifth line we add 882 to 792, giving 1674. We thus output the fifth line once and carry forward a remainder of 531 to the sixth line, which is also output once.

The pattern is that approximately one line in three is missed out, when the remainder drops below 351. However, after line 13 is missed, the remainder has reached 1062, sufficient to allow all three of the following lines to be output. Once again, the method shows it can make the best use of available resources.

Sample code for a vertical scaler is shown in Listing 5.8. As well as the `height`, it is passed the resolution, which should be 1 for normal resolution or

```
int vscale (long int height, char res)
{
    static int remainder ;
    if (res==0)
    {
      remainder=0 ;
      return(0) ;
    }
count=remainder+height ;
remainder=count%(1143*res) ;
count=count/(1143*res) ;
return (int)(count) ;
}
```

Listing 5.8: Scaling vertical lines with remainder adjustment

2 for fine resolution. If **res** is 0, this is taken as being a special case and the remainder is initialized to zero.

The method of borrowing dots to avoid losing data in short runs has no obvious parallel when scaling vertically, but it is just as essential that we avoid losing information when scaling down. We don't want detail on lines that aren't output to be lost. For instance, the lower-case letters c and e can easily be confused if the scan line with the horizontal stroke in the e happens to be missed out. The missing line cannot always be reconstructed from the context. For instance, if a fax consisted of hexadecimal codes, confusing either of these pairs would be quite a serious matter. This brings us to the second section of the scan line output routines, which is the code for generating data for output.

Generating data for output

All our output devices accept bit-mapped graphics output. Devices that can output single scan lines at a time all accept a stream of bytes which are read from left to right, as written. Table 5.3 shows how the first three bytes of such a bit-stream are used to encode the first 24 bits of each line.

Table 5.3: Mapping bits and bytes to dots for single scan line output

	Byte 1	Byte 2	Byte 3
MSB	dot 1	dot 9	dot 17
bit 2	dot 2	dot 10	dot 18
bit 3	dot 3	dot 11	dot 19
bit 4	dot 4	dot 12	dot 20
bit 5	dot 5	dot 13	dot 21
bit 6	dot 6	dot 14	dot 22
bit 7	dot 7	dot 15	dot 23
LSB	dot 8	dot 16	dot 24

The code for mapping out run lengths for devices working like this is basically the same as that for mapping the contents of the reference line when doing 2-D coding; the address and size of the array and run-lengths are the only real differences. Most page-based laser printers, and most screens, accept this type of single-line graphics data. The only addition to the code we used when mapping run lengths into a reference line is that we keep track of the final black bit on each line; this enables us to save time when reproducing images by not outputting trailing white space.

A modification to this mechanism is necessary for devices that have to output multiple lines of graphics at one time. The original of this type is the Epson 9-pin dot matrix printer, which outputs data in slices of 8 lines. (The ninth pin is unused in graphics modes.) While a stream of bytes is again output, each byte is used to encode the same bit for all 8 lines in the slice. The printer cannot print one line at a time, and has to print the whole slice. Table 5.4 shows how the first byte maps to the first bit of all the lines, the second byte maps to the second bit, and so on.

Table 5.4:   Mapping bits and bytes to dots for 8 scan line output

	Byte 1	Byte 2	Byte 3
MSB	line 1 dot 1	line 1 dot 2	line 1 dot 3
bit 2	line 2 dot 1	line 2 dot 2	line 2 dot 3
bit 3	line 3 dot 1	line 3 dot 2	line 3 dot 3
bit 4	line 4 dot 1	line 4 dot 2	line 4 dot 3
bit 5	line 5 dot 1	line 5 dot 2	line 5 dot 3
bit 6	line 6 dot 1	line 6 dot 2	line 6 dot 3
bit 7	line 7 dot 1	line 7 dot 2	line 7 dot 3
LSB	line 8 dot 1	line 8 dot 2	line 8 dot 3

Clearly, the code use for mapping a run-length for a single-line based output device is not suitable for use when the destination device outputs multiple lines in slices. In a sense, a device such as a laser printer, which takes single-line input, is analogous to a serial transmission, as it effectively handles single bits strung together in a sequence. On the other hand, a printer that outputs in slices is more like a parallel transmission, which manages whole bytes at a time but can't transmit a single bit at all.

For all output, an array is dynamically created by multiplying the number of lines in a slice by the number of dots across each page to obtain the bit count, and dividing by 8 when allocating the memory in bytes. Obviously, a printer driver for something like a Canon Bubblejet, which prints 48 lines in each slice, would require 48 times as much memory as a printer of the same resolution that accepts single lines. In fact, at 360 dpi the Canon Bubblejet has a resolution better than many laser printers. An 8.5" line at 360 dpi requires 3060 bits; 48 such lines require an array of 18360 bytes. If memory is tight, and 18K isn't going to be available, then the slice would have to be output in portions.

It is possible to develop a generic routine for outputting multiple lines of data which works for all cases where the slice is exactly divisible by 8. It is therefore suitable for basic 9 pin dot matrix printers, 24 pin near-letter quality printers and 48 nozzle inkjets. The difference is that while a single byte is sufficient when only 8 pins need to be controlled, three bytes per dot are needed for 24 pin printers, and six bytes per dot are needed for 48 nozzle inkjets that work on the same principle. Table 5.5 shows how the first three bytes are used to encode a single bit from each line of a 24-line slice.

Table 5.5:   Mapping bits and bytes to dots for 24 scan line output

	Byte 1	Byte 2	Byte 3
MSB	line 1 dot 1	line 9 dot 1	line 17 dot 1
bit 2	line 2 dot 1	line 10 dot 1	line 18 dot 1
bit 3	line 3 dot 1	line 11 dot 1	line 19 dot 1
bit 4	line 4 dot 1	line 12 dot 1	line 20 dot 1
bit 5	line 5 dot 1	line 13 dot 1	line 21 dot 1
bit 6	line 6 dot 1	line 14 dot 1	line 22 dot 1
bit 7	line 7 dot 1	line 15 dot 1	line 23 dot 1
LSB	line 8 dot 1	line 16 dot 1	line 24 dot 1

A general-purpose bit-slice output driver needs to know the number of scan lines in a slice. Dividing this by 8 gives the number of bytes in a slice. It also needs to keep a count of the number of scan lines left in each slice; for the first line in each slice this is always set to the total number of lines in the slice, and it is decremented each time a scan line is completed. When this count is decremented to 0, the slice can be output and the count is reset to the maximum again.

Once the number of bytes in a slice is known, the technique for skipping over white runs in a line is simply a matter of multiplying the number of bytes in a slice by the run length, and adding that number to the current index, because each slice skipped is equivalent to a bit skipped.

Mapping black runs in is more complicated. First we find out which bit needs to be ORed into each slice by generating a bit index `bit_inx` from the modulo division of the number of scan lines left in the slice by 8 (the number of bits in a byte). We then use `bit_inx` as a basis for shifting a single bit to generate the correct byte for setting a specific bit via a logical OR.

The next step is to find which byte in each slice needs to have the bit set. For this, we find the number of bytes left in each slice by dividing the number of lines left in each slice by the number of bits in a byte, and rounding up to the next boundary. We generate a slice index `slice_inx` by subtracting this from the total number of bytes in the slice.

We then loop for each bit in the run, and OR in the correct bit to the byte in the output array indexed by **byte_inx+slice_inx**. Before each repetition, we add the number of bytes in each slice to the **byte_inx**.

Note that Listing 5.9 also keeps track of the final black bit on each line; as described earlier, this helps us avoid printing trailing white space on each line.

```
if (color==WHITE) byte_inx+=( *runlength) ;
else
{
    c_bit = 1 ;
    if (bit_inx == 0) bit_inx = 8 ;
    if (bit_inx > 1) c_bit <<= (bit_inx-1) ;
    slice_inx= -((scanlines_left+7)/8) ;
    for (;;)
    {
      if (last_black_byte<=i) last_black_byte=(i+1) ;
      faxline[byte_inx+slice_inx]|=c_bit ; byte_inx+=  ;
      if (!(--runlength)) return(1) ;
    }
}
```

Listing 5.9: Universal routine for outputting n-bit slices

Catering for lost lines and inserted lines

We return to the subject of entire scan lines that risk being lost when scaling a fax image down to fit a lower resolution device. The way these are handled is straightforward. We avoid losing any lines by holding the data in an array which is initialized to white space at the start of each page, and re-initialized only when the array is output. If a call to **vscale** returns 0 and the line isn't output, the data is simply left in the array. If the output device is a printer which outputs slices consisting of multiple lines, the count of the number of lines left in the slice is not decremented if **vscale** returns 0.

When the next scan line is decoded, the black runs that are output are ORed into the array, while the white runs simply adjust the index. This means that if a line isn't output, the data it contains remains to be combined with that from the next scan line. Surplus lines are never simply ignored; they are combined with valid lines instead.

This technique assumes that black data is more significant than white. This assumption is usually justified, and is widely made in the fax business; it is well-known that white on a black background reproduces less well when transmitted than black on a white background.

When a fax is output to a higher resolution device, we have to cater for inserted lines when result **vscale** returns a number greater than 1. For single-line output devices, we just use the value returned by **vscale** as the control variable for a loop which calls our **output** routine.

For multi-line printers that output slices, we have to duplicate lines by isolating individual bits and ORing them in the next bit position. This may be in the next byte of the slice. As with all bit-manipulation, this is a fairly mechanical chore. The only complication is that the count of the number of lines left in the slice has to be decremented each time a line is duplicated. If the count is decremented to 0, the slice is printed. While the array holding the slice has to be re-initialized to white space, the very last line in the slice may now have to be replicated as the first line in the next slice.

As a sample which is typical of the sort of bit manipulation required, we present Listing 5.10. We take the last byte of the slice and shift it left seven times, moving the least significant bit to the most significant position and moving 0 bits in from the left. We then replace it as the first byte of the same slice, with all the remaining bytes being reset to white space. Of course, if **number_of_line** which was returned by **vscale** is 1, we can initialize the whole slice to 0, as the last line doesn't have to be replicated.

```
for (byte_inx=0 ; byte_inx<last_black_byte ; byte_inx+= )
{
    if (number_of_lines==1) faxline[byte_inx]=0 ;
    else faxline[byte_inx]=(faxline[byte_inx+( -1)]<<7) ;
    for (slice_inx=1 ; slice_inx<  ; slice_inx++)
    faxline[byte_inx+slice_inx]=0 ;
}
```

Listing 5.10: Replication of the last scan line in a slice to allow for scaling across slice boundaries

Catering for different devices

We conclude our presentation of decoding faxes with a few words on the physical output of the data to three types of device.

Dumping an uncompressed image to disk

The simplest output device isn't really a device at all; it is a disk dump of a decoded scan line to disk. This can be performed inside a wrapper for an uncompressed TIFF image, and is a convenient way of exporting received fax images to other programs. Uncompressed TIFF is a widely accepted format for this purpose.

Displaying an image on screen

The next simplest device is the option to display an image on a screen. It is worth noting that all the bits in the fax data are complemented before being written to the screen, because a totally blank screen is composed entirely of black space, and any data written to it appears as white dots. In contrast, a sheet of paper is composed entirely of white space, and data printed on it appears as black dots. In TIFF terms, the two have opposite Photometric Interpretations.

Outputting to printers

Most printers are controlled by escape sequences. The sequences for printers supported in the code on the accompanying disk can be found in the OUTPUT.DAT file. The data is always structured in the form of a count byte, followed by the commands themselves. The reason why a length byte is always needed (and why null-terminated strings are no good for this purpose) is because many printers can contain null bytes as part of a command. For example, most dot-matrix printers include their own 16-bit length word with graphics data, and this can easily contain a null.

All printers require an initialization function, which must contain whatever commands are needed as preliminaries before setting the printer up in the required graphics mode. Generally, the simpler the printer, the fewer commands are needed to complete this step. This function usually includes an operating system call to attach the printer and place it in raw mode (as opposed to cooked mode) so that characters such as ASCII EOF codes (0x1a) that occur as part of the data may be streamed to the printer transparently.

The line output function is broadly similar for all printers. The exact escape sequence varies, but the basic pattern is that of a sequence with an embedded count, followed by the image data. The count may be of bytes (in single-line printers). Details of whether anything needs to be sent after the data also vary; typically, multi-line printers follow a slice with a carriage return and line feed, while single-line printers either implicitly move one dot vertically, or need an explicit escape sequence. The process of actually generating the output data was where most differences between single- and multi-line printers were handled. Those now remaining are, by comparison, trivial.

Finally, a deinitialization function is needed for the end of each page. A form feed is an invariable part of this, and if the initialization sequence required placing the printer in a unique graphics mode (as with Hewlett-Packard laser printers), then the form feed is preceded by a return to normal ASCII operation.

6

The Structure of a
Fax Session

Introduction

This chapter examines exactly how data is transmitted from one fax to another. Like the way in which images are constructed and interpreted, this is also governed by one of the ITU recommendations. While T.4 covers how fax data is formatted, it is recommendation T.30 that determines how the data is sent.

If all you want to do is learn how to drive an EIA class 2 fax modem, only the next two pages (covering how a fax session is structured) are essential reading. The key feature of class 2 is that the modem itself takes care of most of the T.30 protocol, and you will normally only need to understand the basics.

If you want to program any class 1 fax modem, or need to use advanced features available on some class 2 modems (such as facilities for debugging fax sessions), then the whole of the chapter is required reading.

The five fax phases

The process of sending a fax is described in terms of these five phases, which in the simplest case of a single page fax would follow each other consecutively:

- *Phase A* handles call establishment and always begins the session. For a caller it consists of dialling the correct phone number and identifying the answerer as another fax machine. For the answerer, the phase consists of detecting the telephone ringing, picking up the line and identifying oneself as a fax machine. Phase A can be carried out either automatically or manually at each end.

- *Phase B* handles the pre-message procedure, and consists of a series of negotiations. The two fax machines compare facilities and options, decide what they have in common, assess the quality of the phone line and choose which parameters to use for the next phase. While this always follows a successful phase A, parameters can be negotiated more than once in a session. After transmission of fax data, either the sender or receiver can request re-entry to phase B so that previously agreed parameters can be renegotiated.

- *Phase C* comprises transmission of a page of T.4 fax data. While this always follows a successful phase B, a fax often has more than one portion of data (e.g. a fax with more than one page). In such circumstances, a batch of repeated instances of phase C (with no re-entry to phase B requested) will always be governed by the parameters negotiated during the single phase B that preceded the whole batch.

- *Phase D* always follows phase C. It consists of post-message exchanges, such as confirmations by the receiver that data has been received, and instructions by the transmitter as to what is coming next. Depending of what these exchanges consist of, phase D may be followed by any of phases B, C or E.

- Phase E terminates the fax session and hangs up the telephone line.

Figure 6.1 shows the basic relationship between these five phases diagrammatically.

Introduction to HDLC

Unfortunately, it isn't entirely practical to describe the working of the phases of a fax session immediately. The exchange of control information between fax machines is handled by a fairly complex communications protocol known as *HDLC*. This is an essential component in the workings of the T.30 session protocol, so before we go any further we explain briefly how it works.

Basic concepts

HDLC (*High-level Data Link Control*) is yet another international standard, this time under the jurisdiction of the ISO. HDLC and its many variants, such as SDLC (IBM), ADCCP (ANSI), LAP-B (used in X.25), LAP-D (in ISDN), LAP-M (part of the V.42 specification) and LLC (used in LANs) lie at the basis of many modern communication systems. It is one of the few items of computer technology whose formal correctness has been proved.

HDLC is a bit-oriented protocol consisting of frames of indeterminate length, which are in turn are composed of fields. Each frame must begin and end with a

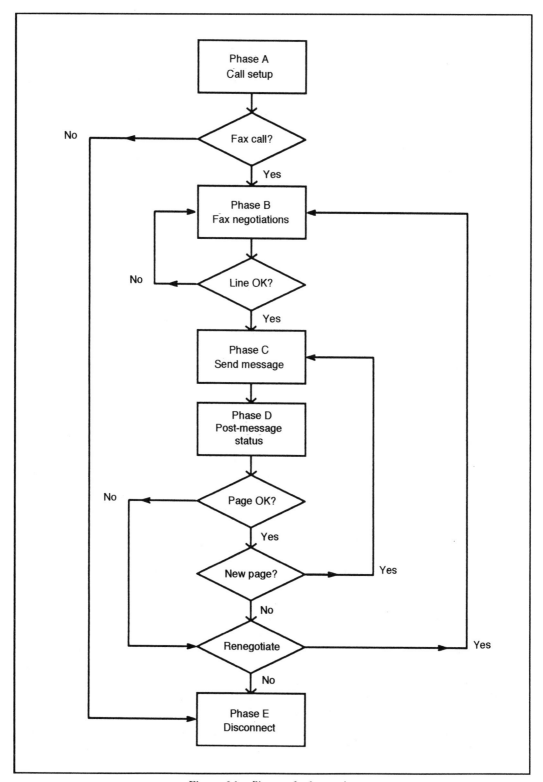

Figure 6.1: Phases of a fax session

unique bit pattern known as the flag sequence, consisting of the sequence 01111110. It is not allowed to occur anywhere within the frame.

Bit stuffing is used to avoid sending a bit pattern identical to that of a flag sequence as part of the data. Also known as *zero-bit insertion*, it consists of the sender inserting an extra 0 bit after any consecutive sequence of five 1 bits that may occur anywhere between the two flags. As we are dealing with bit patterns rather than characters, this stuffing is even performed across character boundaries in cases where bits formatted into ASCII or other character patterns are being sent. The prime requirement is that the flag sequence of six 1 bits must never be sent inside a frame. When the receiver detects a 0 followed by five 1 bits, it checks the next bit. If it is a 0 then it is discarded and the frame continues to be assembled with the next bit. This is called *zero-bit deletion*.

If six 1 bits have been detected, then whatever else may come in, the frame is finished. However, the receiver continues to check the succeeding bits. If the next bit after reception of 0111111 is 0, then a normal end of frame flag has been sent, but if the seventh bit is also a 1 (if the sequence 01111111 has been received), the whole frame has to be discarded. A sequence of between seven and 14 consecutive 1 bits is a signal for a *frame abort*. If the sequence of 1 bits carries on until 15 consecutive 1 bits have been received, that is taken as a signal for an *emergency link disconnect*.

Frame format

A frame sequence in T.30 begins with an initial *preamble*, consisting of flags sent at 300 bps for 1 s ±15%. This is intended both as a means of identifying the modulation scheme to be used, and as a way of synchronizing the receiver and transmitter. This preamble isn't used only at the start of a session, but is needed each time the line turns around. As fax sessions always operate in half duplex mode, this happens quite often. At least 5 seconds out of any fax session is taken up with these preambles, and the overhead is not insignificant.

All HDLC frames have the same basic format. Immediately following the opening flag sequence is an *address field*, used mostly to hold the destination of a frame. As the destination of a fax sent over the PSTN is fixed by the telephone number, this address field is fixed at 11111111.

The address is followed by a *control field*, used to tell the recipient what sort of frame is being sent, whether it is to be followed by more frames, and whether a reply is expected. All fax frames are what is known in HDLC parlance as *unnumbered frames*, so the control field always has the form 1100x000. The fifth bit from the left in the control octet, marked with an x, is known as the *P/F* bit (poll/final), and is always set to indicate to a receiver that a reply is expected (poll) or that the frame is the last of a series (final). The control field is thus set to 11001000 in a final frame, and 11000000 for intermediate ones.

Both of these fields are *octets*, which simply means they are eight bits wide. They are followed by an indeterminate length *information field*.

In the fax implementation of HDLC, the information field is usually divided into two distinct portions. The first 8 bits are always present, and are collectively known as the Facsimile Control Field (*FCF*). Some FCFs are followed by a Facsimile Information Field (*FIF*), though the majority of frames consist solely of an FCF.

One important restriction on a frame is that its transmission must not take longer than 3 s ± 15%. For a receiver this means that if more than 3.45 s elapse between receipt of the opening and closing flags, the whole frame must be discarded. For a transmitter this means no transmitted frame should take longer that 2.55 s to send. Between these limits, the legality of a frame is unspecified, and whether it is accepted or rejected depends upon the implementation.

The frame data ends with a *frame check sequence (FCS)*, calculated using the bits in the frame starting from that one after the end of the opening flag and ending with the bit immediately before the start of the FCF field itself. The FCS is worked out after zero-bit deletion has been performed, so stuffed bits are excluded, and it is transmitted as a 16-bit binary number with the most significant bit sent first.

The recipient of a frame ignores all incoming bits while searching for an initial flag sequence. This phase is often known as *hunt mode*. All the bits received after this are buffered while a continual watch is kept for the closing flag sequence. Once this is detected, the FCS is calculated for all the data stored from the end of the initial flag up to a point 16 bits before the start of the final flag. The calculated FCS is compared with the figure contained in those last 16 bits of the frame, and if the two match the frame is assumed correct; if the frame check sequences are different, it is discarded.

Figure 6.2 summarizes the structure of a typical HDLC frame as seen by a fax machine.

Flag	Address	Control	FCF	Data	CRC-16	Flag
01111110	11111111	11001000	110000001	000000000111000010111101	0001000001001011	01111110

Figure 6.2: Structure of an HDLC frame

Bit numbering in octets

There is a definite terminological problem when referring to those bits that make up an HDLC octet. It may seem ludicrous, but in the absence of further information a reference to bit 2 of an otherwise unidentified eight bit sequence could apply to any one of four possible candidates.

Normally, an eight-bit byte has a least significant bit (LSB), which is the bit set in the representation of a binary 1, and we follow the usual arithmetic convention of placing this as the right-most bit (00000001). An eight-bit byte also has a most significant bit (MSB) which is at the other end, and is the bit set in the binary number 127 (10000000). The LSB and MSB are also known as the *low bit* and the *high bit*, respectively.

There are three conventions that use a similar LSB $\rightarrow$ MSB directional approach for referring the individual bits of a byte:

- The most common is to call the LSB bit 0, and the MSB bit 7.

- A rather confusing alternative is to refer to the bits by the hexadecimal value that the entire byte would have if just one particular bit was set. Using this convention, the byte has the pattern 00000001 if only the LSB is set, so the LSB becomes bit 01h. As the MSB is set in the pattern 10000000, the MSB is known as bit 80h.

- The third approach simply refers to the bits in a byte from right to left as bits 1 through 8. This corresponds to the order in which the bits in a byte are sent and received using asynchronous protocols.

However, as can be seen from Figure 6.3, when they are received using bit-oriented synchronous protocols, bits are almost invariably numbered from left to right; this is the convention adopted here. Bits are numbered in this way when received via HDLC because the data in a frame is essentially unstructured. The bits that make up the octet are neither numbers nor asynchronous data bytes, but simply bits following each other. The leftmost bit is the first bit sent in any HDLC octet, and is the first bit received. Whatís more, as thereís no need for the data to be sent packaged conveniently into lumps of 8, it could quite legitimately be the only bit there is. Thus, it would make no sense to refer to it as anything other than bit 1.

MSB							LSB	
7	6	5	4	3	2	1	0	Convention 1
80h	40h	20h	10h	08h	04h	02h	01h	Convention 2
8	7	6	5	4	3	2	1	Convention 3
1	2	3	4	5	6	7	8	Synchronous

Figure 6.3: Four possible methods of referring to bits

Identification of fax frames

Describing HDLC in words is more daunting than analysing the frames in practice when a class 1 fax modem is being used. As it includes synchronous to asynchronous conversion facilities, a properly functioning modem will deliver a frame to your applications software after checking the FCS and indicating whether the frame is good. All the data inside the frame is supplied neatly bundled into byte sized chunks which are all 8 bits wide. We continue to refer to these as octets to reinforce the fact that they aren't necessarily characters, and ought to be viewed as simple bit patterns.

Since flags are not generally forwarded (they delineate the frame rather than being a part of it), the first octet is always going to be the fixed fax address of 11111111, which we can safely ignore. The second octet is the control field, which simply tells us whether another frame is due to follow or whether the sender is expecting a reply, which we can look at later if we need to. Consequently, the first octet we really need to examine is the third in the frame. This is the FCF, which tells us what the purpose of the frame is and what other information it might carry.

Frames sent as part of a fax session fall naturally into a number of discrete groups according to both their meaning and structure. The first group consists of initial identification and information frames used in negotiations; the second group consists of command frames which control a session; and the third group consists of response frames which provide a means of monitoring a session.

The circumstances under which these frames are sent, together with their order and timings, are what the T.30 fax session protocol is all about. We are now ready to start looking at it in detail.

Summary of the phases of a fax session

Phase A

Phase A is slightly more complicated than simply dialling, as it includes what the ITU term *call events*. The sound of these rather than the names are what will be familiar to most people. The two tones that are used in Phase A are the calling tone (*CNG*) and the called station identification (*CED*). The CNG tone is the intermittent bleep heard when you pick up the phone and there's a fax at the other end. The frequency is 1100 Hz, and it remains on for 1/2 s and goes off for 3 s before repeating. It plays no part in the transmission itself, and is used for all fax transmissions in groups 1, 2 or 3. The tone is simply to identify the call as coming from a non-human, and is supposed to be mandatory for all automatic fax machines. Its use is optional for manually operated machines, as the operator should be able to give a suitable verbal message.

The CED tone is the high-pitched whine heard when you dial and get through to a fax machine by mistake. This has a frequency of 2100 Hz and should be on continually for a period of between 2.6 and 4.0 secs, starting from 1.8 to 2.5 s after the call has been answered. Like the CNG tone, the CED plays no part in the communication itself, and is there so that a human listener knows they have connected to a fax and not a person. The CED is always followed by an interval of 75 ± 20 ms.

Once the call has been answered (or immediately after blind dialling in situations where it is not possible to detect whether a call has been answered), the caller sends the CNG tone described above. In the gaps between the tones it listens for the preamble signal to be sent from the answering station, after which it expects to receive an *initial identification* signal. It repeats the CNG tone if it doesn't receive these, and continues with this until a timeout expires. The length of the timeout, T1, is part of the T.30 recommendation, and is set to 35 ± 5 s.

Meanwhile, at the other end of the telephone link, the answering station picks up the line and sends out its CED tone. It follows this blindly with its preamble and the initial identification signal expected by the caller. It then listens for a suitable response indicating that phase B has begun. If it hears nothing within 3 s ± 15%, it repeats the cycle and carries on until its own timer T1 expires.

So although the CNG and CED tones are the most obvious part of phase A, they play no part in any fax negotiating. The first exchange of information is when the calling modem detects the initial identification and sends a suitable response.

Initial identification

This term refers to the first negotiating frames which the answerer speculatively transmits as part of phase A. They include both fax capability and identification information. While it is the caller that initiates a session, it is the answerer who sends these first frames. In all subsequent stages, the session is always controlled by the transmitter, with the receiver in an essentially passive role. Only the transmitter ever issues command frames, and only the receiver ever issues response frames. Until the first negotiating exchange, there is no way of knowing which station will be receiving and which will be sending. Though it is the originating station that usually wants to transmit a fax, the fax session protocol allows for *polling*, the common term for a caller telling an answering station that it is prepared to receive any faxes the answerer has for transmission.

Up to three frames are sent as part of the initial identification by the answering station in phase A. Their FCFs all start with the bits 0000:

● The only compulsory frame is that carrying the Digital Identification Signal (*DIS*), which gives the caller details of what capabilities of the answerer has. Its FCF has the format 00000001. We've seen that there are a number of options possible in how the T.4 recommendation is implemented, including such essentials as the ability to transmit or receive, communication speeds, fax resolution and paper size. The answerer's capabilities are carried in a number of bitfields in the FIF accompanying this frame.

The DIS is always the last of the frames sent in an identification sequence, so it is essential that the control field in this frame has the format 11001000. Bit 5 in the control octet is always set to indicate that a frame is the last of a series, and that a reply is expected. Conversely, the optional frames possible in the sequence before the DIS always have the control field set to 11000000.

The DIS is usually preceded by the optional Called Subscriber Identification frame (*CSI*). Its FCF has the format 00000010. The CSI frame conveys the identity of a fax machine by sending its telephone number as 20 ASCII characters in the facsimile information field. The format of the data is modified, as it is sent backwards, that is, from right to left. The T.30 specification states that the LSB of the least significant digit should be transmitted first.

Fax machines that keep a log of all calls received usually record the information contained in this frame. The official T.30 documentation specifies that the CSI frame must contain a valid international telephone number, including only the + sign, spaces and digits. It gives the table shown in Table 6.1, showing the representation of these allowable characters.

Table 6.1: Character codes for the T.30 text information frames

Character	MSB->LSB
+	00101011
0	00110000
1	00110001
2	00110010
3	00110011
4	00110100
5	00110101
6	00110110
7	00110111
8	00111000
9	00111001
Space	00100000

Alert readers will notice that the above codes are identical to the standard ASCII codes for the same characters. Many fax machines now also include ASCII letters as part of the CSI frame, though this probably isn't a good idea as the use of letters prevents the ID from being recognized by a machine that only supports digits. However, where letters are supported, they too are sent backwards in the same way as the ASCII codes for the officially supported digits listed above.

- The DIS can also be optionally preceded by a Non-Standard Facilities frame (*NSF*). Its FCF has the format 00000100. Whether driven by the needs of users to send more detailed documents at higher speeds and with better security, or by manufacturers continually looking for improvements that might give them a competitive edge over their rivals, improvements in the performance of fax machines are likely at any time. Being aware of the fact that telecommunications technology advances so rapidly, the ITU made sure that the T.30 standard was flexible enough to incorporate customized enhancements by using the NSF frame.

The NSF frame must include a FIF of at least two octets, the first consisting of a recognized ITU country code. The second octet of a non-standard FIF is reserved for a country to allocate codes among its domestic manufacturers.

The third and any subsequent octets are entirely customizable, and if their contents are recognized by the caller, there is no real limit to what could be added to the fax session protocol. It is not unusual for a number of NSF frames to be sent, indicating that a number of different manufacturers' extensions are supported by a particular fax machine, and that a fax can emulate one of a number of models.

The only official restriction on what can be included in the NSF is that, like all frames, it must not take longer than 3 s ± 15% to transmit. In practice, NSFs are considerably shorter than the theoretical maximum of 90 octets. If the NSF isn't recognized by the caller then it simply ignores it.

Once the DIS frame has been sent by the answering station it waits for a command from the caller. If this doesn't arrive within 3 s ± 15% it returns to the start of phase A and repeats the cycle beginning with the CED. Once a command acknowledging the DIS has been received, however, phase B of the session is deemed to have begun.

Phase B

Phase B consists almost entirely of an exchange of HDLC frames which together determine the parameters governing the message transmission itself. The frames the answerer sends as its initial identification sequence are the first of these, and the originating station can respond in a number of different ways. The most common reply to an identification sequence is for the originator to send a *command to receive*. By issuing this, a station claims the role of transmitter and the consequent right to control the rest of the session.

The possibility of polling means the command to receive is not issued uniquely by a caller in response to an initial identification sequence. It could equally be issued by an answerer in response to a poll command. For some reason I've never been able to follow, the FCFs for virtually all the frames in a T.30 session have the first bit set to indicate which of these possibilities has occurred. If the first bit is a 0, then a frame is being issued by the same station that sent the DIS; if this bit is a 1, a frame is being issued by the station that responded to the DIS. The T.30 recommendation denotes this bit of the frame control field with an X in the first position.

The command to receive, like the initial identification sequence, can contain three types of frames. All begin with the four bits X100:

- The only compulsory frame is the one carrying the Digital Command Signal (*DCS*). Its FCF has the format X1000001, and it is always the last frame sent as part of the command to receive, so its control field will always have bit five set to 1.

The DCS frame corresponds to the DIS frame sent as part of the initial identification. It includes a FIF with a very similar format. What differences there are arise from the fact that while the DIS frame conveys the capabilities of a station, the essentially passive role of a receiver means that the selection of which of the available options are to be used has to be left to the transmitter. For example, the DIS frame might indicate that a station has the capability of receiving at any fax speed from 2400 bps to 9600 bps, but the decision as to which of those speeds will be used is taken by the transmitter. It notifies the receiver as part of the FIF field of the DCS frame in the command to receive. The details of the FIF bitfield for the DIS and DCS frames are all listed in Table 6.2.

- The DCS is usually preceded by the optional Transmitting Subscriber Identification frame (*TSI*). Its FCF has the format X1000010. The format of the facsimile information field of the TSI is the same as that of the CSI frame, consisting of a 20 character ASCII coded telephone number.

- The DCS can also be optionally preceded by a Non-Standard Facilities Setup frame (*NSS*). Its FCF has the format X1000100, and like the NSF frame, it is manufacturer or user dependent and must include a FIF of at least two octets, of which the first must be a recognized ITU country code. This frame should only be sent if the initial identification sequence included an appropriate NSF frame, and it should never be spontaneously generated.

DIS/DTC/DCS Facsimile Information Bit Fields

We now summarize the information conveyed in the facsimile information field (FIF) of the following closely related frames:

- The Digital Identification Signal (*DIS*) sent by an answering station as part of the initial identification sequence. To anticipate slightly, the facsimile information field for this frame has exactly the same format as that used in the Digital Transmit Command (*DTC*) sent as part of a polling sequence. You will therefore see occasional references to this frame along with the DIS.

- The Digital Command Signal (DCS) sent as part of the command to receive prior to all transmissions

Table 6.2 covers the 24 bits in the FIFs of the above frames common to all group 3 fax machines. Although these FIFs have been extended many times by the ITU, support for most optional extensions is by no means common. Many fax machines, fax modems and fax software offer what are essentially the same facilities, all of which are based on the original bit field definitions. Any fax machine implementing FIFs of a longer length is downwardly compatible with the original three octets, as they are all able to recognize that if bit 24 in a DIS is set to 0,

Table 6.2: Meanings of bitfields in DIS/DTC/DCS frames

Bit	DIS/DTC	DCS
1	0 (Ability to send Group 1)	0
2	0 (Ability to receive Group 1)	0 (Command to Receive Group 1)
3	0 (Group 1 IOC=176)	0 (Group 1 IOC=176)
4	0 (Ability to send Group 2)	0
5	0 (Ability to receive Group 2)	0 (Command to Receive Group 2)
6	0 (reserved for Group 2)	
7	0 (reserved for Group 2)	
8	0 (reserved for Group 2)	
9	1 Ability to transmit Group 3	
10	1 Ability to receive Group 3	Command to Receive Group 3
11	data rate table	data rate table
12	data rate table	data rate table
13	data rate table	data rate table
14	data rate table	data rate table
15	Ability to use fine resolution	Command to use fine resolution
16	Ability to use 2-D coding	Command to use 2-D coding
17	page width table	page width table
18	page width table	page width table
19	page length table	page length table
20	page length table	page length table
21	minimum scan line time table	minimum scan line time table
22	minimum scan line time table	minimum scan line time table
23	minimum scan line time table	minimum scan line time table
24	extend field	extend field

then no extensions beyond bit 24 are supported. If a fax machine which has only basic capabilities connects to a machine that can handle a number of extensions, and receives a DIS frame that has bit 24 set and one or more extra octets following the third and last compulsory one, it simply ignores the extensions and responds with a basic 24-bit DCS frame.

The convention used for the bitfields in our table is that a value of 0 in a DIS/DTC frame is interpreted as meaning that an option is unavailable, while a 1 indicates that the station does have a specific capability. As DCS fields are basically commands sent from the transmitter to the receiver rather than options, a 1 is interpreted as a selection that will be used, while a 0 means that an option is not selected.

The meanings of the bit fields in the table are pretty obvious. All fax modems, which are incapable of supporting group 1 or 2 transmissions should ensure that bit fields 1–8 inclusive are always set to 0 as they have no application to group 3 (digital) fax.

One bitfield that needs a little reflection is bit 9, by which an answering machine indicates that it has the ability to transmit. This bit is always set in the DCS frame; after all, there's no point in issuing a command to receive if you haven't the ability to send anything. However, when this bit is set in a DIS frame, it means a fax machine has documents available for polling. A system that only wants to receive should always reset this bit to 0.

Bits 11–14 and 17–23 should be interpreted by consulting the four tables 6.3, 6.4, 6.5 and 6.6.

Table 6.3 shows how the values of bits 11, 12, 13 and 14 are used to indicate the speed capabilities of a fax machine and the eventual selection by the transmitting station. This is where there is the greatest difference between the meaning of bits in the DIS/DTC capability frames and those of the DCS command frame. Some knowledge of the speeds used by the various modulation schemes is essential in knowing how they work. The modulation schemes themselves operate transparently to the user.

Table 6.3: Data rate table for bitfields 11–14 in DIS/DTC/DCS frames

Bits 11-14	DIS/DTC speed capability	DCS rate selected
0000	V.27 ter fallback	2400 bps V.27 ter
0100	V.27 ter	4800 bps V.27 ter
1000	V.29	9600 bps V.29
1100	V.27 ter V.29	7200 bps V.29
0010		14400 bps V.33
0110		12000 bps V.33
1010		
1110	V.27 ter V.29 V.33	
0001		14400 bps V.17
0101		12000 bps V.17
1001		9600 bps V.17
1101	V.27 ter V.29 V.33 V.17	7200 bps V.17
0011		
0111		
1011		
1111		

Versions of the T.30 specification with support for speeds in excess of 9600 bps are relatively new. Until 1992, bits 13 and 14 were unused but reserved. Although we stated that all extensions would be left until later, we have made an exception in the case of higher fax speeds, because the bits used to define these speeds were already reserved in the original version of T.30, and their introduction is an inherently simple addition which offers increased reliability and speed with little extra complexity. (The discussion of modem modulations and speeds that follows supplements our earlier discussion of fax modulation schemes, which readers should refer to if they are unclear on any of the details.)

- V.27 ter supports transmission at 4800 bps with a fallback to 2400 bps. This was the original modulation scheme selected by the ITU in their quest to send a typical A4 page in a minute. 2400 bps transmission using V.27 ter is the most resilient way of sending a fax under bad line conditions, and is the lowest speed common to all fax machines. It conveniently has a special FIF bit capability coding all of its own. The technology used in V.27 ter synchronous transmission is of the same type, and dates from the same era, as that used in V.22 1200 bps asynchronous modems, and is based on DPSK (differential phase-shift keying). However, most V.22 data modems were capable of asynchronous full duplex communications only, and therefore were unable to support fax operations.

- V.29 is now the most common modulation scheme used for fax, and supports 9600 bps transmission with a fallback to 7200 bps. (Note that 4800 bps transmission at V.29 is not supported as it offers no advantages over the mandatory V.27 ter 4800 bps transmission rate.) As with V.27 ter, the technology used in V.29 synchronous transmission mirrors that of the V.22 bis 2400 bps full duplex asynchronous modems developed at the same time, both based on quadrature amplitude modulation. V.22 bis modems were originally incapable of supporting fax operations for the same reason as V.22 modems didn't support fax, as they couldn't operate synchronously. The introduction of error correction and data compression methods such as MNP5 and V.42/V.42 bis required that asynchronous modems must also be able to operate in synchronous mode. The development of integrated chipsets to handle this requirement probably explains why 2400 bps data modems with error correction and data compression facilities can so easily be supplied with 9600 bps fax support at virtually no extra cost, and their introduction from 1990 onwards led to the sudden growth in the availability and popularity of fax modems as we know them today.

- V.33 supports 14400 bps transmission with a fallback to 12000 bps. It dates from the same time as V.32 9600 bps asynchronous modems, and one V.32 option uses the same trellis coding as V.33. However, V.33 was designed for leased lines rather than public telephone networks, and though covered in the T.30 specification, it isn't seen on normal fax units designed to connect to ordinary phone lines.

- V.17 supports 14400 bps transmission with fallbacks to 12000, 9600 and 7200 bps. It is basically a half duplex fax version of the V.32 bis full duplex standard, and is often supplied as part of the same integrated chipset in the same way that facilities for V.22 bis data and V.29 fax are often supplied together. Where two modems support V.17, this scheme should be used in preference to the same data rate using V.29. This is because 9600/7200 bps transmission using V.17 trellis coding is more reliable than 9600/7200 bps transmissions using V.29 QAM.

Apart from the special case of V.27 ter fallback, it is assumed that, where a DIS or DTC frame includes a capability for any modulation scheme, all the speeds included in the scheme are covered. This means that every fax system with V.17 capability must offer it at all speeds from 14400 bps down to 7200 bps.

Table 6.4 shows how the values of bits 17 and 18 are used to indicate the page width capabilities of a fax machine and the eventual selection by the transmitting station. Table 6.5 shows how the values of bits 19 and 20 are used to indicate the page length capabilities of a fax machine and the eventual selection by the transmitting station.

Table 6.4: Page width table for bitfields 17–18 in DIS/DTC/DCS frames

17-18	DIS/DTC page width capability	DCS page width selected
00	1728 dots in a 215mm line	1728 dots in a 215mm line
01	1728 dots in a 215mm line	
or	2048 dots in a 255mm line	
or	2432 dots in a 303mm line	2432 dots in a 303mm line
10	1728 dots in a 215mm line	
or	2048 dots in a 255mm line	2048 dots in a 255mm line
11	invalid but interpreted as 01	invalid but interpreted as 01

Table 6.5: Page length table for bitfields 19–20 in DIS/DTC/DCS frames

19-20	DIS/DTC page length capabilities	DCS page length selected
00	A4	A4
01	unlimited	unlimited
10	A4 and B4	B4
11	invalid	invalid

Table 6.6 shows how values of bits 21 to 23 are used to indicate the minimum scan line required by the receiver, together with the eventual selection by the transmitting station. Though the way this figure is negotiated is part of the T.30 recommendation, the available options are themselves defined as part of T.4. This field is unlike others in the DIS/DTC/DCS FIF set, as it has no real meaning for the transmitter, which simply does what it is told.

Table 6.6: Scan line time table for bitfields 21–23 in DIS/DTC/DCS frames

21-23	minimum scan line time required, scan line time used
000	20 ms both normal and fine resolution, 20 ms
100	40 ms both normal and fine resolution, 40 ms
010	10 ms both normal and fine resolution, 10 ms
001	5 ms both normal and fine resolution, 5 ms
110	10 ms normal resolution and 5 ms fine resolution
011	20 ms normal resolution and 10 ms fine resolution
101	40 ms normal resolution and 20 ms fine resolution
111	0 ms both normal and fine resolution, 0 ms

Polling

We have already made extensive reference to the existence of polling as a standard facility available in the T.30 recommendation, though it is not implemented by all fax machines and is ignored by many class 2 fax modems. Polling is initiated by responding to the initial identification sequence with a *command to send*, often known as the polling command, and is only possible if the remote machine had bit 9 if its DIS frame set to indicate it has the facility of being polled. This means it is capable of transmitting faxes. If bit 10 is also set, the remote station has the ability to receive too. It isn't unusual to find that fax machines set up to be polled answer calls by sending out a DIS frame with bit 9 set and bit 10 clear, forcing a caller either to hang up or poll.

There are up to three frames in this command, and their FCFs all start with the four bits 1000:

- The compulsory polling frame is that one carrying the Digital Transmit Command (*DTC*). This is the converse of the DIS frame, and it gives the answerer details of what a caller's capabilities might be. Its FCF has the format 10000001. As it is the last frame in the command, this is indicated in the control field by setting bit five to 1, whereas all the optional frames in the command would have bit five in their control frames reset to 0. The FIF has the same format as that of the DIS – refer to Table 6.2 for more details.

- The DTC is usually preceded by the optional Calling Subscriber Identification frame (*CIG*). Its FCF has the format 10000010. The format of the facsimile information field of the CIG is the same as that of the CSI frame.

- The DTC can also be optionally preceded by a Non-Standard facilities Command frame (*NSC*). Its FCF has the format 10000100, and it must include an FIF of at least two octets, of which the first must consist of a recognized ITU country code. Like the NSF frame, it is manufacturer or user dependent, but while the NSF frame is sent unsolicited in the hope that the recipient will be able to comply, and can quite safely be ignored, the NSC frame should only be sent in response to the receipt of a compatible NSF, and its appearance in any other context is a sign that something is amiss.

Other possible responses to an initial identification sequence

While the sequences ending in the DTC and DCS frames are the most common responses to the DIS frame, they are not the only ones possible.

One obvious possibility is cancelling the entire session, and the answerer should be able to cater for the caller hanging up at this stage. While this is a common occurrence when a human has called a fax machine in error, it may also be a result of a decision by the originator. For instance, the CSI frame might have been used as some sort of security check. Alternatively, the DIS frame might have revealed that the answerer didn't have the capability to receive the document that was being sent.

The simplest of the alternative responses is that a DIS can itself be used to respond to a DIS. The use of a second DIS as a response to an initial DIS is effectively a method of an originator ceding control of the session to the answering station, and is obviously dependent on the answerer being able to transmit as well as receive. However, the T.30 recommendation indicates that if a DIS is received that is identical to one that was sent, it should be ignored.

Another possibility is that the use of non-standard procedures by both sides could take the session out of the realms of T.30 recommendation altogether. This is something that would start with the NSS command, but where it would end up is a completely open question.

The main area that needs to be catered for is where the DIS or one of the preceding frames is either corrupted because of line noise (when the FCS will be incorrect), or doesn't appear to make sense, or wasn't received at all as the other station wasn't ready for it. We have already identified one way of handling this, for the unique situation where an initial identification signal is being sent after a CED tone to a station that may still be in phase A. In this case, the CED tone is resent, followed by the initial identification sequence, and this continues until timer T1 expires.

However, there is a more general set of error handling procedures which comes into effect in similar situations once phase B has been firmly established at both stations. Such procedures are not confined to problems with identification sequences, but can be used as needed to resolve problems with any frames.

Error handling procedures

The main source of errors we need to consider is frame corruption caused by transmission errors on the telephone system. There are a number of ways in which this type of error can disrupt a session. One particularly hazardous phenomenon is when a lost flag stops a frame from being recognized. This can often be detected only by using a timeout mechanism. An easier error to detect is when the contents of a frame are corrupted, as in this case the FCS doesn't match the contents. Other internal consistency validations, such as checking if the last frame in a command has bit 5 of the control octet set, can also reveal that errors have occurred.

However, a second source of errors is caused by the mechanical or electrical limitations of the equipment being used, ranging from unavoidable but recoverable timing violations to unrecoverable errors and serious data loss. These occur whether stand-alone fax machines or computers and fax modems are used. From a sender's viewpoint, it makes no difference whether a delay in responding to a command is the fault of slow printing or paper feed on a fax machine or of a slow computer system. Though some specific problems are unique to particular systems, such as buffer overruns and serial port errors, which only affect computers, the protocol for handling errors needs to be something that is common to all fax installations.

There are two special frames used in error handling, neither of which have an accompanying FIF, and both of which have the first four bits of the FCF set to X101:

- The Command Repeat frame (*CRP*) can be sent as a response to an otherwise valid frame that appears to have been sent in error. A CRP frame facsimile control field has the format X1011000. If this frame is received at any point when a response is expected, the station receiving it should retransmit its last command entirely. A CRP is a valid response to an initial identification signal that arrives corrupted.

    The CRP frame is a T.30 option, and not all machines implement it. Those machines that don't do so recover from frame transmission errors by using the requirement that all responses to commands must be received within 3 s ± 15%. If no response is received, the sender should retransmit the command.

    Whether a command is retransmitted as a result of a timeout or receipt of a CRP, no frame may be sent more than three times in succession. If a valid response is not detected after the third attempt to send a frame, the sender should disconnect.

- The Disconnect command (*DCN*) should be sent when a command has been repeated three times and either failed to generate a valid response or has been met with a CRP on each occasion. The DCN frame facsimile control field has the format X1011111, and no response is expected. Following either transmission or reception of a DCN, a station should disconnect the line.

We have already mentioned one basic rule in T.30, which is that once the first part of phase B has been completed, and the roles of two fax stations as transmitter and receiver have been established, it is the transmitter that controls the session. Only the transmitter may send commands; the receiver must respond to all commands, and the receiver is only allowed to send responses when it has received a command. Unsolicited frames, whether commands or merely queries, may not be sent.

The DCN frame is an exception to this, as disconnecting is an action a receiver can take unilaterally. A receiver has to disconnect when its synchronization with the transmitter has broken down, which could occur when a command is received that the receiver cannot possibly respond to, and is repeated if queried with a CRP frame or by permitting a timeout. Another possible reason for disconnecting is when a 3 s transmitter timeout has been permitted, but no command is resent. Disconcertingly, some fax systems will disconnect for this reason even when a perfectly good command is received and was not replied to because the receiver wasn't ready.

Disconnection also occurs if no commands are received for a long enough time. A receiver determines when this has happened by keeping a timer, T2, set initially to 6 ± 1 s and reset every time an HDLC flag arrives. If T2 times out when a command is expected, the receiver has to abort the session. For this reason, a transmitter that has to perform an operation that might take some time, but still wants the receiver to remain on-line, is advised to send continual flag octets to keep T2 from expiring and taking the session with it.

Apart from the transmitter timeout of 3 s with no response (which the receiver can also enforce) and the receiver timeout of 6 s with no command, there is one other timing consideration used both to determine when errors have occurred and to limit their effects – the rule that no frame is allowed to exceed 3 s ± 15%. It is needed because if there were no maximum frame size, the loss of one HDLC closing flag could prove fatal. With no way of discarding incomplete frames, when the command came to be retransmitted there would be no way of distinguishing its opening flag from the lost closing flag of the previous attempt at transmission.

Negotiating session parameters

The negotiation process consists of the transmitter working out the best set of parameters common to both stations and then notifying the receiver of the results. What constitutes a 'best' set is an open question, and it is up to the programmer of fax software or the designer of a fax machine how much latitude the user is allowed in determining this.

The general presumption is that where an additional option is available and has not been specifically disabled, then it ought to be used. So where two fax machines have both normal and fine resolution capabilities, fine resolution will be used. Where both 1-D and 2-D coding are permitted, 2-D coding will be used. For all connections, the highest practical transmission speed will be used. However, the highest speed common to any two machines isn't necessarily the best to use for any particular connection. It is generally accepted that while speed of transmission is a good thing, it should never be sought at the expense of reliability, and that if there is a trade-off to be made, the speed ought to be reduced to gain a more reliable connection.

Some telephone connections are better than others, and some call destinations are always more prone to transmission problems. A call between two telephones on the same modern all-digital exchange always gives a more reliable connection than one made between two old Strowger exchanges, and a fax machine that can whiz pages through in around 20 seconds using 14400 V.17 connections needs some way of telling when a transmission has to crawl along at 2 minutes per page using 2400 bps. Since negotiations between fax machines take place using a highly reliable 300 bps V.21 connection, agreement on a common speed via the initial DIS/DTC/DCS frames is no indication of whether sending data 12 times faster using a completely different modulation scheme is going to give satisfactory results.

The practicality of a particular speed is determined by the transmitting station sending a so-called Training Check Frame (TCF). This isn't a proper HDLC frame, and consists simply of series of 0 bits sent for 1.5 s ± 10%. The sequence of events at the transmitter is as follows:

- The transmitter sends a DCS frame that includes the desired transmission speeds and modulation scheme in bits 11–14.

- Transmissions at 300 bps via the V.21 channel 2 modulation scheme used for negotiating then cease.

- The transmitter observes the compulsory delay for a period of 75 ± 20 ms between all consecutive transmissions where there is any change in the modulation scheme.

- The TCF is sent using the modulation scheme and high speed specified in the DCS.

- The transmitter then stops high speed transmissions and starts to listen for a response from the receiver to the TCF using V.21.

Meanwhile, the concurrent sequence of events at the receiver is as follows:

- Following successful receipt of the DCS frame, the receiver identifies the required modulation scheme from bits 11–14.

- The receiver switches from V.21 reception and awaits the TCF at the high speed negotiated.

- Knowing that the TCF data is supposed to consist of 0 bits and last for at least 1.35 s, the receiver evaluates the quality of line from the data actually received.

- Once high speed TCF transmissions have stopped, the receiver returns to V.21 and transmits one of two possible pre-message response frames at 300 bps. All pre-message response frames have FCFs beginning with bits X010 (remembering that X is set to 1 by the station receiving the last DIS).

If the TCF indicates the line is good enough to receive a fax at the required speed, the receiver sends a Confirmation to Receive Frame (*CRF*). The FCF of this frame has the format X0100001, and there is no accompanying FIF. Once a CRF is transmitted, the receiver returns to the higher speed negotiated to await the fax data, while a transmitter receiving a CFR switches back to the higher speed and begins to send the fax. The CRF indicates the end of phase B, while the transmission of fax data is the province of phase C.

On the other hand, if the TCF indicates to the receiver that the line cannot support a particular speed, it sends back a Failure To Train frame (*FTT*) instead. The FCF of this frame has the format X0100010, and like the CRF frame, there is no accompanying FIF. Following an FTT frame, the receiver remains at V.21 and awaits a command, while the transmitter composes and usually sends another DCS frame indicating what it intends to try next. The only other possible response is to send a disconnect command and hang up the line.

These details leave open a couple of questions on which the T.30 recommendation is deliberately vague. The first concerns the exact procedure the receiver ought to adopt in evaluating the content of the TCF, while the second is the exact strategy a transmitter ought to adopt if it receives the FTT frame.

Clearly, a receiver that detects a continual sequence of 0 bits for at least 1.35 s knows that the TCF is good. If there is no detectable pattern of 0 bits, the TCF is clearly bad. However, what a receiver should do if it detects 1.65 s of 0 bits with a single 1 bit in the middle is unspecified. What a receiver should do if only 1.3 s of 0 bits are detected is equally unclear. To the extent that the decision as to the amount of corruption in a fax which is acceptable is a subjective decision, these decisions are rightly left to implementors to decide.

My own opinion is that where the TCF is composed mostly of 0 bits with a single short burst of garbage, there will often be no benefit in sending the FTT frame in the hope that a lower speed will make this go away. With a minimum scan line time of 20 ms, a single error lasting no longer than 20 ms would be unable to affect more than two lines whatever speed is used. However, if a single error of this kind occurring in a training check lasting 1.5 s is typical of the frequency of error bursts on the line, the number of lines affected could be affected by the speed used. At 9600 bps an average page would take 30 s, and in the worst case 40 lines out of the 1143 on the complete page would be corrupted with over 96% of the fax being correctly reproduced. However, at 4800 bps a single page would take 60 s to send, and an error burst every 1.5 s would result in twice as many lines being corrupted simply because the fax took twice as long to be transmitted.

So while it is true that, in general, dropping the speed results in a more reliable connection, this doesn't apply to certain types of line noise, such as that caused by random clicks on exchange equipment. If you receive a TCF with random 1 bits all over it, you should certainly respond with a FTT frame, but it's worth checking to see if a frame is mostly correct, as in this case dropping down to a lower speed might cause more errors.

The strategy that a transmitter adopts on receiving an FTT frame is also unspecified, and needs to be worked out, as the T.30 recommendation is silent on the subject. The obvious plan is to drop the speed a notch for each failure. However, just to ensure that the first failure at the highest speed isn't due to a one-off error burst, and really is a problem with line quality, it could be reasonably tried twice just to make sure a line problem does exist. If we start with two failures at V.29 9600 bps, we would then drop down successively to V.29 7200 bps, V.27 ter 4800 bps and, finally, to V.27 ter 2400 bps, attempting to obtain a CFR frame from the receiver each time. If we started with two failures at V.17 14400 bps, we would drop down successively to V.17 12000 bps and then V.17 9600 bps, on the grounds that V.17 modulation is more reliable than V.29 at the same speed. It may therefore seem perverse to try V.29 9600 if V.17 failed at that speed, but I'd recommend doing it anyway, just in case the problem is that one of the stations isn't handling V.17 properly. If V.29 9600 fails, we then try V.17 7200 bps followed by V.29 7200 bps for the sake of completeness, finishing as before with V.27 ter 4800 bps and V.27 ter 2400 bps.

1	V.29 9600 bps (1)
2	V.29 9600 bps (2)
3	V.29 7200 bps
4	V.27 ter 4800 bps
5	V.27 ter 2400 bps

Figure 6.4: Fallback on a V.29 and V.27 ter modem

1	V.17 14400 bps (1)
2	V.17 14400 bps (2)
3	V.17 12000 bps
4	V.17 9600 bps
5	V.29 9600 bps
6	V.17 7200 bps
7	V.29 7200 bps
8	V.27 ter 4800 bps
9	V.27 ter 2400 bps

Figure 6.5: Fallback on a V.17, V.29 and V.27 ter modem

Phase C

Phase C of a fax transmission begins as soon as the receiver has notified the sender that a training sequence has been successful. Both stations return to the higher speed negotiated during Phase B. In cases where V.17 modulation is used, there is a slight modification to this pattern, as the TCF is sent using V.17 with a long training sequence, while phase C data is sent using V.17 with a short training sequence. Confusingly, there are two meanings of 'training' here; 'long training' and 'short training' are low-level phenomena concerned with timing constraints on the speed with which the data link itself can be established; while the sort of training referred to in 'training check frame' concerns the integrity of the data sent over the connection once it has become established.

During phase C, fax data formatted according to the T.4 recommendation is transmitted in a continual stream. A delay is almost inevitable before this begins, as the fax machine or software first has to wait for the connection at the higher speed to become established. The latest version of T.30 includes a recommendation that there should be a wait of 75 ± 20 ms after receiving any T.30 signals before sending data at any of the higher speeds. This recommendation wasn't present in the earlier versions, so it is presumably added for greater reliability. It is also usual for delays to be introduced for various mundane mechanical reasons, such as the need to get the scanning machinery started in the case of some fax machines, or the wait while a fax image is loaded in from disk in the case of software driving a fax modem. There is no risk of such delays corrupting the start of the data, as the possibility of a delay in phase C data before data transmission begins is catered for in the T.4 specification by the requirement that any data that may be sent before the first EOL code is to be ignored. What happens during this period varies with the manufacturer, but the usual practice is to send a continual stream of 1 bits. To send garbage or white noise is unsafe because there is no way of guaranteeing that an EOL code won't slip in by mistake. Sending a continual stream of 0 fill bits is even less safe, as almost any type of line noise is certain to result in a spurious start of fax EOL being sent.

However, once the T.4 data begins to be transmitted synchronously, no further delays are allowed, and there are no gaps or pauses permitted in the data. Of course, fill bits may have to be added before any EOL codes as needed to bring each line up to the minimum scan line transmission time negotiated during phase B, but these have to be inserted in the data stream by the transmitting station. It is also worth remembering that, apart from a range of minimum scan line transmission times, T.4 recommendations also state there is a maximum scan line transmission time of 5 s. If a scan line lasts longer, the receiver should disconnect.

When the data at the end of the T.4 page is exhausted, phase C concludes with a return-to-control (*RTC*) sequence consisting of six EOL codes. This is also part of the T.4 recommendation. The RTC should not be regarded as being a part of the page data itself, but is sent afterwards to indicate that the page data is finished. This distinction becomes important if T.4 pages are pre-prepared by software for later transmission through a fax modem, as some modems require that T.4 data should not usually include the RTC sequence at the end of a page, but should leave the modem firmware to insert it instead.

Phase D

Phase D begins for the transmitter once the RTC sequence has been sent, and for the receiver once the RTC sequence has been received. Both stations now need to return to the low speed modulation scheme previously used for negotiating in Phase B. However, before the transmitter does this, it needs to observe another compulsory delay for a period of 75 ± 20 ms, similar to the delay required before the TCF was transmitted in phase B. The reason is that, once again, there are going to be consecutive transmissions with a change in the modulation scheme.

Once communications have been re-established at the negotiating speed, one of three possible post-message commands will be sent from the transmitter to the receiver. All post-message command frames have FCFs beginning with bits X111 (remembering that X is set to 1 by the station receiving the last DIS). None have any accompanying FIF:

- The End Of Message frame (*EOM*) instructs the receiver that the page is finished and the transmitter wants the session to return to the start of Phase B again. In other words, the receiver will need to transmit another DIS so that the parameters for the session may be renegotiated. There are any number of reasons why the transmitter may want to renegotiate, but the most common is that it needs to change the resolution for the next T.4 fax image from normal to fine, or vice versa.

  A number of comments in T.30, as well as the EIA class 2 specification, appear to indicate that the EOM can be used as a kind of end-of-document marker in a multi-part transmission, followed by the transmitter returning to phase B to renegotiate exactly the same parameters. In this case, the EOM frame apparently serves to notify the receiver that a new document is about to be sent and the next page received is not simply the next page of the previous document. It seems likely that this is a misinterpretation of the specification, and that the word 'document' is used in T.30 in a technical sense, and simply refers to a sequence of pages with a common page format.

  Other possibilities which may precipitate a return to phase B include attempting a change in the dimensions of the page, or even a change from transmitting a message to polling for one. The facsimile control field of the EOM frame has the bit pattern X1110001.

- The MultiPage Signal (*MPS*) instructs the receiver that the page is finished and the transmitter wants to return to the beginning of Phase C, i.e. the next page of the document follows using exactly the same negotiated parameters as the page just sent. The facsimile control field of the MPS frame has the bit pattern X1110010.

- The End Of Procedure frame (*EOP*) instructs the receiver that the page, document and session are finished and that the transmitter now wants to disconnect. The facsimile control field of the EOP frame is the bit pattern X1110100.

The T.30 recommendation includes an option for these three post-message commands to be sent in an alternative form, indicating a Procedure Interrupt Request (*PRI-Q*) that asks an operator to intervene. In this form, the fifth bit from the left is set to a 1. Thus a *PRI-EOM* has the format X1111001, a *PRI-MPS* is X1111010 and a *PRI-EOP* would be X1111100. All successful procedure interrupts would be followed with a return to the beginning of phase B when the fax session resumed. These PRI-Q frames were primarily of historical interest, and date from the time when line noise was a serious problem, and visual inspection of a received fax with subsequent verbal confirmation to proceed was the most practical method of sending documents.

Having said that, one never knows when an apparently redundant feature will prove useful, and some manufacturers are reintroducing machines with procedure interrupt capability in their continuing efforts to maximize sales by product differentiation. It isn't clear yet whether this will catch on. Though most fax sessions now are completely automated, this is largely because users don't often know how to operate a fax beyond inserting a documents and punching in the phone number; it is always possible that an easy to use feature will prove so successful that it will become generally available.

Transmitter procedure interrupts

The standard is rather vague both on the method that ought to be used to trigger a PRI-Q command, and the procedure to be followed after a PRI-Q frame is sent. Not only is a receiver allowed to ignore the procedure interrupt request if no operator intervention is possible, but it is also permitted for a receiver to have no capability for procedure interrupt at all. The T.30 recommendation states that the fifth bit of a PRI-Q signal need not be utilized provided this does not result in errors, but omits to indicate how this can be guaranteed.

The information on how transmitters should generate and handle procedure interrupts is given in a small flowchart buried in the corner of the second transmission page of the main fax session flowchart figure 5.2/T.30. The complete flowchart extends over six successive pages of the T.30 recommendation, and though it has a lot of useful reference information, it is a complex figure which is virtually useless unless you know what you're looking for.

Proper handling of procedure interrupts requires a fax machine or fax modem capable of automatically switching the line to and from a handset. Many fax modems do have dual sockets labelled PHONE and LINE with inbuilt barge-in protection, so lifting the handset during a call doesn't affect the fax. This doesn't necessarily mean these devices permit procedure interrupts to be handled correctly. On the other hand, a fax modem without a socket for a telephone, or a dedicated fax machine without a handset, will almost certainly have problems in handling procedure interrupts.

The classical position for fax machines is that procedure interrupt is always triggered by a human operator, and always requires a human operator at the remote end for successful completion. The fact that fax modems and the computers that control them have far more processing power than dedicated fax machines means this position may not correspond to reality in future; the manufacture of multi-function modems capable of doing more than fax transmission and reception might result in the development of software using the procedure interrupt facility to switch between fax and some other type of modem operation rather than between fax and voice. There is some support for this approach in the EIA specifications for fax modems, but as far as I know, nobody has developed any software which takes advantage of interrupting a fax session in this way. You should, nevertheless, bear in mind that the following description of the procedure interrupt would not apply in all its detail to such developments.

We begin with a procedure interrupt on a transmitting fax, typically triggered by a human pressing a suitably labelled button on the fax machine. At the next opportunity, a PRI-Q interrupt command is generated. If a receiver cannot handle interrupts, it ignores the PRI-Q bit and carries on as if the normal form of the command had been received. A receiver that can handle a procedure interrupt doesn't respond immediately, but instead attempts to notify its own operator. If the remote operator is available, they lift their own handset and press their own interrupt button. The receiving fax then acknowledges the PRI-Q frame by sending its own special procedure interrupt response frame; we'll cover these PIP and PIN frames shortly. The transmitter confirms that it has received the interrupt response by resending the PRI-Q a second time, and alerts the operator to pick up the handset. Once this is done, the call is switched to the handset and the fax is no longer connected. The receiver does the same once the second PRI-Q is received. After the voice conversation is over, the fax session can be resumed at the start of phase B again by pressing the manual send or receive buttons on the fax.

So, under this interpretation, every successful PRI-Q command is sent twice; once to ask for the interrupts and a second time to acknowledge the interrupt response. This ought to work properly with all machines implementing procedure interrupts, and is illustrated in Figure 6.6.

The T.30 recommendation defines a specific timeout known as T3, a period of 10 ± 5 s during which the station receiving the PRI-Q frame or a PIN/PIP response attempts to locate a human (presumably via bells and flashing lights). If no human responds, the receiver should proceed as if the PRI-Q bit had not been set. Meanwhile, the transmitter sending the PRI-Q should resend the command if no response is received within 3 s $\pm$ 15%. If the command is sent three times without a response being received, the transmitter should disconnect. Note that this means a receiver unable to locate a human should play safe by responding and ignoring the PRI-Q bit within 7.5 s of receipt.

Phase D responses

After any of the post-message command frames has been sent, the transmitter waits for confirmation from the receiver before continuing. Predictably enough, the reply to a post-message command under these circumstances is a post-message response. All post-message response frames have FCFs beginning with the bits X011, and none are accompanied by a FIF.

The usual response to a post-message command is to send a message confirmation frame (*MCF*) whose FCF consists of the bit pattern X0110001. Following this, the session continues with the appropriate frame as instructed by the command being acknowledged.

If the post-message command was an EOM, indicating the transmitter wishes to return to the start of phase B, the receiver must acknowledge this by following its MCF either by resending its identification sequence or, at least, resending its DIS. However, note that the MCF should always be sent as a final frame even though it will be followed by another frame in this situation; a return to phase B marks a clear discontinuity in the fax session.

There may be occasions when the receiver is aware that the data which had been received during phase C was not entirely satisfactory. While the T.30 standard recognizes that this is possible, and makes reference to copy quality checking, it merely states that 'some algorithm' should be used for evaluation. The most obvious method for checking reception quality is to check the data

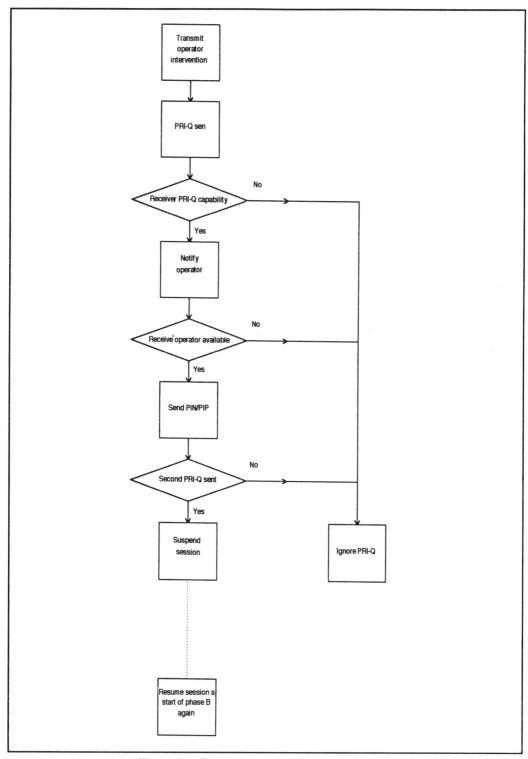

Figure 6.6: Transmitter generated procedure interrupt

received for conformity with the T.4 standard for fax data. Cases where lines lacked the correct number of dots, or where portions of the data made no sense, could be detected fairly easily 'on the fly' with sufficient processing power. In such a situation, the receiver has the option of requesting a retrain instead of simply confirming a message. Where a retrain is requested, the transmitter doesn't have to renegotiate all the parameters, but simply returns to the end of phase B and sends another TCF (training check frame). The receiver then re-evaluates the line quality and either sends a CFR (confirmation to receive), in which case the session continues at the start of phase C again, or the receiver sends an FTT (failure to train), which effectively tells the transmitter it must drop the speed.

There are two types of retraining requests: a Retrain Positive frame (*RTP*), whose FCF consists of the bit pattern X0110011, tells the transmitter that although a retrain is being requested, the page just received is of a satisfactory quality. On the other hand, a Retrain Negative frame (*RTN*), whose FCF consists of the bit pattern X0110010, tells the transmitter that the last page wasn't satisfactory.

After having sent an MPS, a transmitter receiving an RTP should simply comply with the request and carry on with the next page of the fax as usual. Though the T.30 recommendation doesn't explicitly make the point, it would seem to be a reasonable strategy for a transmitter receiving an RTN in the same circumstances to recommence the session after retraining by resending the last page if it has the capability to do so. However, if the transmitter cannot manage this, then at least it can note in its log which page of this particular fax will need to be resent manually.

The situation in which a retrain request is received after a transmitter has sent an EOP to indicate that it wants to disconnect is slightly more problematic. Although the protocol permits sending either of the retraining requests as a response to any of the post-message command frames, it doesn't make any sense to send a retrain positive request after an EOP. A retrain negative does make sense, as it is the only method of indicating that a page has failed to arrive properly, and will at least ensure that whoever sent the fax will see that a part wasn't received. However, asking for a retrain positive just before a disconnection when a page has been satisfactorily received is a complete waste of time. I suggest the only sensible course is to ensure that a retrain positive request is only sent by a receiver as a response to an MPS command.

Receiver procedure interrupts

The final response frames are the two procedure interrupt responses. These can be sent in response to a PRI-Q frame as an indication that the receiver agrees to the interrupt request. This was described above. However, the receiver can also ask for an interrupt spontaneously, in the same way that it can request a retrain on its own initiative. Like a PRI-Q post-message command from a transmitter, a procedure interrupt request from a receiver is a statement that operator intervention is required. If operator intervention is not possible, the session should resume at the start of phase B.

The circumstances under which a procedure interrupt might be requested by a receiver are not made clear. The obvious situation is where an operator notices that the fax being sent is the wrong document, or is being fed in back to front so the pages are all blank. In this case, frantic jabbing at the interrupt button on the fax is quite justified. Another likely candidate for the interrupt (which in this case could be triggered automatically) is where a machine has run out of paper while receiving. As with a transmitter originated procedure interrupt, the PIN or PIP frame is sent at the next available opportunity.

There are two types of procedure interrupt responses. A Procedure Interrupt Positive frame (*PIP*), whose FCF consists of the bit pattern X0110101, tells the transmitter that although a procedure interrupt is being requested, the page just received is of a satisfactory quality. On the other hand, a Procedure Interrupt Negative frame (*PIN*) whose FCF consists of the bit pattern X0110100, tells the transmitter that the last page wasn't satisfactory. Neither of these signals is met often. T.30 states that while all transmitters should be able to recognize PIP and PIN responses, the ability to send them is optional for receivers. Recognizing PIP and PIN doesn't mean a transmitter has to accede to the interrupt request; it merely means that if one is received, it shouldn't generate an error.

A receiver which spontaneously sends a PIP or a PIN frame continues the session by simply waiting for the next command from the transmitter. If the transmitter is capable of handling procedure interrupts, it sends a PRI-Q version of its last command, and the session continues as if the interrupt request had originated with the transmitter; this means that the PIN or PIP is sent twice, once to request the procedure interrupt, and again as an acknowledgement of the PRI-Q. The PRI-Q is also sent twice, once as a response to the first PIN or PIP, and again as a confirmation of the second.

If the receiver sends a PIP or PIN and the transmitter cannot handle procedure interrupts, it is not permitted simply to ignore the response and retransmit the original command (although this appears to be an option in the 5.2/T.30 flowcharts). A transmitter unable to handle procedure interrupts should resume the session at the start of phase B. This means that, effectively, PIN and PIP are treated as RTN and RTP.

Phase E

This is initiated by the transmitter sending a DCN frame, to which no response is expected. The transmitter then closes the connection and hangs up the phone. If the DCN was detected by the receiver then it also would hang up, but a DCN frame can also be sent when giving up on the session because of an irrecoverable error. In this case, the receiver may well not see the disconnect frame and would have to rely on a timeout instead.

Table 6.7 summarizes the T.30 frames covered so far. Where the first bit of the FCF is listed as an x, this should be set to 1 if we received the last DIS or reset to 0 if we sent the last DIS.

Table 6.7: FIF definitions for the basic T.30 frames

Initial identification		
NSF	answering nonstandard facilities	00000100
CSI	answering station identity follows	00000010
DIS	answering capability follows	00000001

Phase B commands		
NSS	transmitter nonstandard facilities	x1000100
TSI	transmitter station identity follows	x1000010
DCS	transmitter parameters and TCF follow	x1000001
NSC	polling nonstandard facilities	10000100
CIG	polling station identity follows	10000010
DTC	polling capability follows	10000001

Phase B responses		
CFR	confirmation to received	x0100001
FTT	failure to train	x0100010

Phase D commands		
EOM	end of document	x1110001
MPS	end of page	x1110010
EOP	end of transmission	x1110100
PRI-EOP	end of transmission interrupt request	x1111100
PRI-EOM	end of document interrupt request	x1111001
PRI-MPS	end of page interrupt request	x1111010

Phase D responses		
MCF	message confirmation	x0110001
RTP	Retrain positive	x0110011
RTN	Retrain negative	x0110010
PIP	Procedure interrupt positive	x0110101
PIN	Procedure interrupt negative	x0110100

Phase E command		
DCN	disconnect	x1011111

Optional error notification		
CRP	command repeat	x1011000

7

Fax Session: Advanced Topics

Introduction

The fax session protocol devised by the ITU has been subject to a process of continual development since it was first published. In this chapter we examine the various enhancements that have been made to it.

As all such enhancements are optional, so is most of this chapter. However, some of the topics, like error correction and binary file transfer, have generated much interest recently as manufacturers of fax machines and software have been marketing them as desirable features. Those wishing to know about this technology will find that the chapter discusses the issues and techniques needed to understand and implement leading-edge fax systems. It isn't light reading, and the bulk of our account is tied closely to the standards documents.

It isn't difficult to design and code file transfer protocol and error checking mechanisms, nor is it difficult to send and receive files successfully between two stations running identical implementations of a specification. What is far more difficult is to do the same thing with a second station running the same protocol, but developed by someone with whom you have no contact and will almost certainly never meet.

Implementation of protocols like Kermit and Zmodem is kids' stuff by comparison, simply because they have a specification backed up by freely available sample code. If there is ambiguity in the specification, you can look at the code and see what actually happens. This luxury is denied to us here in this case as our two fax machines have no common code to refer to. The standard really is all there is, and that's why we have to go into it in so much detail.

Extensions to the fax specifications

There have been a number of extensions to the original group 3 protocol since it was established in 1980. While the T.4 recommendation has always included a number of options such as 2-D coding and variations in paper sizes and transmission speeds, it made few concessions to ensure the integrity of the image being transmitted. Any substantial revision to T.4 is also likely to lead to revisions in the T.30 negotiating mechanism, as additional options will usually require more bit fields to be added to the DIS/DCS/DTC facsimile information field. The extend field in bit 24 is designed specifically to allow for this; when bit 24 is set, an additional octet follows the three compulsory ones. The last field in this fourth octet, bit 32, is another extend field, which, when set, indicates that a fifth octet follows. This process continues, with each additional octet ending in yet another extend field.

At the time of writing, the most recent revision of T.30 is dated 03/93, with its last extend field being bit 72. The original three octets have been extended to nine, with the process continuing at what seems to be an ever increasing pace. You may be advised that a serious fax programmer must obtain the latest T.4 and T.30 versions from which to work. While it is true that having up-to-date versions does no harm, and is undoubtedly a requirement for anyone who wants to write software with every refinement and capability possible, the fact is that all versions of ITU standards are downwards compatible, so there is no need to obtain the latest versions if you don't intend to use the additional features they document. Fax machines tend to have a fairly long life, and though technology is always offering improved dial-up services, these are always incremental. Ground-breaking advances in communications are never at the expense of backward compatibility, as incompatibility makes communications impossible.

Though none of the T.30 fax recommendation extensions are compulsory, some are so easy to implement that it make no sense not to support them. The most obvious instance is the ease with which V.17 modulation, with its higher speed capability, can be slipped into the basic three octet negotiating format. However, provided bit 24 is set to 1, the first two additional octets represented by bits 25 to 40 in the table below include a number of options that are becoming increasingly common. They are worth the extra effort involved in implementing them. Equally, some options were included for reasons that have more to do with political infighting in the relevant study groups than with their technical merit, and there is little chance of coming across them.

The current extensions to T.30 fall into five main groups:

- 1988 fax extensions; bits 25–40
- 1993 fax extensions; bits 41–50
- Data file extensions; bits 51–58
- Text transmission extensions; bits 59–64
- Open Document/ISDN extensions; bits 64–72.

The last three of these groups extend the boundaries of the recommendation far beyond the simple sending and receiving of facsimile messages on normal telephone line. This chapter begins with the first two groups, which are firmly based in the traditional fax world of image transfer over phone lines.

We start with a group consisting of those features whose implementation depends upon the two extra octets added to the specification in 1988. The workings of the fields in question are well established, and are listed in Table 7.1.

Bit 24 extend field

The basic T.30 session protocol uses only three octets, with bit 24 set to 0. A calling fax machine that receives a DIS frame with the bit 24 extend field set to zero should respond with only the basic three octets, and should set its own bit 24 to zero. The answering station has indicated that it does not understand any field beyond bit 24. If it understood any of the extended options, but wanted to indicate it didn't use any, it would have sent additional octets set to 0. As it didn't signal that it understood any additional fields, none should be offered.

2400 bps handshaking

Fax negotiations normally proceed at a rate of 300 bits/s. It is theoretically possible to shorten the transmission time by a few seconds if these negotiations can be handled at a faster speed. There is an option to permit 2400 bps handshaking to achieve this, but it is rarely used.

Table 7.1: First group of fax extensions

Bit	DIS/DTC Capabilities	DCS parameters
24	Extend field=1	Extend field=1
25	2400 bps handshaking	2400 bps handshaking
26	Uncompressed mode	Uncompressed mode
27	Error correction mode	Error correction mode
28	0	0=256 octet ECM frames
		1=64 octet ECM frames
29	Error limiting mode	Error limiting mode
30	Reserved for G4 on PSTN	Reserved for G4 on PSTN
31	T.6 coding	T.6 coding
32	Extend field	Extend field
33-37	Width extensions	Width extensions
38-39	Reserved for width expansion	Reserved for width expansion
40	Extend field	Extend field

Uncompressed mode

Uncompressed mode is an extension to the 2-D coding scheme, covered extensively in Chapter 3. It is another option not often seen on fax machines, probably as it is seldom needed and is computationally expensive to implement.

Error correction

Error correction mode (ECM) is an increasingly popular option. We have already mentioned that a major weakness in the basic T.4 and T.30 recommendations is they make few allowances for errors occurring in transmission and reception. It is true that the division of an image into lines which have a fixed structure, beginning with a white run length and ending with a unique EOL code, serves both as a means of limiting and detecting errors. It is also true that the ability of a receiver to request retraining and report that specific pages have not been received properly provides much needed feedback to the sender about how a message is getting through. Nevertheless, the way in which basic fax standards handle data corruption generated by line noise can only be described as crude. The usual error detection method is visual inspection of the page received by a human operator, and the only really effective method of error correction is to resend whole pages. It is not difficult to design a better system for transmitting clean copy, and this is precisely what was included in the 1988 revision.

Whether error corrected fax is something the public is looking for is another matter. The quality of many faxes sent over the telephone is pretty disgraceful, and this is not because the telephone lines are bad. The phone lines are better now than they have ever been, and it is often possible to transmit faxes with near 100% accuracy without using error correction. The shabby appearance of many faxed documents is simply down to the poor maintenance of fax machines, particularly the scanning portion of the system, where lack of regular cleaning is the cause of streaky faxes with black dots scattered partly at random, and partly where lots of dust gets pushed around towards the edges of the paper.

While data communications normally have to be 100% reliable to be of any use, the redundancy in standard pages of faxed text is sufficiently great for quite bad faxes to remain reasonably legible. Many fax users appear quite happy with dirty faxes provided they are able to read them. The realization that error correction will simply enable the dirt on scanner heads to be sent more accurately over a phone line is enough to make anyone wonder whether there is much point in introducing error corrected faxes. Nevertheless, we devote a large part of this chapter to explaining how it works. The reason is not simply because the standard is defined and is being used for sending faxes, but also because it is essential to many of the more recent enhancements to fax technology. For instance, faxes sent using group 4 T.6 compression over normal phone lines, and computer files sent using the T.434 binary file transfer protocol, both rely on error correction mode as an underlying protocol.

We now look at error correction mode at each relevant phase of the fax session in some detail. What follows is inevitably highly technical, and is mostly going to be of interest to those with EIA class 1 modems where ECM has to be implemented by the application software. It can be skipped by users of class 2 modems. In general, these either don't provide error correction or do so transparently.

Phase B in ECM

A fax station indicates that it has the capability to receive documents using ECM as part of the phase B negotiation procedure by setting bit 27 of the DIS information field to 1, and a transmitter indicates that it is using ECM by setting the same bit in the DCS frame. All other negotiations proceed during phase B in exactly the same manner as they would with a conventional transmission, with a training check frame being used to ascertain the line quality. It is only once phase C begins, and the connection switches to a higher speed, that error correction mode looks significantly different, as simply sending T.4 encoded data down a phone line is no longer what is required.

However, it is worth noting that returning to the start of phase B is likely to be more common when ECM is available than it is when machines don't have this facility. This isn't just because fax systems with ECM are more up-market, and tend to have more facilities, but because ECM gives fax systems a far greater range of possibilities.

Leaving aside any necessity for retraining if a bad page is received, there are only three reasons why phase B would need to be re-entered in standard mode: if the paper size is changed; a change from 1-D to 2-D coding; and (and the only one I've ever seen in practice) when the vertical resolution needs to be changed. The most obvious addition to these is that ECM requires a return to phase B to change the frame size (as defined via bit 28 in the DCS FIF listed in Table 7.1). ECM is also used for the transfer of various types of non-fax image, which can be intermixed in the same session as normal fax documents, and the renegotiation of parameters each time phase B is entered is an essential ingredient in making this work. We concentrate here on the use of ECM to transfer uncorrupted fax images.

Phase C in ECM

Before any image data is sent in phase C, a synchronization sequence is sent consisting of continuous flags (011111101111110 ...) for 200±100 ms. This is needed because error corrected transmission consists of sequences of HDLC frames, and random data should not be sent. Continuous flag transmission enables a receiver to lock onto the first non-flag bit sequence and know that marks the start of the first frame. The continual flags play the same role here that continual 1 bits play before the EOL, which marks the start of data in standard uncorrected mode. In both cases, the need is for a receiver to know unambiguously the point at which significant data begins to be received.

A T.4 fax image is divided, for error correction purposes, into frames containing either 64 or 256 octets. It is unlikely that a T.4 fax image will be an exact multiple of these numbers, so the last frame is allowed to be a short one. A receiver must be able to handle both frame sizes, with the transmitter indicating which it will use by setting bit 28 of the DCS information field. The frame size cannot be changed while a DCS information field remains in force. While none of the recommendations explicitly say that the choice of frame size is to be made on the basis of telephone line quality, it generally makes sense for this to be the main criterion.

The FCD frame

The HDLC frames used to send the T.4 data have an identical construction to that discussed in Chapter 6, and are known as Facsimile Coded Data (*FCD*) frames; they all have FCFs with the format 01100000. The Facsimile Information field (*FIF*) consists of the segment of T.4 data being sent. The run-length codes, the EOL sequence at the end of a line and the RTC sequence at the end of a page, together with any 2-D tag bits at the start of a line, are all included in the FIF as part of the data to be sent. Minimum scan line times don't apply to data sent in error corrected mode. The transmitter must set bits 21–23 in the DCS to indicate a 0 ms scan line time, and the receiver must be able to accept any data sent with a minimum scan line time of 0 ms. Though fill bits are not mandatory, the transmitter can still use 0 fill bits at the end of lines to align data on octet boundaries. Pad bits (which are ignored and can therefore be either 0 or 1) can also be used after an RTC to align the last frame on an octet boundary.

Frame sequencing

Inserted between the FCF and the FIF is an octet defined as an 8-bit binary number, which allows received frames to be placed in their proper sequence, which enables missing frames to be detected. Being 8 bits wide, this number can range from 0 to 255, with the first frame always being 0. The 1988 T.4 recommendation, detailing one part of the error correction scheme, referred to this frame number as being part of the FIF, which could therefore be either 65 or 257 octets long. However, the 1988 T.30 recommendation makes continual references to an FCF being 16 bits long in error correcting mode, with the first octet containing the frame identifier and the second containing the frame position.

It does no harm to think of the frame sequence octet as a simple binary representation of a decimal number inserted between the FCF and the FIF, but belonging with neither. This is possibly a more realistic view, because it is unlike the rest of the HDLC frame in one crucial respect. While the other bits in the FCD frame are sent in the same left to right order as they appear on a printed page, the frame number octet is sent from right to left, as if it really were a decimal number sent as one asynchronous byte with its least significant bit transmitted first.

The RCP sequence

As an octet is a sequence of eight bits, we so far have quite a familiar concept to anyone thinking in terms of error corrected asynchronous file transfer protocols that could be constructed on the basis of a file being divided into blocks with lengths of either 64 or 256 characters. However, the blocks in error corrected fax transmissions consist of a number of frames rather than a number of characters or bytes. The usual block size is 256 frames, which makes sense given that the frame number has to be contained in one octet and so can't be used to identify more than 256 frames. One transmitted block is also referred to as a partial page, and consists of a series of consecutive FCD frames, followed by a sequence of three consecutive Return to Control for partial Page (*RCP*) frames. The RCP has no facsimile information field, and its HDLC information field consists of an FCF with the format 01100001. The reason for the transmission of the RCP frame three times is simply that it is an application of the maxim that 'what I tell you three times is true'. The recommendation explicitly states that for a receiver, the receipt of one valid RCP frame is sufficient to indicate the end of a block.

Both the FCD and the RCP frames are sent with an HDLC control field set to 11000000, so at least one further frame must be transmitted as the receiver will only be able to respond once a frame arrives with the P/F bit set. Once an RCP sequence is detected, the two stations have 50 ms to end phase C and return to a special error corrected version of phase D. At this point, error corrected mode starts to get complicated, which is quite logical considering that the most complicated part of any error correction scheme isn't the setting up, transmission or even error detection, but the mechanism for responding, retransmitting and retrying once data has been sent.

It is worth noting that the overhead for transmitting a frame using error correction is a minimum of two flags, one address octet, one control octet, one FCF, a frame sequence octet and two frame check octets. Note that the same flag can be taken as both the end of one frame and the start of the next. Each block additionally has an overhead of three RCP frames that aren't required in normal transmissions.

In terms of characters, and assuming no errors, the 7 extra characters mean that a fax sent using 64 octets per frame would take over 10% longer to send with error correction than without, while a fax sent using 256 octets per frame would take less than 2.5% longer. However, an error rate of around 1 in 2000 bits would be sufficient to ensure that a fax consisting of 256 octet frames (which contain

Opening flag 01111110	Address 11111111	Control 11000000	FCD-FCF 01100000	Frame sequence octet 00000000 to 11111111	FIF of 64 or 256 octets	16-bit FCS	Closing Flag 01111110

Opening flag 01111110	Address 11111111	Control 1000000	RCP-FCF 01100001	16-bit FCS	Closing Flag 01111110

Figure 7.1: Phase C in error-corrected mode

2048 bits) would have serious problems with error correction, as every frame-would probably be corrupted. The same error rate when using 64 octet frames (which contain 512 bits) would affect only one frame in four.

It is tempting to conclude that 64 octet frames are safer, but the error rate on normal telephone lines is considerably less than 1 bit in 2000 and in any case, it tends to run in bursts rather that randomly hitting single bits. Most ECM frames sent on modern phone lines don't need any retransmissions. 256 octet frames are a more efficient choice under these conditions.

Phase D commands in ECM

Standard uncorrected mode presented us with three possibilities for post-message commands in phase D, depending on whether the transmitter wanted to return to phase B and renegotiate parameters, return to phase C to send another page of the document using the same parameters, or go on to phase E for a normal disconnection. Further, each of these possibilities could be combined with a PRI-Q request for operator intervention.

Error corrected communication has a different set of post-message commands, all of which have FCFs that begin with the four bits X111, where X is set to 1 if the station received the last DIS. The post-message commands available in non-error corrected mode should not be sent. None of their FCFs are valid; a receiver must either not respond and allow a timeout, or query them with a CRP frame, and if the transmitter persists the receiver has to disconnect.

In contrast, the receiver has available a superset of the standard post-message responses. The additional responses available when ECM is being used all have FCFs beginning with the four bits X011.

The Partial Page Signal (*PPS*) command always follows the RCP at the end of a normally transmitted block. Unusually, a PPS frame carries two FCF octets. Though this could be thought of as one 16 bit FCF, it makes more sense to think of it as two separate FCFs: the first FCF octet for a PPS frame has the format X1111101, which identifies the frame as a PPS; the second FCF follows the first, and identifies the type of PPS being sent. Unless the block that has just been sent is the last one of a page, this second FCF is filled with 0 bits, and the frame is referred to as a PPS-NULL. If the frame just sent is the last on a page (in which case it should have ended with a T.4 RTC or T.6 EOFB sequence), the second octet is formatted in the same way and with the same meaning, as the post-page messages for non error-corrected transmission. Table 7.2 shows the formats the second FCF octet in a PPS frame can take.

The two FCFs are followed by an FIF of three octets containing counters, formatted identically to the frame sequence octet contained in the FCD frames. The first octet contains the page counter, set to 0 at the start of a call and incremented for each page transmitted. The second octet contains a block counter, set to 0 at the start of each page and incremented for each partial page transmitted. So there is no limit to the size of a fax, and so that unlimited page lengths can be sent, both these octets are sent as modulo 256 numbers (so they wrap round to 0

Table 7.2: Possible FCF2 octets in PPS frames

NULL	partial page boundary	00000000
EOM	end of document	11110001
MPS	end of page	11110010
EOP	end of transmission	11110100
PRI-EOP	end of transmission	11111100
PRI-EOM	end of document	11111001
PRI-MPS	end of page	11111010

when incremented past 255). The third octet is the frame counter, which has the same value as the frame sequence octet of the last FCD in the block just sent; it thus can't have a value greater than 255. It holds the total number of frames in the block minus 1 (as the first frame is numbered as frame 0).

Opening Flag 01111110	Address 11111111	Control 11001000	PPS-FCF1 x1111101	EOP-FCF2 1111010	Page Count 0 to 255	Block Count 0 to 255	Frame Count 0 to 255	16-bit FCS	Closing Flag 01111110

Figure 7.2: Example of a PPS-EOP

Phase D responses in ECM

After the transmitter has sent the block as a series of FCD frames, followed by the RCP frames (all at high speed) and returned to negotiating speed to send the PPS, it waits for a response from the receiver. If the block has been received properly, the receiver sends a standard message confirmation frame (*MCF*), whose FCF consists of the bit pattern X0110001. This is identical to how a receiver behaves in standard non-error corrected mode. The transmitter may then return to the start of phase C and continue with either the rest of the page if a PPS-NULL had been sent, or the next page if PPS-MPS had been sent. In the case of a PPS-EOM, both stations return to phase B, while in the case of a PPS-EOP, the transmitter sends a DCN frame and disconnects. The PRI-Q frames are interpreted in a similar way to interrupt requests in standard mode, as no procedure interrupts are permitted between partial pages.

If the block was not received correctly, the receiver sends a Partial Page Request (*PPR*) frame, whose FCF has the format X0111101. A PPR is always sent with a fixed 256 bit FIF consisting of a map of all the FCD frames in the block that has just been received, with a frame that has being incorrectly received being marked by a 1 in the map. The first bit sent in the FIF corresponds to FCD frame 0, with the last bit of the 32 octets in the FCF corresponding to FCD frame 255. Where a partial page has fewer than 256 frames, the extra bits in the PPR must still be present, and should be set to 1.

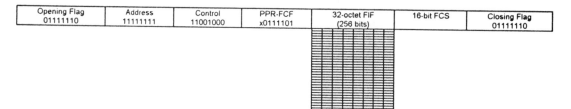

Opening Flag 01111110	Address 11111111	Control 11001000	PPR-FCF x0111101	32-octet FIF (256 bits)	16-bit FCS	Closing Flag 01111110

Figure 7.3: Example of a PPR

Predictably enough, a transmitter that has received a PPR frame returns to Phase C and the high transmission speed, and resends the synchronization sequence of training flags. It then retransmits only those FCD frames that were identified as having errors, complete with their original frame sequence numbers between the FCF and FIF, which allows the receiver to insert them in their proper place. The retransmitted frames are followed by the RCP frames (all at high speed) and then by a return to negotiating speed to send a PPS identical to that transmitted the first time the block was sent. The receiver updates its map of correct frames and sends either an MCF if a clean copy of each frame in the block has now been received, or an updated PPR to show the blocks which still have uncorrected errors. This sequence can be repeated up the three times.

Once the transmitter has received a fourth PPR in succession it has to make a decision on the best course of action. The annexes describing how ECM operates don't give a guide on what criteria should govern this decision. However, a mix of common sense and experience of local line conditions, combined with the degree of importance attached to a perfect transmission, ought to be sufficient to allow rules to be drawn up. For example, suppose the first attempt to send a 256-frame block resulted in 50 frames containing errors, and that this was reduced to only two frames with errors by the time the fourth PPR was received. The transmitter might decide on the 'one more time' approach and resend the frames once more, or it might decide that 2 frames in error out of 256 makes little difference. The point to remember about error detection and correction is that it has to be worthwhile. On the other hand, when error correction is used for fax data that uses T.6 MMR encoding, perfect transmission is essential as the lack of EOL codes means that synchronization after an error burst is possible.

If the transmitter does want to carry on correcting, it tells the receiver by sending a Continue To Correct (CTC) frame, which has an FCF of X1001000 accompanied by an FIF consisting of the first two octets of a standard DCS frame FIF. The receiver discards all except bits 11–14, which it used as the speed to move to for the next attempt at correction. The transmitter can carry on with the same speed as before, or drop down a level, or even move down to 2400 bps. The receiver responds to the CTC frame with a Continue To correct Response (CTR) frame, which consists of an FCF with the format X0100011.

On the other hand, the transmitter may decide to cease correcting and move to the next block. It indicates this by sending an End Of Retransmission (*EOR*) frame, which is like the earlier PPS frame in that it has two FCF octets. The first FCF identified the frame and has the format X1110011, while the second has the same format as the second FCF of the PPS frame preceding it, as shown in Table 7.3.

Table 7.3: Possible FCF2 octets in EOR frames

NULL	partial page boundary	00000000
EOM	end of document	11110001
MPS	end of page	11110010
EOP	end of transmission	11110100
PRI-EOP	end of transmission	11111100
PRI-EOM	end of document	11111001
PRI-MPS	end of page	11111010

There is no FIF accompanying the EOR frame. There is also no requirement for the second FCF of the PPS and any subsequent EOR frames to be identical, and a switch from MPS to PRI-MPS could in some circumstances make sense. The receiver responds to the EOR frame with its own End of Retransmission Response (*ERR*) frame, which has no FIF and an FCF format of X0111000. The acronym ERR is unfortunate, as the frame is actually a confirmation response indicating the acceptance of transmitted errors notification. It plays exactly the same role in the flow of the session as the MCF frame, and releases the transmitter, which can proceed with the next block in the case of an EOR-NULL, the next page in the case of an EOR-MPS, a return to phase B in the case of an EOR-EOM, and with a phase E disconnection in the case of an EOR-EOP.

It is worth noting that the EOR option is specifically prohibited when error correction is used for file transfers rather than for faxed images; it is also prohibited when a group 4 image coded according to the T.6 recommendation is being transferred.

Flow control in ECM

In standard uncorrected mode, there is no real flow control between the receiver and transmitter. A receiver indicates the minimum scan length time it takes for a line to be printed, the transmitter simply pads out its EOL with 0 bits to make up short lines to the minimum time, while the receiver effectively prints out 'on the fly'. At the end of a page of data, a receiver needs to reply to a post message frame within 3 s to avoid a timeout, and within 9 s to avoid a disconnection. A fax machine that cannot process a scan line inside 40 ms or cannot feed and cut a page inside 9 s simply doesn't make it to the market.

When using ECM the situation is different, as the receiver must able to store correctly received frames in memory. While it may be possible to print out text as it is received, there is no guarantee this will work all the time, because if the first frame of a block is corrupted, but all the others are fine, the receiver obviously can't print out the contents of any of the good blocks until the first bad block has been retransmitted. As the first bad frame received effectively blocks further printing, all correct frames received in the rest of the block will have to be stored. The memory needed for this is not trivial.

If we are using large frames, each will contain 256 octets. Each block contains 256 large frames, so a complete block requires $256^2 = 65536 = 64K$ 8-bit bytes of storage. A worst case scenario is that described above, where the first frame of a block is corrupted but the rest are received correctly. Just after the first frame has been successfully resent, the receiver will (for a moment) have a full 64K of data stored and ready to be printed, but won't have printed anything yet.

Implementation of a simple flow control procedure, where a receiver can tell a transmitter to stop sending and wait, gets around this problem. In error corrected mode a receiver which hasn't finished processing any partial page responds to a PPS with a Receive Not Ready (*RNR*) frame. Though this can be done at any point within the usual time period allowed for a response (3 s ± 15%), it is sensible to leave it for as long as possible. On receiving the RNR the transmitter then sends a Receive Ready (*RR*) frame telling the receiver to respond with its status. Neither of these flow control frames have any facsimile information field attached. The RNR frame FCF has the format X0110111, while the FCF of the RR frame has the format X1110110.

The first time a transmitter sees the RNR response, it starts a timer T5, set to 60 ± 5 s, and the RNR-RR sequence can continue until this timer expires, at which point the transmitter must send a DCN frame and disconnect. Obviously, the receiver should be able to respond with MCF well before this (or respond with an ERR if the transmitter sent an EOR frame to stop correcting). The procedure interrupt frames PIP or PIN also serve as adequate responses to an RR enquiry, but we have already said that these aren't generally used in automated fax operations.

The ability of two fax stations to synchronize transmission of pages using flow control frames makes the scan line printing times used in non-error corrected modes something of an irrelevancy. In fact, setting this to 0 reduces transmission times. This is why all machines capable of ECM must be able to accept data without any fill bits before an EOL, which is what a scan line time of 0 means.

Table 7.4 summarizes the initial FCFs of the frames used in the optional error correcting fax mode.

Error Limiting Mode

The details and deficiencies of error limiting mode can be found in the NOTES directory on the accompanying disk. ELM is essentially a modification to the

Table 7.4: Facsimile control fields of the ECM frames

Phase C context frames		
FCD	01100000	Facsimile Coded Data (T.4 data sent as the FIF in an HDLC frame at high speed)
RCP	01100001	Return to Control for Partial page (end of data and return to low speed - sent 3 times)

Phase D commands		
PPS	X1111101	Partial Page Signal (followed by a second FCF)
CTC	X1001000	Continue to Correct (sent with revised transmission speed)
EOR	X1110011	End Of Retransmission (followed by a second FCF)
RR	X1110110	Receive Ready (flow control status command)

Phase D responses		
PPR	X0111101	Partial Page Request (indicating frames with errors)
CTR	X0100011	Continue to correct Response
ERR	X0111000	End Of retransmission Response
RNR	X0110111	Receive Not Ready (flow control not ready response)

coding method used in generating a fax, and apart from the setting of bit 29 in a DIS/DCS frame, a T.30 session sending an image using error limiting mode looks identical to the same image being sent in normal mode. In this respect it is different to error correction mode. It seems that while the latter was proposed by United Kingdom representatives at the ITU, with support from American and European delegations, the error limiting mode was proposed as an alternative by the then-USSR. The adoption of both schemes as optional extensions to T.4 was a compromise which was perhaps typical of the times.

Width extensions

The presence of a fifth octet in the DIS/DTC/DCS frames is indicated if the last bit in the fourth octet (bit 32) is set to 1. It is devoted to extending the possible specifications for the width of a faxed page. We have already described how the basic T.30 recommendation indicates the width of a fax is by setting bits 17 and 18 in the DIS/DTC/DCS. The meaning of these bits is as given in Table 7.5.

However, the T.4 recommendation had been extended to enable sizes smaller than A4 to be sent. It is possible to send a 151 mm line encoded as either 1216 or 1728 dots, and a 107 mm line as either 864 or 1728 dots. These possibilities are clearly not catered for in the available options for bits 17–18, and there is no room there for any expansion. Consequently, bits 34–39 are reserved for encoding

Table 7.5: Widths specified in bits 17–18 of the negotiating frames

17-18	DIS/DTC page width capabilities	DCS page width selected
00	1728 dots in a 215mm line	1728 dots in a 215mm line
01	1728 dots in a 215mm line 2048 dots in a 255mm line 2432 dots in a 303mm line	2432 dots in a 303mm line
10	1728 dots in a 215mm line 2048 dots in a 255mm line	2048 dots in a 255mm line
11	invalid but interpreted as 01	invalid but interpreted as 01

other possibilities for fax recording widths, with bit 33 being set to 1 to indicate that these bits are to be used rather than the original bits 17–18. At the time of writing, only bits 34–37 are used, and Table 7.6 shows how they are defined.

Table 7.6: Widths specified in bits 33–34 of the negotiating frames

DIS/DTC width capabilities	DCS page width selected	
33 = 0	bits 17-18 valid	width indicated by bits 17-18
33 = 1	bits 17-18 invalid	width indicated by bits 34-37
34	1216 dots on a 151 mm line	middle 1216 dots from a 1728 dot line
35	864 dots on a 107 mm line	middle 864 dots from a 1728 dot line
36	1728 dots on a 151 mm line	invalid
37	1728 dots on a 107 mm line	invalid

The last bit of the fifth octet, bit 40, indicates the presence of at least one further octet. We can assume that the pattern established with the first extend field in bit 24, and continued with extend field in bits 32 and 40, will carry on, and all further octets will carry an extend bit in the eighth position. Bits 48, 56, 64 and 72 are already defined this way in the latest T.30 draft recommendation, and all fax software ought to be prepared to accept DIS negotiating frames of an indeterminate length. There are already reports that some older fax modems and fax software have only allowed for five octets in a DIS frame, and are unable to talk to the latest generation of fax machines that extend their capability frames beyond bit 40.

More recent extensions to the fax specifications

The remaining extensions to the T.30 fax specification weren't added until the 1993 revision. The group we are concerned with next depend upon bits 41 to 50 for their implementation, and comprise what are currently the final additions to the purely fax aspects of T.30. Most remaining extensions are concerned with non-facsimile applications, which we deal with shortly, but all the 1993 additions are still quite new, and it is unclear exactly how widely they will be implemented. At the time of writing, I was unable to locate a fax machine that implemented any of the new features in accordance with the specification. Because this meant it would not have been possible to test any program code for the 1993 additions to T.30, we have not included any, either in the book or on the companion disk.

Table 7.7: Second group of fax extensions

Bit	DIS/DTC	DCS
41	R8 x 15.4 lines/mm	R8 x 15.4 lines/mm
42	300 dpi	300 dpi
43	R16 x 15.4 lines/mm	R16 x 15.4 lines/mm
	or 400 dpi	or 400 dpi
44	Inches preferred	0=metric 1=inches
45	Metric preferred	
46	Scan line time for 15.4 lines/mm	
	0 = same time as for 7.7 lines/mm	
	0 = Ω the time for 7.7 lines/mm	
47	Selective polling capability	0
48	Extend field	Extend field
49	Subaddressing capability	0
50	Password capability	0

Changes to fax resolution

Bits 41–46 substantially extend the resolution available to faxed documents. The basic horizontal resolution used by all faxes has traditionally been approximately 8 dots/mm, while the basic vertical resolution has been either 3.85 lines/mm for normal resolution or 7.7 lines/mm for fine resolution. The old horizontal resolution of 8 lines/mm is referred to as R8 in this version of the standard. A new set of options doubles this to approximately 16 dots/mm, and is referred to as R16. The exact resolutions specified are double those of the R8 set, and are specified as 3456 dots/215 mm for A4 pages, 4096 dots/255 mm for B4 pages and 4864 dots/303 mm for A3 pages.

As well as doubling the horizontal resolution, vertical resolution can also be doubled to a superfine 15.4 lines/mm. It's worth noting that a normal resolution A4 page that used to be scanned as a 1728×1143 bitmap can now be scanned as a 3456×4574 bitmap.

The point is that while the speed at which faxes can be sent has increased six-fold (from 2400 to 14400 bps), there has been an eightfold increase in the possible size of an uncompressed bitmap (from 241K to 1930K). This is another example of the way in which the beneficial effects of an advance in technology in one area is more than swallowed up by the increased demands which are made on it by advances in other areas.

There are also new options in bits 41–46 to bring the scanned image more into line with computer technology, which has remained stubbornly attached to measurements based on inches rather than millimetres. Resolutions for scanners and printers tend to be measured in hundreds of dots per inch, with the most common resolution being 300 dpi both horizontally and vertically. As there are approximately 25.4 mm to 1 inch, this translates to the uncomfortable figure of 11.8 dots/mm. It falls halfway between the vertical resolutions of 7.7 and 15.4 lines/mm and the horizontal resolutions of 8 or 16 lines/mm. The lack of a convenient match has made faxing between computers and computer-based faxes more awkward than it should be, and the availability of a signal for genuine 300 dpi compatibility resolves this problem.

There is a much better natural fit for 200 and 400 dpi computer resolutions, which at 7.87 and 15.74 dots/mm correspond quite well to the old fine fax resolution of 8 and 7.7 lines/mm, and the new superfine resolution of 16 and 15.4 dots/mm, respectively. However, bits 44 and 45 are used to specify whether the metric (millimetre based) or imperial (inch based) dimensions are to be used, and the value of these bits modifies the meaning of bits 15 and 43 in a DIS capability frame as shown in Table 7.8.

Table 7.8: Resolutions specified in bits 15, 43 and 44–45 of the negotiating frames

Bits 44-45	Bit 15	Bit 43
00	(invalid)	(invalid)
10	200 dpi	400 dpi
01	R8 x 7.7 lines/mm	R16 x 15.4 lines/mm
11	either 200 dpi	either 400 dpi
	or R8 x 7.7	or R16 x 15.4

Bit 44 is set in the corresponding DCS frame to indicate whether the metric or inch-based version is to be used. Note that a resolution of 200 dpi is actually considered to be equivalent to the old fine resolution of 1728 dots/line with 7.7 lines/mm, while 400 dpi is considered to be equivalent to the superfine 3456 dots/line with 15.74 lines/mm. While the interconnection of these supposedly equivalent inch and metric based resolutions can cause more than the previously allowable 1% distortion, the ITU now recognize this is the unavoidable price to be paid for the fact that computers and facsimile use different units of measurement.

Bit 46 is used to indicate whether a scan line time of 10, 20 or 40 ms set in bits 21–23 for a resolution of 7.7 lines/mm can be halved if a resolution of 15.4 line/mm is used instead, or whether a line takes the same time to print irrespective of the vertical resolution. (Bits 21–23 contain a similar set of options for matching scan line times at 3.85 and 7.7 lines/mm.)

Additional frames

While many of the additional features that may be needed for specific fax requirements can be added by judicious use of the non-standard frames NSF, NSC and NSS, there are three extensions to the frames that can be sent as part of the phase negotiations which have now been included in the T.30 specification as optional features. All three extension frames use an FIF with the same format as that of the CSI/CIG/TSI frames. Like the ID frames, the additional frames discussed here are only supposed to include the + sign, spaces and the digits 0 1 2 3 4 5 6 7 8 9. It's difficult to see what harm could be done by ignoring this restriction and using any ASCII character. The argument that the field wouldn't be recognized by a machine that only understands numbers doesn't make sense when all three frames already need to rely on agreed formats if they are to be of any use. In particular, extending the range of characters available for use in security fields can only increase their usefulness.

Selective polling

The first of the additional frames is the Selective Polling (*SEP*) frame, sent as part of the command to send, before all DTC polling requests, and having an FCF with the format 10000101 and an FIF containing a reversed ASCII string. The SEP frame identifies a particular document to be sent when one fax system polls another. There is no provision in the original T.30 specification for the selection of one document from among a number of candidates for sending, and the SEP frame remedies this.

The SEP frame can only be used if bit 47 in the DIS is set. The answering machine sends a DIS to the caller, which responds with a SEP frame as one of the optional frames sent with any CIG and NSC frames before the DTC frame which tells the answering station it is being polled. The SEP frame cannot be sent under any other circumstances. While the lack of document selection in the original polling method was certainly a deficiency, the problem with using a SEP frame to identify the document is that it requires callers to know in advance what documents are going to be available. Modern interactive computer-based fax-on-demand systems that use touch-tones to allow callers to select the document they wish to receive are a more flexible way of solving this problem, and don't require the caller to invest in new fax machines if they want to be able to poll selectively.

Subaddressing

The second frame added to the T.30 recommendation is the Subaddress (*SUB*) frame, sent as part of the command to receive before any transmission, and being one of the optional frames preceding the DCS. It has an FCF with the format X1000011, where the X is set to 1 if the station was the one that received the last

DIS. The SUB frame can only be used if bit 49 in the DIS or DTC is present and set. The actual subaddress is contained in the FIF as a 20 digit reverse ASCII string. As its name implies, this frame is intended for situations where one central receiving point controls access to a number of different destinations. The most obvious application is where a single fax board on a computer network needs to ensure that an incoming fax is automatically routed to the private directory of the intended recipient.

Passwords

The last of this group of additions to the T.30 specification are the two Password (*PWD*) frames. These can only be used if bit 50 in the DIS is present and set. The first of these is the polling password frame, which can appear as part of the command to send in the same context as the SEP frame. It is another of the optional frames sent before the DTC frame which tells the answering station it is being polled. In this context, the password frame has an FCF with the format 10000011, and the password is contained as a 20 character reversed ASCII string. The polling password frame provides additional security for confidential documents. Polling fax systems that were restricted to earlier versions of T.30 could use only the CIG field for security, which was not at all secure, as the CIG field was defined as being identical to the TSI and CSI fields, and was supposed to consist of the telephone number of the fax machine. Having a number that could either be looked up in a phone book or that would be automatically given to any caller provided little security, as it is only too easy to program a fax machine with someone else's phone number. A separate password field is clearly a far more secure method of restricting access to pollable documents.

The second of the password frames is the transmission password frame, which can appear as part of the command to receive and is sent before any transmission. Like the SUB frame, it is one of the optional frames that can precede the DCS. The FCF of this PWD frame has the format X1000101, and the password itself is contained as a 20 character reversed ASCII string. The T.30 recommendation doesn't say what the purpose of password-protecting callers to a receiving fax station might be. One obvious use is to prevent the reception of junk faxes, and ensure that all faxes received on a particular machine are from authorized users.

Application and implementation of additional frames

The exact way the PWD, SUB and SEP frames are interpreted and used by machines that receive them is not spelled out in T.30. Presumably, some sort of database of passwords, callers and destinations for PWD frames is necessary, with separate databases for subaddresses and pollable pages for the SUB and SEP frames. For instance, an answering station specifically set up for polling would need a database enabling specific pages to be matched to SEP numbers, and would also allow caller passwords to be maintained and altered. There is also no procedure given for a situation where SEP frames were sent that matched no local document, or PWD frames arrived that had faulty passwords, or SUB frames arrived with addresses for which there was no match.

While the T.30 recommendation doesn't actually say that fax systems implementing these additional frames ought to be based around computer systems, it is difficult to see how dedicated fax machines would be able to provide the flexibility and capacity needed for this and still compete on price and, more importantly, on ease of use.

Though the facilities provided on many recent fax machines (and fax software) for the permanent storage and quick dialling of commonly used numbers can easily be enhanced to store commonly used subaddresses as well, this does nothing for less frequently dialled numbers, and in the case of a fax machine, requires more familiarity with how it is programmed than the casual user has. There is no point in permanent storage of passwords either on fax machines or in fax software, as this is akin to sticking computer access passwords on the front of a terminal or workstation. Not only does it provide no security, it also provides an illusion of security to systems managers, which is even worse. Permanent storage of PWD frames offers no advantage over existing TSI and CID frames.

There is a world of difference between loading a document into a fax and dialling a phone number, and loading the same document into a fax and having to enter a 20 digit secure password and a 20 digit subaddress in addition. The first operation is just as easy as making a phone call, and is surely one of the reasons fax has become so popular. The second operation is troublesome and error-prone, and seems to be something that, by destroying ease of use, also restricts the mass appeal of fax.

The basic point at issue is that if the growing complexity of the T.30 options means you need a computer at each end to take full advantage of all the features, it becomes reasonable to ask what the point might be of sending a document as a fax instead of as a computer file. If the answer is that there is none at all, this wouldn't imply that fax is a technology doomed to become outdated. It would simply mean that providing standards giving advanced electronic mail features to fax machines is a dead end. Ultimately, this is something that only time and the market will decide.

Non-fax extensions

The rest of this chapter is devoted to more of the recent enhancements made to the T.30 recommendations that enable fax technology to encompass areas previously the sole province of e-mail.

The remaining 1993 additions to T.30 are not concerned with enhancing the speed or resolution capabilities of standard fax, but are best seen as extending the scope of the fax session protocol itself. Almost certainly, the growth in the fax modem market and the more widespread use of computer faxing has been one of the main triggers for these developments.

Common to most of the latest features is the requirement that an applications programmer is acquainted with a far wider range of source documents. Until now, the basic T.4 and T.30 recommendations have been sufficient for implementing virtually all the features we've covered. This reliance on a small number of fairly self-contained documents is about to change dramatically. Among the recommendations that contain substantial portions of the requirements for the remaining features of T.30 are:

- T.50   7-bit International Alphabet No. 5 (ASCII) character set

- T.51   8-bit character set

- T.434   Binary File Transfer (BFT)

- X.208   Specification of Abstract Syntax Notation One (ASN.1)

- X.209   Basic Encoding Rules for ASN.1

- T.505   Processable Mode 26

- F.551   Telematic File Transfer (TFT)

Others are merely touched upon, such as the T.611 recommendation for a Programmable Communications Interface (APPLI/COM). The fact that there are so many documents to bring together is no accident. Since 1988, the ITU, together with the ISO, have been trying to integrate all the so-called 'telematic services' into a common set of recommendations which interwork in standardized and predictable ways. While the need for backwards compatibility has meant this has had no effect on how previous versions of the fax standards worked, the most recent additions are heavily influenced by the concepts developed as a result of this policy. Rather than attempt to make each recommendation into a stand-alone specification, the most recent standards contain a list of 'normative references', comprising all the related standards. Sometimes, documents listed are non-essential, but it is increasingly true that you have to understand more than one recommendation at a time.

Too many standards

In general, it seems that implementing many of the more recent ITU recommendations has increasingly become a sort of treasure hunt, with each basic recommendation containing references to other documents, which in turn refer to a third layer, and so on. This creates problems, as it is increasingly difficult to assemble all the documents together. It isn't unusual to find them written in different styles, and inconsistencies do creep in. Above all, the fact that standards documents are neither designed for teaching and education nor include any mechanisms for verification of a correct implementation causes no end of problems. Designers and programmers do get specifications incorrect and even standards committees can misinterpret or ignore the work of other standards bodies. Despite my best efforts, it is quite possible that I've misinterpreted some of the specifications myself.

Even so, I've tried to integrate all these various documents in the rest of this chapter. This isn't completely seamless, and it isn't intended to be. In some places I've taken the option of simply summarizing those portions of the specifications I think are useful, and referred directly back to the source documents for more details.

It would have been possible to double the size of this portion of the book by going into all the details of all the recommendations, and this would have severely distorted the contents. As near as I can tell, the latest features catered for in T.30 are implemented on 0% of all stand-alone group 3 fax machines. While this figure is an approximation at the time of writing, and may change in future, it makes little sense to attempt an authoritative presentation under these circumstances. Clearly, I haven't been able to verify the operation of any of these facilities.

Certain software packages do support various forms of binary file transfer, but even this cannot be said to be standardized, as benchmark machines for testing compliance with the specifications are not available. Consequently, the information following on the remaining bitfields is given primarily for completeness, and I can't claim any special expertise on their use. While I have tried to give sufficient detail to enable anyone implementing the remaining features to be able to do so, I cannot be certain that my account will always be adequate.

Most of the remaining groups of fields in the current T.30 specification are not used for sending facsimile images, but have been added to T.30 so that fax and other types of communications and computer technology can coexist more peacefully.

Data file transfer

The first group we consider are those fields catering for data file transfer. Anything that isn't an image which can be decoded using T.4 or T.6 coding mechanisms, and isn't identified as a character message (covered in the next set of T.30 extensions), is known as a data file. The ability to handle data files is indicated in the first instance by a receiver setting bit 51 of the DIS or DTC, showing it has this general capability. However, it also needs to set bits 53, 54, 55 and 57 (defined below) to indicate what sort of data files can be accepted. Bit 51 is a general indication that the information contained in the HDLC frames isn't an image and shouldn't be printed. Bit 51 is confusingly referred to in T.30 as the 'capability to emit data file'. In fact, it need only be set by a receiver, not by a transmitter, in which case a file is being received rather than 'emitted'. The full set of fields relating to data file transfer is given in Table 7.9.

Defining a standard method of transferring general computer files using a fax modem happened in the late 1980s. The reasoning behind it was that fax technology offered a standard for fast, affordable and reliable half duplex 9600 bps communications at a time when the fastest full duplex asynchronous data modem standard was V.22 bis, operating at 2400 bps. As a fax modem was synchronous

Table 7.9: File transfer modes specified in bits 51–58 of the negotiating frames

Bit	DIS/DTC	DCS
51	Data file capability	Unused
52	Reserved (FSI)	Reserved (FSI)
53	BFT capability	BFT transmission
54	DTM capability	DTM transmission
55	EDI capability	EDI transmission
56	Extend field	Extend field
57	BTM capability	BTM transmission
58	Reserved	Reserved

and didn't need to send start and stop bits, a 9600 bps fax modem was capable of throughput as fast as a hypothetical 12000 bps data modem, and it could send a file five times faster than the then state-of-the-art data modem. The fact that both V.29 fax and V.22 data modulations use the same basic QAM technology made possible the production of units with both capabilities.

Even when the V.32 bis 9600 bps standard for asynchronous modems was approved, they were quite expensive; modems of the 9624 family that offered 2400 bps for data with 9600 bps for fax were more economical. It made sense to use the fax channel for file transfer, thereby offering equivalent performance to V.32 modems at a far lower price.

While this remains true to some extent at the time of writing, 9624 family fax modems are no longer state-of-the-art, and are not that much cheaper than V.32 bis modems which also offer V.17 fax capability. Trellis encoding technology lies behind the current generation of both fax and data modems, and it seems likely that future developments in communications will result in fax and data standards running in parallel. The price/performance advantage that lay behind the original proposals for sending computer files using fax modems is now largely a thing of the past.

However, the concept of transferring data files over a fax link has received new impetus from the latest generation of fax communications software. Many of these programs, in theory, enable the same document to be sent to a fax machine in fax form, or to a computer in binary form, without the person doing the transmitting needing to know what the capabilities of the remote system might be.

Problems with machine-independent fax protocols

In practice, things are seldom so simple, as the data file transfer procedures are negotiated through the T.30 mechanism in much the same way as the speed or resolution. The problem is that, while a fax parameter such as the resolution is a known quantity, the content of a binary file is not. A computer which announces it has the capability of receiving a binary file could end up receiving something like a Postscript document instead of a fax. While this is fine if the computer can handle Postscript, if it does not it would clearly have been better off with the fax image. Even if the file format is acceptable, the receiving computer may not have a printer with the correct capabilities, such as the fonts required to print the document out.

The reason for the immense growth in the number of installed fax machines is because fax is a universal medium, and requires the very minimum amount of technical expertise on the part of sender and receiver. In addition, the sender knows that once a fax has been sent, there will be a copy of the faxed document on someone's desk almost immediately. A computer fax system based on fax modem technology does need to be set up properly, but just as a normal fax machine needs to be properly fed with paper, the onus is on the receiver to make sure that this is done. Once a computer is set up to handle fax traffic properly, it is if anything more reliable than a fax machine.

All the mechanisms for sending original computer files in exactly the same way as faxes depart from the universality of fax as a transmission and reception medium, as there is always the possibility that a computer file will be sent to someone unable to handle it. There are obvious commercial advantages to manufacturers of word processing and other packages in sending such files, as there is a much greater chance that a group of users who regularly transmit documents will need to standardize on one particular package.

The point is, there isn't a well-defined standard for moving word processing files from one machine to another. Even plain ASCII-based text isn't always transportable, as both the character sets and the conventions for end of line characters vary widely between machines. Conversion presents no problems for recipients who know what they are doing, but there is a clear difference between this type of document transfer and sending someone a fax. Solutions that integrate faxing and file transfer too seamlessly run the risk of sending someone a file that they can't handle.

There is no doubt that sending a document from one computer to another as a fax makes little sense. A perfectly good file containing genuine text, which can be edited and printed out using standard software and peripherals, is sent as a fairly low resolution graphics image. The disadvantages of this are obvious. It is bad enough that the resolution (and therefore the clarity) of a fax is less than that of even the cheapest printer. To add insult to injury, this is generally accompanied by a massive increase in size, generally about tenfold. The received file cannot be edited without either retyping the text or running the image through optical character recognition software. Even then, any automatic structures that may be included in the original file, such as outlining and indexing, will be lost. Sending the original file is quicker and more flexible.

However, using software specifically designed for file transfer between computers can often be better than sending a data file using fax protocols. A good file transfer protocol is much quicker than fax, as it doesn't involve 300 bps negotiations and complex training procedures.

The ability to transfer data files isn't handled by any but the most complex fax software, and many fax modems can't handle the error correction protocols used. In contrast, computer file transfer protocols such as XModem, YModem, Kermit and ZModem are widely available, highly reliable, and work on any data modem whether or not it has fax capability. While fax phone numbers and data modem phone numbers can't be randomly mixed for any broadcast calls, we have already seen that the lack of standardization in data file formats makes a random grouping of computer fax phone numbers of dubious value in any case.

The integration of data file transfer with electronic mail is possibly a better long term bet for developers than integrating data file transfer with fax. While this latter option may once have seemed the way forward, the erosion of the performance differential between data transfer modulations and fax modulations, combined with the lack of standardization of word processing file format between computers, makes the enterprise less attractive as a generalized document transport mechanism than it seemed a few years ago. Despite this, data file transfer standards do exist, and are being used, and the computer industry has always been unpredictable. Therefore, this chapter includes an introductory outline of the relevant T.30 mechanisms.

Error correction for data files

All the facilities in these groups depend upon ECM for correct functioning. Data files are sent in exactly the same way as error-corrected faxes, and bit 27 must be set in the negotiating frames for data file transfer to be considered. Just as a request to send T.6 data by setting bit 31 is ignored unless the error correction bit 27 is also set, so these fields too depend upon the availability and selection of ECM at both ends of the link.

There are a number of preliminary points that need to be spelled out regarding the compulsory use of ECM in any sort of file transfer context:

- The basic structure of frames and block is identical to that used for fax. However, though the syntax of various frame exchanges may look the same, the meanings are different. For example, the MPS multi-page signal is interpreted as a multi-file signal.

- Frame sizes of both 64 and 246 octets are supported, but the type of padding out between the end of data and the RTC permitted in fax messages is not allowed in file transfers. The last frame of the last short block must not be padded at all; the last octet sent in the last frame of the last block must be the last byte of the file. The ability to send short frames, optional for fax machines, is mandatory for file transfers.

- The first bit transferred in any octet corresponds to the least significant bit of the corresponding byte of the file. In other words, there is no need to invert each byte before sending it or after it has been received.

These rules apply to all file transfers using ECM.

Type of data file transfer

In this section, we cover the various options given in Table 7.9, which lists the types of data file defined in T.30.

Basic Transfer Mode

The simplest is Basic Transfer Mode (*BTM*), which requires bit 57 to be set. This is just what it says; the file being transferred has no information sent with it. Apart from knowing it is not an image, nothing else can be inferred. There is some similarity with the Xmodem protocol here, which also transfers the content of computer files with no additional information, not even a name. The context of the transfer is often enough to let a user know what a file happens to be. The TSI can often identify a transmitter sufficiently for the file type to be known.

Document Transfer Mode

Document Transfer Mode (*DTM*) is indicated by setting bit 54. This is a fairly simple method of including a file description along with the file contents. The description is sent before the file content, and consists of a series of lines of ASCII text terminated by a blank line. The file description is quite firmly structured. Each item of information, known as a field, must occur on a line by itself, and must be preceded by at least one additional line stating what field is going to be sent. This line contains field identification information in both numeric code form and also as an ASCII string delimited by colons. More information on DTM descriptors can be found in the NOTES directory on the accompanying disk.

EDIFACT transfer

EDIFACT Transfer (EDT) is indicated by setting bit 55. An EDIFACT (Electronic Data Interchange For Administration Commerce and Transport) file is structured according to ISO/IEC 9735 rules, and is outside the scope of this book.

Additional ECM rules for BTM, DTM and EDIFACT

All three data file transfer modes covered so far require some fairly sensible modifications to the ECM specification. These rules (which don't all apply to BFT mode) are as follows:

- EOR signals are not allowed. In a situation where a fourth PPR has been received, the transmitter must either send a CTC and drop to a lower modem speed or abort the transfer by sending a DCN frame and disconnecting.

- In the case of a failure, the whole file should be retransmitted. Partial file transmissions are not allowed, and there is no scope for resuming an interrupted transfer from the last point where the connection was known to be good.

● Blocks are separated by PPS-NULL frames. PPS-MPS frames can be sent to concatenate multiple files for sending in one session. In other words, a single file is the logical equivalent of a single page of a fax. The PPS-EOP frames are used to end a session once the last block of the last file has been transmitted.

● The PPS-EOM command can be used to switch transfer modes during a session, enabling BTM, DTM and EDT transfers to take place during one call. Though it isn't specifically permitted, I can't see any reason why the return to phase B shouldn't be done to negotiate sending a fax rather than a different kind of data file. Presumably there is nothing to stop this happening the other way round.

Binary File Transfer

Binary File Transfer (*BFT*) is available in a number of commercial software packages, and is becoming more widely available, but this is probably for marketing reasons of product differentiation. The trend is to provide more and more features, be they useful or not. We have already questioned the benefits of transferring computer files using BFT as a matter of principle, and one purely practical consideration which mitigates against the take-up of BFT is Microsoft's decision to implement a non-standard protocol in their own Microsoft at Work fax package bundled with Windows for Workgroups.

BFT is similar in concept to DTM, but includes a more rigorous file description primarily designed for automatic processing. It is indicated by setting bit 53 in the negotiating frames. There is a separate standard for BFT, contained in Recommendation T.434. This is possibly the most technically difficult of all the standards referred to in this book, and is not for the faint-hearted. Full details of the BFT protocol have been included in the NOTES directory on the accompanying disk.

Character capability: character mode and mixed mode

The fourth group of fields added to the latest version of T.30 allow for the transfer of files containing printable characters rather than fax image data or binary data. Like fax image data, character files can be printed out, but the amount of data being transferred typically takes up 90% less space than a comparable image, and thus can be sent in one-tenth of the time.

Unlike the ability to do data file transfers, the ability to handle characters directly is quite easy to build into any fax machine. The only modification needed is to give the fax the ability to generate a pattern of dots for a particular character from a character generator table. Compared with what lies behind decoding T.4 image data, this technology is quite straightforward. It is within the capability of even the humblest of dot matrix printers, and given that many fax manufacturers are already heavily represented in the printer market, fax machines with added character generators are something I would not be surprised to see quite soon.

Any receiving fax that has character capability should indicate this by setting bit 59 in its DIS/DTC frame. Like the corresponding data file bit 51, this is referred to as showing a 'capability to emit character file'. Presumably, this means the receiver has the capability to emit the characters to a printer. Another similarity with data file transfer is that bit 27, indicating ECM capability, must also be set. Additionally, the type of character transfer that can be accepted should be indicated by setting bits 60 and 62 as appropriate.

This portion of the T.30 specification is rather vague, and contains nothing on the relationships between bits 59, 60 and 62. Most of the information on character modes is actually included in annexes D and E to recommendation T.4, which covers only the format and not the negotiating procedure, for which readers are referred back to T.30. The bit definitions given there are as shown in Table 7.10.

The presence of the unused bits 61 and 632 may be an indication that these modes are not yet fully specified. (On the other hand, bits 13 and 14 were set aside as spare bits reserved for future modulation schemes years before V.17 modems appeared, which surely constitutes evidence that SG8 is quite capable of planning some distance ahead.)

Table 7.10: Third group of fax extensions: character capabilities

Bit Number	DIS/DTC	DCS
59	Capability to emit character file capability	unused
60	Character mode	Character mode
61	Reserved for control document	Reserved for control document
62	Mixed Mode	Mixed Mode
63	Reserved for future negotiation mechanism for character file transmission	
64	Extend field	Extend field

Character mode

The T.4 recommendation for character mode is quite detailed. The page format is fixed at a maximum of 55 lines of 77 characters each. Shorter lines and pages are, of course, permitted. The size of the character matrix is specified as being 20 x 16, which should include any inter-character and inter-line gaps. The character sizes are designed to be printed out at 6 lines/inch. Characters are assumed to be printed left to right and top to bottom. There is a default top margin of 130 blank dot lines, which is a little over eight character lines, and a left margin of 104 dots (a little over five characters). The first dot in the first printed character is therefore the 105th dot on the 131st line.

As a full line of 77 characters, each of which is 20 dots wide, would take up 1540 dots, a left margin of 104 dots leaves a maximum right-hand margin of 84 dots. Most lines will of course be shorter. The margins are very generous compared to the guaranteed reproducible area as defined in the T.4 recommendation.

Table 7.11 shows the character set. As you can see, it is more or less basic ASCII plus extensions above decimal code 127. The symbols in positions 2/4, A/4 and A/4 (which include currency symbols) are liable to change from country to country. Note that the 16 characters in column C, from ASCII 192, are diacritical marks such as accents, and are non-spacing. They overprint characters rather than being characters in their own right.

There are also six control codes, defined as shown in Table 7.12.

The end of line sequence is defined as being a CR LF pair. Each page should begin with a page initiator sequence of CR FF, and this should also appear at the end of the last page. The tab stops are fixed at five character intervals, beginning with the fifth character of each line.

Table 7.11: Character set for fax character mode

	0	1	2	3	4	5	6	7	8	9	A	B	C	D	E	F
0				0	@	P	`	p			NBSP	°	Do not use	–		K
1			!	1	A	Q	a	q				±	`	¹	Æ	æ
2			"	2	B	R	b	r			¢	²	′	®	Đ	đ
3			#	3	C	S	c	s			£	³	^	©	a̲	∂̄
4			$	4	D	T	d	t			$	X	~	TM	Ħ	ħ
5			%	5	E	U	e	u			¥	µ	‾	♪		ı
6			&	6	F	V	f	v			#	¶	˘	¬	IJ	ij
7			'	7	G	W	g	w			§	•	˙	¦	L•	l•
8			(	8	H	X	h	x			¤	÷	¨		Ł	ł
9			)	9	I	Y	i	y			'	'			Ø	ø
A			*	:	J	Z	j	z			"	"	°		Œ	œ
B			+	;	K	[	k	{			«	»	¸		Ǫ	ß
C			,	<	L	\	l	¦			←	¼	Non space u/line	⅛	Þ	þ
D			-	=	M	]	m	}			↑	½	"	⅜	Ŧ	ŧ
E			.	>	N	^	n	~			→	¾	¸	⅝	ˌ	
F			/	?	O	_	o				↓	¿	ˇ	⅞	`n	SHY

Table 7.12: Control sequences for fax character mode

Abbreviation	Name	Decimal	Hex
LF	Line feed	10	0a
FF	Form feed	12	0b
CR	Carriage return	13	0d
HT	Horizontal tab	9	09
SS2	Single shift two	25	19
CSI	Control sequence identifier	155	9b

The SS2 character is used as a lead-in to a single box-drawing character, all of which therefore have to be transmitted as 16 bit codes. The set of box graphics is shown in Table 7.13.

Table 7.13: Box drawing characters for fax charactetr mode (these require an SS2 lead-in)

A number of codes for font effects are also defined. Their use is optional, and if not implemented they may be ignored (Table 7.14).

Table 7.14: Control sequences for font effects in fax character mode

Attribute	Sequence	Hex
Normal	CSI 0 m	9b 30 6d
Bold	CSI 1 m	9b 31 6d
Italic	CSI 3 m	9b 33 6d
Underline next character	CSI 4 m	9b 34 6d

ECM in character mode

The structure of the ECM HDLC frames in character mode is identical to that of the frames used when sending T.4 images. Like images (but unlike data files), character mode transmissions can pad out the last frame of the last short block with null 0 octets, which must follow the final CR LF. As with data files, the use of the EOR is prohibited.

The post-message frames are exactly what you might expect. Pages are separated by PPS-MPS commands, and documents within the same session are separated by PPS-EOM commands. The last page of the last block ends with a PPS-EOP.

The specification also states that the PPS-NULL is used to separate ECM blocks in the same page. You may recall that the smallest block possible contains 256 frames of 64 octets, which is 16384 characters. As a page starts with a CR FF and follows with a maximum of 55 lines, each of which can have up to 77 characters plus a CR LF pair, a full page of text would appear never to have more than 4391 characters, and should therefore never need a PPS-NULL. You could artificially force such a page to extend beyond a block by adding vast quantities of diacritical marks or control codes for attributes, such as underlining every character. A line with 77 underlined characters would take up 310 octets, and 55 such lines would need more than one block of 64 octet frames. Sending such a page would require PPS-NULL commands. In normal operation, such heavy use of control codes is not to be expected, so pages should not span more than one block, and the PPS-NULL command should not be encountered in pure character mode.

Mixed mode

Mixed mode (*MM*) permits normal facsimile transmission to be intermixed on the same page as character mode transmissions. A mixed mode fax page is considered to be divided into an indeterminate number of horizontal slices, each of which contain either fax or character data, but not both. We have already come across *FCD* (Facsimile Coded Data); the corresponding acronym for Character Coded Data is *CCD*.

As is usual with all these extensions to T.30, mixed mode requires the use of ECM, with each slice being sent as a partial page. As with other non-fax ECM transfers, the use of the EOR command is forbidden. Three failures to send a clean block (when a fourth PPR is received) must always be handled via the CTC continue to correct command.

So that different types of partial paged can be distinguished, they are sent with different facsimile control fields. An FCD partial page uses the standard facsimile coded data FCF of 01100000, while a CCD partial page uses the character coded data FCF of 01100010. The FCF of 01100001 is used for the three RCP frames following the data, no matter what type of data the last partial page contained.

FCD partial pages follow the usual encoding rules for whatever coding method has been negotiated. However, there are two main differences: every FCD slice must contain an integral multiple of 16 scan lines, presumably making it easier to write code implementing the 16×20 character generator for the CCD slices; and every FCD slice must end with a facsimile slice terminator code (*FSTC*), defined as six EOL+1 codes. This applies no matter what coding scheme is used for the data. Annex E to the T.4 recommendation states that in the case of T.6 data, the FSTC follows the EOFB sequence. Normal RTC codes (6 x EOL) that occur at the end of a page are presumably not used if the final slice on a page happens to be an FCD block, but this is not explicitly stated.

CCD partial pages follow the usual coding rules for character mode in the same way that FCD slices are generated according to the normal T.4 requirement. The most important clarification is that the last line of every slice must always end with a CR LF sequence. A slice line is (obviously) equivalent in height to 16 FCD scan lines.

If the first slice on a page happens to be a CCD partial page, there are two special considerations. The first is that it must begin with a CR FF page initiator. The second is a recommendation that six blank lines be sent before the character data to allow for the fact that the first 130 scan lines might not be reproduced correctly.

Group 3 and Group 4 fax machines

The final group of fields added to T.30 effectively narrow the differences between group 3 and group 4 fax machines, in terms of both capabilities and performance. The fields defined are as given in Table 7.15.

Table 7.15: Group fields

Bit	DIS/DTC	DCS
65	Processable mode	Processable mode
66	Digital Network Capability	Digital Network Capability
67 = 0	half duplex	half duplex
67 = 1	full duplex	full duplex
68-71	reserved	
72	Extend field	Extend field

So far we've said relatively little about group 4 machines, except that they function over ISDN instead of PSTN, and that they employ T.6 rather than T.4 image encoding techniques. This has probably given the impression that group 4 is nothing but a faster and more accurate version of group 3. The reality is that it offers far more.

The specification for group 4 was developed around 1980, at the same time as the specification for Teletex. Both the PC and modem industries were in their infancy, and the fax revolution was some distance away, and largely unanticipated. The main method of non-voice business telecommunications was the telex network, which used a five-bit digital code and communicated at a rate of 50 bps, or 10 characters per second. Teletex was seen as a future candidate to replace the ageing telex network for routine business transactions, and group 4 fax was designed with an optional capability to communicate directly with the Teletex network. The relevant specifications are scattered among a bewildering variety of T recommendations; the most notable of these are F.200, T.60, T.61 and T.62. Few would deny that Teletex has been a notable failure. As we pointed out in Chapter 1, an international standard isn't always guaranteed commercial success.

Group 4 machines are divided by the ITU into three basic classes, somewhat confusing given that the EIA group 3 fax modems are divided into classes on a different basis. A class 1 group 4 fax machine needed only the ability to send and receive fax images. A class 2 group 4 fax had to add the ability to receive both Teletex and mixed mode documents. Similar to the new group 3 mixed mode, these contained both Teletex character and image data. A class 3 group 4 fax had to add the capability to send Teletex and mixed mode as well as being able to receive them. There were also other differences between the group 4 classes which are not relevant to this discussion, such as their different resolution capabilities.

Processable mode 26
Group 4 fax has always been heavily influenced by newer concepts that the ITU and ISO have been developing together, such as Document Architecture Transfer and Manipulation (*DATAM*) and File Transfer Access and Management (*FTAM*). The definitive statements of these ideas are now grouped together under the title Open Document Architecture (*ODA*). More details can be found in recommendation T.411/ISO 8613-1 Annex F 'Open document architecture and interchange format: Introduction and general principles' and subsequent recommendations. T.412 deals with document structures, T.414 with document profiles, and T.415 covers interchange formats. The internal architectures of ODA compliant documents can be found in T.416, which contains the character content architecture, while T.417 has the raster graphics content architecture and T.418 lays out the geometric graphics content architectures.

Processable mode 26, or *PM 26*, which is defined by setting bit 65 in the negotiating frames, is very much in the ODA mould, and the specification can be found in recommendation T.505. There are a number of similarities with the T.434 BFT standard (notably the use of ASN.1), which is similarly aligned with the ISO FTAM specifications. PM 26 is effectively one of a number of methods for transferring a document in all three possible permutations of formatted form (like a fax or a printout) which shows what it looks like, and processable form (like a postscript program) which shows how it is composed. The formatted form gives the content and the processable version enables the file to be edited and altered.

While the graphics format can be any of the usual T.4 or T.30 coding methods, the character format seems to be an attempt at defining a universal word processing interchange format. In the commercial world, this type of architecture is the province of companies such as Adobe, and it is worth noting that the ITU themselves use the proprietary industry standard Postscript format for electronic distribution of their own documents rather than any of their own recommendations.

I find it a little worrying that, apart from the reference to PM 26 in the definition of bit 65 of a DIS/DTC/DCS frame, there is no other reference to PM 26 in T.4 or T.30. In view of this, there seems no point in covering PM 26 in any detail, as it is something that I wouldn't advise anyone to implement unless things change dramatically. It is more cost effective to use a normal file transfer protocol to send a standard word processing file and rely on a suitable file conversion utility to sort out any problems.

Digital network capability

Setting bit 66 in the DIS/DTC/DCS is an indication that the group 3 fax machine has ISDN capability. This is largely outside the scope of this book. However, T.4 points out that there are two different technical means by which group 3 machines operate over ISDN. The first, G3F, is defined in T.4 Annex F. It is derived from the group 4 class 1 specification, with which it is compatible. The second, G3C, is defined in T.30 Annex C. This is based on the normal PSTN T.30 specification. The reason for the existence of two different versions is because, for political reasons, the ITU didn't at first want to encourage the possible convergence of group 3 and group 4 technologies. This left the door open for different manufacturers to adopt proprietary solutions to fill the nascent demand for 64 kbit/s ISDN group 3 compatible machines. Eventually, as two options had been developed, both were standardized. The ITU are currently working on solutions to the resulting incompatibility problems.

Full and half duplex

Full duplex capability is set via bit 67 in the DIS/DTC/DCS frame. It is applicable only to G3C machines, as defined above, that use T.30 protocols to communicate over ISDN connections, though a full duplex modem version is being studied. When a full duplex connection is used for ECM, partial pages are sent consecutively without waiting for an acknowledgement. Any unacknowledged commands that may come in while a partial page is being sent are dealt with at the end of a partial page transmission. Note that this option is not very well specified. The problem is that while sending blocks without waiting for an acknowledgement is a well-established technique for file transfer using a so-called 'sliding window' approach, all such techniques depend for their reliability on a clear understanding of the size of the window in blocks, and on the receiver being able to acknowledge blocks selectively.

However, neither the MCF nor PPR responses contain any indication of the block to which they relate. It follows that a full duplex transmitter must assume all responses are received in sequence and that none have been lost. As an ISDN connection can be assumed error free, this isn't going to cause a problem, but any attempt to extend the standard to allow the use of full duplex with PSTN and standard modems isn't possible without some further refinement of either the response frames themselves, or more likely, of the open-ended nature of the frame stream.

Such a modification could take the form of a stipulation that no more than one unacknowledged frame can be outstanding at any time. If a session times out with no response to a frame being received, this could trigger a return to transmission from the point at which the last MCF was received. The behaviour of the flow control frames and the minimum buffer requirements for a full duplex fax machine also need specifying in more detail. No doubt issues such as these are among those being discussed by the ITU. Until they have completed their work, I would regard the full duplex option as being irrelevant for dial-up fax sessions. Any recommendations should be looked at quite closely, as full duplex operation has the potential to develop into yet another option where two systems with similar implementations can communicate quite easily, but where problems arise as soon as machines appear with interpretations which are incompatible.

8

Introduction to EIA
Fax Modems

Introduction

This chapter should be understood by anyone writing software to control a fax modem. In it we describe the various classes of fax modem, and the differences between them. Most of the chapter is taken up with an account of things that may go wrong when a modem is linked to a computer and placed in fax mode. While we highlight problem areas, and suggest how to avoid common pitfalls, the main benefit to be gained from the chapter is an appreciation of the limitations of existing fax modem standards.

What are fax modems?

Pedants would say that a fax modem is basically anything capable of transferring fax image data over telephone lines using fax session protocols, and thus all working group 3 fax machines have to incorporate a fax modem. Strictly speaking, this is quite true, but we are concerned in this book with a rather narrower set of devices.

We intend to cover intelligent modems, capable of normal data communications as well as sending and receiving to and from other fax systems. They can be controlled through a normal asynchronous communications port by issuing commands generally corresponding to the AT command set. Fax modems can be either external, in which case they are stand-alone boxes connected by cable to an asynchronous serial port on a computer, or internal, consisting of a board that plugs into a slot on the motherboard. Commands controlling the fax side of operations are based on specifications produced by the TIA/EIA TR-29.2 subcommittee on Digital Facsimile Interfaces, which consists largely of US modem manufacturers. We generally refer to it simply as the EIA.

Internal and external modems are functionally identical from the programmer's point of view. Most, if not all, fax modems can also operate as conventional asynchronous data modems. We refer to such modems here as operating in either data or fax mode. Of course, what a modem handles in fax mode has just as much claim to be considered data as anything else. However, the data handled by a modem in fax is sufficiently different to that in other modes to justify this usage. In particular, the bit-oriented nature of fax images makes any attempt to assign specific meanings to any data byte a waste of time. Identical 8-bit patterns mean different things depending on the coding method used and the pattern's position in a sequence. We thus use the term data mode to distinguish the non-fax side of modem communications from the fax aspects.

More terminology

The language used by the EIA is that specific type of English peculiar to standards bodies, which is designed to be unambiguous and exact. However, it occasionally fails to achieve this objective. When this happens it merely manages to confuse, and impose an unnecessary barrier to clear understanding. I find the use of the terms *DTE* and *DCE* by the EIA standards documents to be such a barrier. In practical terms, a DTE is a computer, while a DCE is a modem. I have avoided using the terms DTE and DCE in this and subsequent chapters on EIA standards.

A DCE is generally called a *modem* here, though sometimes it's called a *fax modem*. I prefer to talk about either *computers* or *application software* instead of DTE, as this is more accurate and descriptive. For instance, where the standards talk about the DTE 'responding' to the modem, it is clear something is being described that is really the responsibility of the applications software, and therefore of the programmer. In a book of this type, which doesn't simply present or paraphrase the relevant standards but seeks to offer help to those implementing them, I think this approach makes sense.

Command set standards

There was no official source of standards for commands used to control data modems, so the industry developed around a number of unofficial commercial sources. Hayes, who originated the AT command set, were one of the more significant companies in this area, but they cannot be said to be solely responsible for the current state of the software command set. Other companies have been responsible for enhancing the technology (such as Microcom, originators of the MNP error-correction protocol), while manufacturers such as Rockwell, one of the main suppliers of modem chipsets, have an equally strong influence on the accepted list of commands. In addition, any manufacturer departing from commands supported by major communications software manufacturers will have great difficulty in selling their products.

The situation regarding commands used to control a modem's fax operations has always been different to other modem developments. Unlike functions like data compression and error correction, faxing with a modem is not transparent; it needs active co-ordination between the modem and the software that drives it. This was realized early in the development of the fax modem, and the existence of a standard command set was seen as the best way of turning this apparent liability into an asset.

The work involved in defining the software command set was channelled through a recognized standards body, which is why the *de facto* source of most of the current fax modem software specifications is the EIA.

Fax modem service classes

All EIA fax modem commands are distinguished by the fact that they all begin with the prefix **AT+F**. The most basic of these commands is the simple query

```
AT+FCLASS=?
```

which asks the modem what service classes it supports. For instance, class 0 is defined as indicating asynchronous data and any modem capable of data communications ought to include a 0 in its response. The concept of service classes arose quite early on in the deliberations of the TR 29.2 committee, and needs some explanation, as it is fundamental to understanding the differences between types of fax modems.

Opinion in the TR 29.2 committee was originally split as to what tasks a fax modem ought to undertake. A minimalist group thought the modem should be mean and lean, offering the bare minimum of facilities needed to conduct a fax session. A rather less minimalist group felt that more than the bare minimum of facilities was needed, as many operating systems would be unable to provide the real-time support needed to handle critical parts of the fax session protocol in a satisfactory manner. A third group thought the fax modem should effectively be able to handle virtually all the work involved in sending a fax in much the same way that fax machines do.

The committee agreed to differ, and work began on three different classes of fax modem, reflecting the three different views. The higher the service class, the more work the modem should be capable of. All the EIA specifications for asynchronous fax modem standards contain an outline of these classes:

- Service class 1 aims to insulate an application controlling a modem via an asynchronous serial port from the synchronous nature of fax communications. A class 1 modem will handle the necessary conversions from asynchronous to synchronous, and vice versa, and will take care of the requirements needed to support the HDLC framing used by the fax negotiating procedure. However, the composition and interpretation of all data, whether it is negotiation data or image data, is entirely the responsibility of the controlling computer. While the modem will give some timing support by providing accurate clocking, the onus is on the application software to issue the right commands at the right time.

- Service class 2 provides all the necessary software support to insulate the controlling application from the need to know anything about the fax session protocol. All the application software has to do is to provide a restricted set of options, mostly to do with the construction of the fax image, and then provide or receive data in the proper format. Though the modem takes care of the negotiation, encoding and decoding the fax itself is up to the application software.

- Service class 3 insulates the application software from needing to know anything about either the format of a fax image or the fax session protocol. An application simply provides the modem with data in an agreed format (text or graphics) for sending, and accepts data converted to a similar form on reception.

A hierarchy of classes

The fate of these three proposals demonstrates that the higher the class, the greater the difficulty in getting the standard through the committee.

Class 1

The minimalist Class 1 standard proposal (SP 2188) emerged quite quickly. It consisted of only half-a-dozen commands, was only 20 pages long, and was approved in 1990 as standard ANSI/ TIA/EIA-578. Fax modems conforming Class 1 started selling soon after.

Class 2

The Class 2 proposal (SP 2388) had reached ballot stage by this time, and was due for approval as ANSI/TIA/EIA standard 592. It consisted of over 50 commands and was 90 pages long. It did not gain approval at that time. This didn't stop Class 2 modems based on the unapproved draft of October 1990 from going on sale soon after, and class 2 modems based on the same proposal are still being sold and used today. The 1990 proposal never did get approved, for reasons that are still unclear. While there were certainly deficiencies and ambiguities in the Class 2 specification, they appear to be no worse than those in other documents (such as Class 1). In particular, fax software houses have not found it difficult to write Class 2 compliant software, despite alleged problems with the specification. And while it is true that there are admitted bugs in various modem implementations of Class 2, the fact that a manufacturer has misinterpreted the specification can surely be no reason to deny it approval.

Class 2.0

What was eventually approved as EIA-592 was a revised proposal (SP 2388-B), commonly known as Class 2.0. It contained much of the functionality of the unapproved Class 2 based on earlier drafts of SP 2388, but was incompatible with it. The philosophy behind the unapproved Class 2 and the approved 2.0 is identical, and there is a one-to-one correspondence between many commands. However, the differences are sufficient to ensure that software written for Class 2 modems will not work with a 2.0 modem. There are still far more Class 2 compliant modems than Class 2.0, and the eventual approval of Class 2.0 didn't herald the abandonment of Class 2. There was really no reason why it should, as most existing Class 2 products were working reasonably well, and moving to a different command set could only involve manufacturers in extra costs and expose users to an extra source of teething bugs. While Class 2.0 may be the officially approved version, the draft of Class 2 has the status of an unofficial but more widely adopted standard.

Class 3

The Class 3 standard proposal (SP 2725) never emerged from the committee, and it is probable that it never will.

Class 4

Recent reports indicate that a service Class 4 is being developed. This breaks the original pattern of increasing class number corresponding to an increase in modem functionality, as Class 4 is projected to be similar to Class 1, with modifications intended to relieve the timing burden for multitasking computer systems.

Class 8

The EIA have no commitment to introduce additional service classes in numeric order. Service Class 8 has already been defined for modems with voice capability.

Class 2.1

The latest version of the 2.0 specification warns of the possibility of service Class 2.0 being supplemented by a service Class 2.1 in future.

Class 1.0

There is also a proposal before the ITU for an internationally approved version of Class 1, which includes the requirement for a modem to respond to the **AT+FCLASS=?** query with a 1.0 response. It is probable that this standard will become known as Class 1.0.

The +FCLASS command

Going back to the **AT+FCLASS=?** command, a response of **0,1** indicates that a modem can handle data and fax Class 1, a response of **0,2** indicates that it can handle data and fax Class 2, while a response of **0,1,2** indicates that it can handle data and both Class 1 and 2. A modem that can handle the approved Class 2.0 rather than the unapproved Class 2 responds (predictably enough) with **2.0** instead of a **2**.

The **AT+FCLASS** command can take any one of three forms. The test parameter sequence **AT+FCLASS=?** causes a modem to respond with a list of its capabilities. The read parameter sequence **AT+FCLASS?** causes it to respond with the current setting in force for the service class parameter, while the set parameter sequence **AT+FCLASS=<class>** causes it to adopt the setting instructed by the class parameter. The sequence in Table 8.1 shows how the various forms of this command can be used on one particular modem.

While this may appear straightforward, it isn't as simple as it looks. There are a number of things to note about this sort of modem command sequence, which are typical of the problems programmers trying to write for various implementations of the EIA specifications come across, so we'll discuss this deceptively easy example at some length. We have no choice but to delve into the minutiae of the standards at times, as it is there we find the detail that can make a whole session fail.

Table 8.1: Example +FCLASS dialogue

AT+FCLASS=?		ask the modem what its capabilities are
	0,1,2	the modem can handle data, class 1 fax, class2 fax
	OK	the modem responds OK
AT+FCLASS=1		tell the modem to enter fax class 1
	OK	the modem responds OK
AT+FCLASS?		ask the modem what service class is in force
	1	the modem responds with class 1
	OK	the modem responds OK
AT+FCLASS=2		tell the modem to enter fax class 2
	OK	the modem responds OK
AT+FCLASS?		ask the modem what service class is in force
	2	the modem responds with class 2
	OK	the modem responds OK
AT+FCLASS=0		tell the modem to return to data mode
	OK	the modem responds OK
AT+FCLASS?		ask the modem what service class is in force
	0	the modem responds with class 0 = data mode
	OK	the modem responds OK

General modem defaults

Possibly the point most obvious to anyone who has controlled any type of modem using a Hayes command set is that capturing the above sequence from a modem is only possible if it is echoing commands, is operating with responses enabled, and is in verbal response mode. The standard settings to force this mode are **ATE1Q0V1** (where **E1** sets command echo on, **Q0** sets quiet mode off, and **V1** sets verbal responses on). However, these commands are not themselves part of any of the EIA fax modem standards, which don't mention command echo or quiet mode. Though they do acknowledge that both verbal and numeric result codes may be used, the standards state simply that a user-selectable option should be provided to switch between the two, without saying what it might be.

Luckily, I don't know of any modem implementing any of the **AT+F** fax command sets that can't handle a command sequence such as **ATE1Q0V1**. Certainly, setting **Q0** should be regarded as an essential part of initializing a fax modem, as controlling a fax session is impossible without analysing result codes. I've always been surprised that the commands to set a modem to fax mode don't automatically set **Q0** also. The value of the other parts of the command sequence is more questionable. In fact, setting **E0** for command echo off is generally a better option if you don't need to capture command echoes, since not having to take time to echo user commands back gives a modem a measurable performance boost. The Class 2.0 document is the only one to recommend specifically that any command echo be turned off.

Deciding to opt for numeric or verbal result codes is largely a matter of taste, but the arguments for using verbal codes are stronger for fax than for data use, as so many of the test parameter command sequences return numeric values that may be confused with numeric result codes. This results in an overloading of digit response codes. A response of **0** sometimes means data mode and sometimes that the command has been executed. While it ought to be possible to tell from the context what a **0** means, this is not always going to be the case if it isn't known how the modem behaves normally.

Command and response syntax

The standard for sending AT commands to modems is common to all service classes. Only the low order 7 bits of the characters in a command are evaluated, and none of the commands are case sensitive (with the possible exception of commands that include string literals inside quotation marks, for which some Class 2 modems maintain the case of alpha characters).

Each command must begin with an **AT** sequence, and any characters before the AT should be ignored; though not mentioned in all the standards, some modems require a short period of silence before any **AT** command. The body of the command (such as **+FCLASS**) must consist of ordinary printable ASCII characters, with all spaces being ignored, and a **<cr>** is used as the command terminator. The only other permissible character is the backspace, which can be used in the body of a command to remove the last character entered.

More than one **AT+F** command can be included on a line provided the commands are separated by a semicolon. Only the first of the commands on each line need have the full **AT+F** prefix, as subsequent commands can begin with a simple **+F**. While all commands that take numeric parameters are supposed to evaluate missing parameters as 0, it would not be wise to rely on this. Entering the command **AT+FCLASS=** (leaving out the final 0) should be the equivalent of **AT+FCLASS=0**. On some modems this is the case, but on others an **ERROR** response is the result.

Despite this difference in handling deliberate mistakes, the rules are fairly unambiguous and admit few exceptions. I've not found any modem that departs from the command sending specification in any significant way.

The unanimity regarding command syntax applies to a lesser extent when we come to response syntax. The standards point out that there are two types of response: information text and result codes. Information text has to be preceded and followed by a **<cr><lf>** pair. All information has to appear on a line by itself. The same applies to verbal result codes, while numeric result codes aren't supposed to be preceded by anything and should be followed by a lone **<cr>**. While most modems follow these rules, you occasionally find one that inserts an extra **<lf>** or **<cr>**, or both. These variations are minor, and don't generally cause any problems. The real problems begin not with the syntax of the responses, but with their content.

The OK response

Let us take as an example the issuing of a final **OK** (verbal) or **0** (digit) response after completion of an **AT+FCLASS** command. While this is a common response sequence, not all modems behave in this way. On a strict interpretation of some of the specifications, it could even be seen as being an illegal response. Neither the original Class 1 specification nor the draft Class 2 specification say it should be issued. While the Class 2.0 specification requires the final **OK** response for the **+FCLASS=?** command, this can't be taken just as tidying up a loose end, as the final **OK** isn't mentioned in connection with the **+FCLASS?** read parameter command.

Since the **AT+FCLASS** commands are common to all EIA fax modem standards, it is quite disconcerting to find that the response is capable of such differing interpretations. Considerations such as these, and the fact that a response of 0 could indicate either a service class or a command response, makes me a little uneasy about opting for digit result codes. While the exact sequence of digital responses for any particular modem is no doubt easy to find out and program, software that uses verbal result codes is almost certainly going to be easier to port from one manufacturer's modem to another.

More ambiguous responses

Modems that incorporate components from one of the major chipset manufacturers such as Rockwell will often behave in the same way. But while modems that use Rockwell components are dominant in the middle of the fax modem marketplace, they certainly have no monopoly. Not only are chipsets from other manufacturers widely available, it is not uncommon (especially at the top and bottom ends of the market) for modems using components from the same manufacturer to vary quite widely in their responses.

Some of these interpretations will not only defeat obvious methods of distinguishing a 0 parameter from a 0 meaning an **OK** response, but can also fool modems that are using verbal result codes to avoid ambiguities. For example, suppose you are trying to check what service class a modem is using, and you decide on an algorithm that interprets the first digit received after an **AT+FCLASS?** command as being the service class. Such an approach would be defeated by a modem that returned 001 instead of 1 when in class 1.

Responses to unsupported options

It is bad enough that manufacturers have differing views as to how to present certain numeric responses, and don't always agree on whether a final **OK** or 0 is required after a response. In some situations even more confusion can arise: modems with the Sierra chipset have been known to accept a command to enter the non-existent Class 9!

It is worth noting that there is not necessarily a correlation between how a modem interprets a particular specification and its reliability in performing when it comes to sending and receiving faxes. A modem can behave in an entirely orthodox manner and be so badly built as to be unreliable in pure operational terms.

Parameter delineation

Some of the most serious discrepancies between modems are those affecting responses to test parameter commands. While most modems produce a list of numeric values separated by commas, programmers writing for unknown modems must be prepared to accept alternative presentations of the same information. One reported variation is for a modem to enclose the numbers within a set of brackets, thus:

```
AT+FCLASS=?              ask the modem what its capabilities are
         (0,2)           the modem can handle data and class2 fax
```

Variations of this type are more common among Class 2 than Class 1 modems. Most variations come under the category of differing interpretations of the same specification.

Command timings

We continue with our analysis of the response to an initial **AT+FCLASS** sequence. It isn't a terribly useful command in itself, but it is a fairly useful hook on which hang a more general discussion of a number of important introductory points common to all fax modem command.

Examination of a simple dump of data sent to and from the modem (like the dialogues listed earlier) gives no indication of the timings of the commands and responses. In this connection, the question as to whether a final **OK** or 0 is appended to the modem capabilities assumes a greater significance. It is an important issue, not just because it can lead to a misinterpretation of a numeric response, or even because some modems respond with an **OK** where others respond with an **ERROR**. The most critical aspect of any uncertainty in a response is that if an application doesn't wait for a final **OK** when one is being sent out, the result will almost certainly be a failure in the command sequence. This is because, although AT compatible modems are becoming more intelligent by the year, most of them have one particular blind spot, in that they can't accept commands which come in while they are still issuing responses.

What this in practice means is that even if a programmer issues the command **AT+CLASS=?** and sees the response **0,1**, it isn't necessarily safe to issue the command **AT+FCLASS=1** immediately. If the modem responds with an **OK** after listing its capabilities, the command will almost certainly be ignored. In fact, even if the modem doesn't send out an **OK** after each response, the parameter set command will probably be ignored if the modem hasn't yet sent out its final line feed after a response such as **0,1**. A modem can't understand anything you tell it while it is talking to you.

The logic behind this is that while a modem is responding to a command, any character it receives is taken as an abort signal, and is discarded once the command being executed has been cancelled. Even in data mode, any character sent to a modem after it has begun dialling and before it has sent a **CONNECT** message aborts the dialling attempt.

This inability of modems to tie their shoelaces and chew gum at the same time can sometimes extend further, with many being unable to understand commands that comes in while they are still working out whether the last character received was a command or not. This is why even sending an **<cr>AT<cr>** string to a modem sometimes fails to produce a response. If the A arrives while the modem is still working on the **<cr>**, it doesn't usually get seen. This phenomenon generally only affects well-written communications software capable of streaming characters to the modem in quick succession. It used to be the case that modems needed delays between each character of each command, but this is not true of any fax-capable modem. If they did need such delay they wouldn't be able to send faxes. However, some models produced by major fax modem manufacturers still can't handle an **AT** command immediately preceded by **<cr>**.

Timing rules

There are a couple of unwritten rules that all fax software should consider, as accurate timing is one of the main keys to successful fax modem programming. First, no command should be sent to a modem unless there has been a period of silence from the modem since its last response; second, commands should always be preceded by a similar period of silence on the part of the controlling software. The two periods can run concurrently. The duration of the period is rather more problematic. Clearly it has to be long enough to do its job, which is to ensure the all stray line feeds and **OK** codes have been absorbed. However, the delay shouldn't be too long; apart from inefficiency this introduces, a delay of longer than 20 ms causes problems for software controlling Class 1 modems, as the fax session protocol can require certain actions to be taken within this period of time.

I've experimented with a number of modems, and have found that under normal conditions, a reasonably safe delay for the **AT+F** commands is to wait for the time taken to transmit two characters.

PC programmers in particular should be aware that the clock on the IBM PC is not up to this job, not because the system isn't fast enough, but because the clock ticks 18.2 times each second, and the granularity dictates that all timings are going to be ±55 ms. More accurate timings will clearly be needed. Assuming the standard 10 bits per asynchronous character, if we are talking to a modem at 19200 bps we need to be able to measure delays to an accuracy of around a millisecond. One technique for managing this type of delay without a clock accurate to a millisecond is described in the NOTES directory on the accompanying disk.

More modem defaults

We turn now to the format of the data and the speed at which the modem commands should be transmitted. One of the few things the different classes of fax modem have in common is the fact that, irrespective of how they handle the rest of the fax session, they can all either accept fax data forwarded by the controlling computer for transmission to the remote fax, or accept data from the remote fax and forward it to the controlling computer. The key here is that while the connection between a computer and fax modem is asynchronous, the link between the fax modem and remote fax is synchronous. Further, while the computer and fax modem exchange data in bytes, all fax data (and synchronous communication generally) is bit-oriented.

On transmission the fax modem strips the start and stop bits from each 10-bit byte received from the computer to leave the 8 data dits, which are transmitted to the remote fax. It is clear that as only 8 bits out of every 10 sent over the asynchronous link are transmitted, the computer must send bits to the fax modem at least 25% faster than the fax modem is transmitting them to the remote fax. If the fax is being sent at 9600 bps, the computer must send asynchronous data at least at 12000bps, otherwise the fax transmitter will run out of data and have nothing to send.

We discuss this phenomenon, known as *underrun*, on page 185, when we deal with flow control. The important thing to realize now is that, since it isn't possible with synchronous data to simply stop and wait until a bit is available to be sent, underrun is a serious condition that, at the very least, results in the fax becoming corrupted, and at worst results in the session being terminated.

On reception this procedure is reversed, with each set of 8 bits received from the remote fax having a start and a stop bit added, with the resulting 10-bit byte being forwarded to the computer over the asynchronous link. Bearing in mind there is no provision for flow control when transferring fax data, a fax modem receiving bits at 9600 bps, must be able to forward bytes at least as fast as 12000 bps to the controlling computer. If the link between the modem and computer is slower than this, fax data will back up in the modem's buffer until they overflow. This is an example of *overrun*.

As the 12000 bps speed required for V.29 fax communications is not commonly found as a systems software option, most fax modems are linked to computers at 19200 bps. This speed is also generally adequate for 14400 bps V.17 fax communications, which on the same calculations as used above require an equally esoteric minimum speed of 18000 bps. It should also be apparent that the most efficient of all possible asynchronous data formats is 8 bits with no parity and one stop bit. The 7 bit formats and those using 8 data bits with parity would simply require an even greater speed differential to compensate for the redundant bits, and would involve unnecessary complications in extracting the data bytes from the asynchronous data.

Enforcing fax modem data format and speed restrictions

Each of the EIA standards says that only the low order 7 bits of a command are used when accepting commands from a computer, but that all 8 data bits are needed for transmitting and receiving phase C fax data. The Class 2.0 document points out that in view of this, it is preferable to use 8 bits with no parity for all communications between fax modem and computer to avoid changing data formats in mid-session.

There is no method of changing speed in mid session, as all Hayes compatible modems either use a fixed speed for all communications, or detect the speed being used from the initial **AT** sent at the start of each command. It used to be common for modems to offer an option on the data communications side to automatically switch speed after a carrier has been established to match the speed of the link that has been established, but most manufacturers have dropped this capability in favour of extended flow control options to enable matching of data rates.

Instead, many fax modems enforce specific speed restrictions to avoid problems. As soon as an **AT+FCLASS=** set parameter command is issued to place them in fax mode, they send their **OK** but then insist on communicating at a minimum speed of 19200 bps.

Commands issued at any other speed are rejected. A variation of this is where a modem accepts commands at any speed, but then switches to 19200 bps when sending or receiving phase C data. However, other modems don't worry about the practicality of the speed at which they are talking with a computer, and are quite happy to accept commands and issue responses at silly speeds well below that required, even where they are clearly incompatible with fax requirements. They may even get as far as answering or dialling before the obvious errors they encounter force a disconnection.

The fax modem specifications are of no real help in establishing what the correct procedure is regarding enforcement of speed. While the EIA standards all say the computer to fax modem speed should be greater than the fax modem to fax speed, they do not say whether this should be enforced, let alone how. Thus modems that enforce minimum computer to modem bit rates whenever fax mode is entered should be regarded as exhibiting a sensible piece of defensive design which may not be required by the letter of the specifications, but is certainly within the spirit.

There is one further complication that anyone writing software to control a fax modem should bear in mind. Most fax modem manufacturers have at some time manufactured 9624 type fax modems, which support 2400 bps for data communications and 9600 bps for fax. A number of modems of this generic type didn't merely refuse to accept **AT** commands at any speed below 19200 when in fax mode, they also refused to accept commands at speeds above 2400 bps in data mode. Some even had the peculiarity that, after having been set to fax mode at 2400 bps, they immediately issued their **OK** response to the command at 19200 bps, with the result that any software waiting for an **OK** response would assume there had been an error.

Though later 9624 models, which incorporated V.42 error correction and V.42 bis data compression, had to drop the 2400 bps data speed restriction, a fair number of the older models were (and still are) being sold. They may not be as capable at data communications as current modem models, but they can perform just as well when it comes to sending and receiving faxes as many more recent fax modems.

A universal AT+FCLASS tester

Bearing in mind the above, the strategy for interrogating a fax modem should be to start by issuing an **ATE1Q0V1** command at 2400 bps on the grounds that, since most modems have a power-on default of data mode, that is the speed most likely to get a response from any modem. This is because modems that support higher speeds also support the lower ones, but the reverse is seldom true. If this command gets no **OK** response, it is possible the modem is in fax mode, and only accepts commands at 19200 bps when in fax mode. The next step should be to switch to 19200 bps and issue an **AT+FCLASS=0** command before starting the procedure again and sending an **ATE1Q0V1** command. If the second attempt also fails, user intervention will probably be required to check cable connections and port assignments.

Once an **OK** has been received, the first **AT+FCLASS=?** command to test the capability of the modem can be issued. Remember that a final **OK** may or may not be present. Earlier we discussed the variations in how the parameter list might be presented. This doesn't make things that complex. As a rough guide, if a **1** is included anywhere in the response string then the modem has Class 1 capability, and if a **0** is included it has data capability. If a **2.0** is included, the modem conforms to the approved Class **2.0** version of EIA-592, while a **2** followed by anything apart from a . implies the modem implements the draft Class 2 specification.

Having found what service class the modem supports, should select the one you want and issue the set service class command **AT+FCLASS=** using that class. To cater for modems that issue their **OK** at a different speed, at this point switch to 19200 bps and check that everything has worked by issuing an **AT** and waiting for an **OK**.

If you are more cautious, you may want to send an **AT+FCLASS?** enquiry to check the setting had actually worked. This last check isn't such a bad idea. If you're using an operating system service to alter the speed, it is possible that the routine to do this also momentarily dropped the DTR (Date Terminal Ready) signal to the modem, which can cause modems to reset themselves back to data mode. If this happens, it can be fixed by issuing a commnad to stop the modem resetting itself when DTR drops. The most common command for doing this is **AT&D0**, and this can be included as part of the initial default setting. This would then consist of the **AT&D0E1Q0V1** command.

Listing 8.1 is a fragmentary function that returns an integer showing what fax class a modem has been set to, using (in order of preference) **1** for Class 1, **2** for Class 2 and **20** for Class 2.0, with **0** for failure. It assumes the availablilty of some of the generic functions listed on page 195, notably **reinit** (to change speed), **export** (to send a commnad to the modem), **import** (to receive a response from the modem within a certain number of seconds) and **wait_for** (to receive a specific response from the modem within a certain number of seconds).

You may suppose that we have now covered everything you need to know before you start proper faxing. But while we may have finished dissecting the **AT+FCLASS** sequence, it would be a mistake to assume we are set up and ready to go. There are a few things we still have to check before proceeding, most of which concern sending rather than receiving faxes.

The need for flow control

There is usually a mismatch in speed between the computer to fax modem rate and the fax modem to remote fax rate, but this doesn't have serious implications for receiving faxes, as all the computer has to do is accept any data the modem sends to it, at whatever the agreed speed. Doing this may be a pain, especially on multitasking machines and those running software that demands a little too much from the CPU, but at least we know where we are and what is required. However, when transmitting faxes, some sort of flow control between computer and modem is essential.

This is easily explained. It is generally the case that a computer transmits data to a fax modem much faster than the fax modem is itself transmitting it to the remote fax, even taking account of the fact that start and stop bits are stripped. The best case to consider is if a fax is being transmitted at the fastest V.17 speeds. Under these circumstances, sending 19200 bps to a fax modem which is only

```
int modemset ()
{
    int faxclass ;
    char ourclass [32] ;
    reinit (2400) ;
    export ("ATHE1Q0V1\r") ;
    if (!(wait_for ("OK",1)))
    {
        reinit (19200) ;
        export ("AT+FCLASS=0\r") ;
        wait_for ("OK",1) ;
        reinit (2400) ;
        export ("ATHE1Q0V1\r") ;
        if (!(wait_for ("OK",1))) return (0) ;
    }
    export ("AT+FCLASS=?\r") ;
    import (ourclass,1) ;
    if (strstr(ourclass,"1"))
    {
        faxclass=1 ;
        strcpy (ourclass,"AT+FCLASS=1\r") ;
    }
    else if (strstr(ourclass," 2"))
    {
        faxclass=2 ;
        strcpy (ourclass,"AT+FCLASS=2\r") ;
    }
    else if (strstr(ourclass,"2.0"))
    {
        faxclass=20 ;
        strcpy (ourclass,"AT+FCLASS=2.0\r") ;
    }
    else return (0) ;
    export (ourclass) ;
    wait_for ("OK",1) ;
    reinit (19200) ;
    export ("AT\r") ;
    if (!(wait_for ("OK",1))) faxclass=0 ;
    return (faxclass) ;
}
```

Listing 8.1: Universal +FCLASS setup

sending on bits at the rate of 14400 per second is bound to result in data backing up in the modem buffers. Even though removal of start and stop bits means 19200 bps asynchronous reduces to 15360 bps synchronously, data still accumulates in the modem at the rate of 960 surplus bits per second.

While some fax modems have generous buffers, even the largest one I've come across, which is 16K, eventually overflows after about 2 mins when being sent data at 19200 bps for resending at 14400 bps using V.17. If a connection is made at 9600 bps using V.29, the buffer fills up much faster and overflows in under 23 secs. Bearing in mind that most fax connections use 9600 bps, and that an average page takes 30 s to send, it is clear that flow control cannot be avoided.

In fact, buffer sizes of 16K are unusual. On-board buffers of 1024 bytes are something that manufacturers boast about, and many fax modems have buffer sizes as low as 104 bytes. Though this sounds niggardly, it should be remembered that all a small buffer needs is rather more intensive flow control than a larger buffer. As long as this is possible, there is no discernible performance difference attributable to buffer size. Of course, intensive flow control is often not possible; a subject we return to shortly.

Transmitter underrun

There is one very important difference between transmitting fax data to a fax modem and virtually any other type of transmission through a standard asynchronous port. Though we happen to come across it first in our discussion of flow control, it also affects the design of a number of other fax software components.

In ordinary asynchronous data transmission if, for some reason, a serial port is left with no characters to send, the result is simple inefficiency. If it happens for a period long enough to provoke a timeout in something like a file transfer protocol, more serious problems may arise; but of itself, a transmission line left in what is known as a 'marking state' is left in a neutral condition. It takes the transmission of a start bit to trigger a receiver into sampling the data line for a series of 8 bit intervals, and if there is no start bit then there is no possibility that any data transfer will occur.

In contrast, if a serial port transmitting fax data via any type of fax modem is left with no data to send, the result is invariably a transmission error. The condition of a transmitter with no data to send is known as 'underrun'. This is serious in the case of fax images because the modem is sending data on in the form of a synchronous bit stream, and there is no such thing as a marking state when sending synchronous data. The level of the data line is always being sampled at regular intervals, and the result is always being interpreted as a 0 or 1 bit.

The implications of this difference are quite profound. Most serial port drivers, and indeed, most communications software, assume that the most critical part of any flow control mechanism is stopping the flow of data when it cannot be accepted. Our discussion of the need for flow control between a computer and a fax modem was presented in just these terms a page or two ago; if there is no flow control, the flow of data will inevitably overrun transmitters and buffers and transmission errors will result. However, it is just as critical when sending fax data that the flow of data is resumed promptly once the buffer full condition is rectified. If the delay in resuming transmission is too great, the opposite condition will prevail and the transmitter will underrun. Corruption of the image data is the result of both overrun and underrun; the two cases are just as serious.

Software flow control

The EIA fax modem specifications state that XON/XOFF flow control must be provided. They also state that other types of flow control such as CTS/RTS can also be provided as extra options selectable by manufacturer specific means (though once again they talk standardese and refer to V.24 Circuits 106 and 133 instead).

This recommendation apparently conflicts with advice most modem manufacturers offer users of the data facilities in their modems, which is to use CTS/RTS flow control wherever possible. However, there are good historical and practical reasons for this recommendation, and equally good reasons why the standard objections to XON/XOFF flow control don't apply.

Fax communications take place over half duplex synchronous communications links. The original meanings of the CTS (Clear To Send) and RTS (Request To Send) signals commonly used for hardware flow control are historically linked with half duplex communications. When using half duplex, a modem cannot transmit and receive at the same time, and data flow is only possible in one direction. Without going into detail, a very condensed version of the protocol follows. The ability of a modem to transmit is controlled by the state of its RTS output and its CTS input lines; when the RTS output is switched on, the modem indicates its wish to assume the role of a transmitter. If the device to which it is connected can receive a transmission, it indicates this by raising a signal on the line connected to the transmitter's CTS input. When its CTS input is on, the transmitter is given permission (cleared) to transmit, but while its CTS is off no transmission should occur.

There is clearly a relationship between the use of CTS and RTS for flow control and their use in half duplex communications. But the meanings of the signals are subtly different, and there is no standard governing the way that such flow control works. This is presumably why the EIA standards prefer to talk about V.24 signals rather than CTS and RTS.

In purely practical terms, XON/XOFF flow control has, in theory at least, one great advantage. It is a software only solution that can be implemented on any system without needing to reprogram serial port controllers, or monitor the state of input signals, or worry about the design of the cable connecting your computer to your fax modem and whether or not it routes the signals to the correct pins. The algorithm to be used when transmitting is basically quite simple:

1. If a character needs sending, check the receiver first.

2. If nothing has been received send the character and loop back to step 1.

3. If a character has been received look at it; if it is not an XOFF, loop back to step 1.

4. If an XOFF has been received, wait for an XON to come in at the serial port; then loop to step 1.

The simplicity of this procedure arises because the data flow is entirely unidirectional. When fax data is being transmitted by the computer, the modem isn't meant to do anything but accept the data and send it on, and XOFF is the only character we ever expect it to generate. We're not interested in XON characters unless they follow an XOFF; in fact, as long as the character received after an XOFF isn't another XOFF, we can assume that it is an XON.

As the flow control is unidirectional, it doesn't matter that the fax data being output is likely to contain XON or XOFF characters; and as we only need flow control when transmitting, it doesn't matter that received fax data is equally likely to contain XOFF or XON characters. This disposes of the main objection to software methods of flow control which arises for data communications, which is that it interferes with the transparent transmission of binary files.

Flow control trigger levels

The theory behind XON/XOFF flow control is unfortunately often defeated by practical considerations on specific installations. The problem boils down to one of timing; in particular, the time taken for application software to respond to an XOFF and an XON. It would be unusual, and very bad design, for a modem to assume an immediate response to any flow control signal, irrespective of whether the signal is sending an XOFF or dropping an RTS line. The almost universal practice for any device implementing any sort of flow control is to set buffer thresholds for the triggering of flow control events. As our discussion applies to any form of flow control, we use the term 'stop event' to refer to issuing an XOFF or dropping a control line, while a 'restart event' refers to issuing an XON or raising a control line.

In a real-time communications environment, the trigger levels set for flow control events are far more important in ensuring good communications than the size of the buffers. This doesn't mean large buffers are a waste of time. Especially in multi-tasking or multi-user environments, a decent size buffer is essential for efficiency, as otherwise fast communication can easily slow the system down. But even in such a computer system, the greatest benefit of a large buffer is that it permits generous flow control thresholds. A threshold set too low can render an otherwise large buffer completely useless.

The sorts of thresholds typically used are a trigger level of 80% of capacity to trigger a stop event and a trigger level of 20% of capacity to trigger a restart event. It is usually quite difficult to find what trigger levels are set on Class 1 fax modems, as there is no software command showing these statistics and the modem specification sheets seldom contain this information. However, many Class 2 modems implement the optional read-only command **AT+FBUF?** which returns four numbers. The first is the total size of the buffer, the second is the stop threshold, the third is the restart threshold and the fourth is the number of bytes currently used in the buffer.

The data in Table 8.2 was captured from three different Class 2 fax modems, and shows how widely buffers and thresholds can vary. The acronyms **<bs>**, **<xoft>**, **<xont>**, **<bc>** stand for buffer size, XOFF threshold, XON threshold and buffer byte count, respectively, and are taken from the Class 2 specification. Note that while all the modems respond with the information in the form of a comma delimited list and no parentheses, the second modem doesn't believe in the final **OK**,

The first modem has set its thresholds at around 20% of buffer full and buffer empty levels, respectively. More importantly, the 3K leeway these settings allow means a modem has around 1.8 s to respond to restart events if a fax session is using a 14400 bps V.17 connection, and over 2.7 s at 9600 bps. The figures for the stop events are even more generous, as we'll see below, but these timings ought to be well within the capabilities of even the most sluggish systems.

The second modem clearly shows that the designer of the firmware fell into the trap of thinking the buffer full threshold was more important than the buffer empty threshold, as the modem will issue a stop command when there are still 324 bytes free in the buffer but waits till only 100 bytes are left before issuing a restart command. As we have seen, there is no practical or theoretical justification for having different timings constraints for stop and start events. In fact, if we really did want the timings to be the same, the restart threshold should have been considerably larger than the stop threshold.

Table 8.2: AT+FBUF?

AT+FBUF?		modem 1 : what are buffer sizes ?
	16384,13107,3276,0	<bs>, <xoft>, <xont>, <bc>
	OK	final OK
AT+FBUF?		modem 2 : what are buffer sizes ?
	1024,700,100,0	<bs>, <xoft>, <xont>, <bc>
AT+FBUF?		modem 3 : what are buffer sizes ?
	104,88,16,0	<bs>, <xoft>, <xont>, <bc>
	OK	final OK

Threshold levels and response timings

We have to look in more detail at the arithmetic involved to work out why this is so. Consider a modem in the middle of a 9600 bps fax session. The restart trigger threshold is the easiest to consider. Since this is issued when there are only 100 bytes left in the buffer, this gives the controlling computer 83 ms to respond.

The stop trigger threshold is more complex, as it needs to take account of the bit rate between the modem and computer. Let's assume this is 19200 bps, which is most common. As the stop command is issued when there are 324 bytes left in the buffer, it is tempting to conclude that, as it takes 168 ms to send 324 bytes at 19200 bps, the computer has 168 ms to respond. However, during those 168 ms the modem has carried on transmitting, and will have managed to send further 201 bytes. This means that instead of the buffer being full, it will have 201 bytes free. It will in fact take 449 ms before the buffer really does become full. Generous though this is may be, it would have been better to design the modem differently. There's no virtue in allowing almost half a second to respond to one event when an equally critical one is given a mere 83 ms. A failure to respond to either event would be equally damaging.

The third modem is the one that makes most demands on its host computer. With a buffer of only 104 bytes, the thresholds have been set at a level 16 bytes from the buffer full and buffer empty levels, leaving 72 bytes in the middle. As the buffer is so small in the first place, it isn't worth arguing about this decision. At 9600 bps (I wouldn't recommend V.17 on such a modem even if it were supported), the host computer has 20 ms to respond to stop events and 13 ms to respond to restart events.

There is a specific problem affecting IBM PCs with buffered 16550 UARTs. These serial controllers include 16 byte receive and transmit buffers which aim to reduce the burden on the central processor of servicing interrupts. While use of these FIFO (first in first out) buffers is an undoubted boon when it comes to receiving data, there is a problem with the use of transmit FIFOs. Once characters are sent to the serial port, you can't get them back. The transmit FIFOs can hold 16 bytes, which happens to be the same size as the trigger level of the third modem described above. If this modem asks its computer to stop sending when there are 16 bytes in the transmit FIFO, a much quicker response from the computer is necessary, and it has only around 12 ms to respond.

Obviously, the threshold levels for flow control are one of the main factors distinguishing which modems are most suitable for any particular system. On a busy multitasking system, a modem with a large buffer and generous thresholds will be more reliable than one with a small buffer and mean thresholds. Unfortunately, this is information modem manufacturers do very little to publicise, possibly because buyers probably wouldn't understand the significance of the figures. Using the **AT+FBUF** command is only possible once you have bought the modem and plugged it in, which is rather late in the day to find out how suitable it might be, and in any case it only works on some Class 2 modems.

Using operating system services for flow control

Because of these difficulties, fax modem controlling software needs to ensure it can respond to flow control events in the shortest possible time. A response to any flow control event within 16 ms is essential for compatibility with the largest number of fax modems. As this is a rather tight timing constraint, flow control cannot be handled as part of the main body of the applications software. There are too many things in the average computer system that take control of the CPU away from an application program for periods longer than 16 ms for this approach to be feasible.

There are three main possibilities for handling flow control:

- It can be handled as part of the serial port interrupt routine. This is the preferred approach on systems where an application is permitted to have access to the hardware. For single tasking single-user microcomputer systems, where the fax application software includes its own interrupt handlers for sending and receiving data, this doesn't present significant problems. It's simply a question of correctly coding interrupt driven flow control routines for all transmitted data.

- If the operating system permits it, flow control can be spawned off and made the responsibility of a separate subprocess with a high enough priority to guarantee it access to the serial port at least every 16 ms while transmissions are taking place. Such techniques are currently a little beyond the scope of this book.

- On multi-user and multi-tasking operating systems that discourage or prohibit access to the hardware at the interrupt handler level, and also can't guarantee a process slices of CPU time at regular intervals, the only method of handling flow control is to rely on operating system services. This assumes that some kind of API or a set of system calls is available from within an application program to permits the selection or deselection of flow control mechanisms as needed.

Flow control using the RTS/CTS lines is generally well supported by most computers and operating systems. On many serial port chips (though not on the 8250 family used on the IBM PC), it can be handled by the hardware. On all systems, whether it works on an external modem or not depends upon whether the cable used to connect the serial port to the modem has been made up correctly. The main problem with this type of flow control is that it isn't supported by all modems.

XON/XOFF is the only method of flow control guaranteed to be available on all EIA compliant fax modems. Anyone wanting to use systems software services for this needs to examine the facilities offered quite carefully, as they may not be suitable for the job they have to do. There is more than one variety of XON/XOFF flow control possible. The main distinctions are as follows:

- Unidirectional flow control is when XON/XOFF is active in one direction only. Bidirectional flow control is when XON/XOFF is active in both directions simultaneously.

- Filtered flow control is when XON/XOFF control characters are used as control characters and are always filtered out of a data stream. Transparent flow control is when the XON/XOFF control characters are both acted upon, and also treated as normal data characters and passed through the handler as part of the data.

This gives four possible permutations:

- The ideal situation is where unidirectional transparent XON/XOFF flow control is available. The application software simply switches flow control on when fax data is about to be transmitted and switches it off again afterwards.

- Unidirectional filtered flow control is almost as good, but may present problems with certain Class 2 modems which signal that fax data should be sent by sending a single XON. Where flow control is filtered, this stops the XON from being seen. This is discussed further in Chapter 10, devoted to Class 2 modems, but can usually be avoided by leaving flow control off until the last possible minute.

- Bidirectional transparent flow control should work but will almost certainly have problems with received fax data, which may well contain an XOFF with no subsequent XON. If this occurs, sending the required responses may be inhibited. Software should thus leave flow control off except when transmitting.

- Bidirectional filtered flow control is the one permutation that definitely makes sending faxes impossible, because the systems software will filter out all XON and XOFF codes from the fax data stream, and the fax image will thus be corrupted even before it is sent. If you need to use system software flow control and the only XON/XOFF variant available is bidirectional filtered, you may have a problem. Try not to buy a Class 1 fax modem that can't support CTS/RTS flow control as you will have difficulty sending faxes with this type of modem.

Avoiding flow control events completely

In case you thought it might have been a way round the problem, flow control can't really be avoided by attempting to set an exact speed (like 12000 bps for V.29 9600 bps transmission). Various features of fax data make this almost impossible. These include the padding of lines to minimum scan times and inserting DLE shielding characters in the fax data stream (discussed in more detail in

Chapters 9 and 10). Considerations such as these make it certain that any attempt to work out an exact baud rate at which no flow control is needed will be doomed to failure. The odd page may get through, especially if it is short (like a cover page), but eventually your luck will run out and either underrun or over-run is certain.

However, there is one way of avoiding flow control altogether. Actually, it is cheating a little to say this avoids flow control, as it doesn't quite do so. It avoids provoking flow control events by anticipating their occurrence in the transmit logic of the application software. The technique shown here is available with any Class 2 modems that support the **AT+FBUF** command. A corresponding method for use with Class 2.0 modems, which offer support for embedded buffer credit enquiries, is described in Chapter 11.

As mentioned earlier, the **AT+FBUF?** command returns four numbers, two of which are the stop threshold and the number of bytes currently used in the buffer. The method of avoiding flow control is simple:

1. Issue the **AT+FBUF?** command.

2. Subtract the number of bytes currently used in the buffer from the stop threshold.

3. Send that number of bytes.

4. Repeat the whole process again until no more bytes remain to be sent.

Methods such as these are not hacks or kludges; they are what the **+FBUF** and similar commands are explicitly designed to facilitate. Unfortunately, not all modems support this feature. Of those that do, modems with the small 104 byte buffers will need feeding with characters every 90 ms or so, and modems with the larger 1024 byte buffer will need feeding every 900 ms. There are many systems where access to the processor even every 90 ms cannot be guaranteed, and locking out for periods as long as 900 ms isn't uncommon.

The fact is that not all systems are ideally suited to real-time computer communication. This is sad but true. For these systems, the only hope of sending faxes is to reduce the real-time element in faxing. At the time of writing, the best way of planning this is to use a Class 2 fax modem with a big (>16K) buffer.

Setting up required defaults after setting AT+FCLASS

Once again, defensive programming dictates that your fax software should explicitly set up any defaults it needs after the correct **AT+FCLASS** setting has been made, in the same way as the **ATE1Q0V1** defaults used initially to get into fax mode were also explicitly set.

Flow control

The most important of these settings has to be that for flow control. Some fax modems automatically switch on XON/XOFF flow control whenever fax mode is entered. Others leave the settings for flow control the same as in data mode, which may be for hardware flow control. Whatever your fax modem does, Murphy's Law means you probably want it done differently. While XON/XOFF availability is mandatory, there is no requirement for it to be a default. In fact, if a modem has an option for no flow control, it is quite possible to enter fax mode with nothing selected at all. Unfortunately, this is one of the least standardized of all the **AT** commands. Picking up a number of modem manuals from a shelf reveals four possible commands for setting flow control options. The most common is **AT&K**, presumably because it is the one used by Rockwell. However, **AT&H**, **AT\Q** and **AT&E** also have many adherents.

DTR control

The command **AT&D2**, which forces the modem to hang up if the DTR line goes low, is often also a sensible setting, supported on most modems. It protects against the modem being left off line if the computer hangs and is reset, or is switched off, or if the cable connector falls out. The command **AT&C1**, which forces the modem carrier detect line to track the modem carrier, is sometimes seen in fax modem setup strings, but it isn't terribly informative in fax situations, as the carrier goes on and off as a matter of course. The dreaded **NO CARRIER** message, which indicates that a data call has failed, likewise has an entirely different meaning in fax situations, and is used routinely on Class 1 modems whenever the line is turned around.

Combining AT commands

The extended set of **AT+F** fax commands can be combined with other **AT** commands to a limited extent. However, all **+F** commands must either be the last on a line, or separated from another **+F** command by a semicolon. Beware of making use of the semicolon to chain commands on the same line. While in theory use of the semicolon as a command separator is unique to **AT+** commands, in practice some fax modems recognize the semicolon as a command terminator, but always generate an **ERROR** for a subsequent command. This is probably because the semicolon is overloaded as a command parameter, as it is used in the data mode command sets as the retun-to-command-mode-character.

Though **ATE1Q0V1+FCLASS=1** would be a perfectly legal command on a Class 1 modem, the command **AT+FCLASS&D2** would not. Be warned that although the command **AT&D2FCLASS=1** is legal, the &D2 part is executed before the **+FCLASS** portion. If you want to ensure that a default is set for fax mode, the only way is to use two commands. First issue an **AT+FCLASS=** command, then an **AT&D2** (or whatever else you need to set).

AT commands not included in the fax set

There are three other non-fax AT commands essential to controlling fax modems. These are the commands used for originating, answering and ending calls. All EIA documents state that mechanisms for doing this are outside their scope, but while the classic commands for these functions aren't part of the fax modem standards, there appear to be no existing fax modems that use anything else. The familiar **ATD** command for dialling (with the usual modifiers such as **T** for tone and **P** for pulse) works on all fax modems, as does the **ATA** command for answering. The **ATH** command for ending a call is also universally supported.

It is worth noting that the controversial +++ sequence, used to return to command mode in data calls, is neither needed nor used for fax operations, and the software standard is an entirely public one. As we've seen, fax communications are divided into distinct phases with their own timing constraints, and the mechanism for distinguishing data from commands is quite different from that needed when making conventional datacomms calls.

Different ways of answering calls

While they aren't part of the fax modem standards, a number of the S-registers present on all intelligent modems can affect fax calls. The S0 register can be set to autoanswer calls after a specific number of rings on the phone. For instance, **ATS0=1** will instruct the modem to answer a call on the first ring. However, it is usual in fax programs to leave autoanswer off by setting **ATS0=0**. When the modem indicates the telephone is ringing (by sending either a **RING** verbal message or alternatively a numeric code of 2) the controlling software then answers the telephone by issuing the **ATA** command.

There is one main advantage in handling incoming calls this way. A modem set to autoanswer with **ATS0=n** will always answer the telephone after **n** rings, irrespective of whether the controlling software is ready for it. This can give rise to problems. For example, where an external modem is used it is possible that the computer could be switched off with the modem mistakenly left switched on and in autoanswer mode. Alternatively, no matter what modem is used, there is always the possibility of the computer crashing or being reset without register S0 being reset to 0. Under some operating systems, the fax software could simply be unloaded by a master process without being given the chance to switch autoanswer off.

It is possible to guard against some of these eventualities by telling the modem to reset itself if DTR goes low (as it would if the computer was switched off, for instance). This is handled by the **AT&D** command mentioned earlier. Of course, **ATS0=0** must be the default setting for this to work at all, and it doesn't cover all possible circumstances where autoanswer is undesirable. Therefore, the explicit answering of a call via **ATA** is considered to be the best mechanism.

S-registers used with fax

Another important register is S7. Though not part of the EIA fax standard, the setting in this register will often determine the time that a modem will wait for a

V.21 channel 2 negotiation carrier after being told to dial or to answer a call. Less usefully, and also less commonly, it may also be used to limit the time a modem will take to establish a carrier at a different modulation during a fax session on Class 1 modems. However, its use for this purpose is not guaranteed to operate on all modems, and software should ideally use an internal clock to establish suitable timeouts under such circumstances. S7 is nevertheless quite an important parameter when a remote fax is being dialled, and the common default of 50 s ought to be extended for fax calls.

Various other S register settings affect how fax modems work. For example, register S3, which changes the default values of the CR command terminator, can disrupt communications with the modem, and should not be changed under any circumstances. Similarly, altering registers such as S22 and S23, which change the values of the XOFF stop character and the XON restart character, respectively, can cause fax sessions to fail. If a fax modem does support the **AT&F** command to reset all factory defaults, there is a good argument for using this if it is to be installed in an environment where unauthorized alterations to its configuration may be made.

Functions required for programming fax modems

While it is possible to use a modem having an AT command set in data mode manually, this is not true for the **AT+F** fax command set. Human reflexes are simply not fast enough to handle real-time requirements. A number of generic functions are required for programming any fax modem automatically.

The following low-level I/O functions are suggestions based on those required by the code in the FAX directory on the accompanying disk:

- An **init** function is required to set up a serial port. On a multi-tasking system the port has to be attached to the fax process. On single user systems, the port hardware may have to be programmed directly, with various other systems services needing to be set up by the fax program. This function will need to be passed a specific speed and data format, and is likely to be responsible for setting up flow control options.

- A **deinit** function is required to reverse whatever the **init** function has done. On multi-tasking systems ports need to be detached and operating system resources need to be freed. On single user systems where the hardware has been directly programmed, this function restores the machine to the same state as it was found it in on entry.

- A **reinit** function is also needed to change either the speed of the port or the flow control options during a session. It might be possible to replace this with a call to **init** followed by a call to **deinit**, but it may be that thistakes too long for some of the fax session requirements.

- Character input and output are handled through simple **rxchar** and **txchar** functions, respectively; the latter may also have to handle XON/XOFF output flow control.

- Where the fax software itself provides buffering functions in addition to that provided by the modem, serial port buffer status must be obtainable through **rxstat** and **txstat** functions for receive and transmit buffers, respectively.

- A **dropdtr** function which drops the DTR signal at the serial port is also useful, as it enables the fax software to hang up and reset the modem.

Accurate timing is essential. If the ANSI C clock functions are either not implemented or are inadequate, timing must also be implemented at a low level.

There are a number of generic routines that can be built out of combining the low-level I/O and clock routines with standard C facilities. Three typical ones, which we saw used in Listing 8.2, are outlined here.

Our first function is called import, and is designed to return modem responses. It is called with a pointer to a buffer for the line received, a maximum line length and a timeout. It returns either true with the buffer containing the line (beginning with the first printable character received and ending with either the

```
int import (char *line,unsigned int length,unsigned int
timeout)
{
    unsigned int i ;
    char k ;
    clock_t t ;
    if (length==0) return (0) ;
    for (i=0 ; i<length ; i++) line[i]=0 ;
    t=(clock()+(timeout*CLK_TCK)) ;
    do  {
        while (rxstat()==0) if (t<clock()) return (0) ;
        k=rxchar() ;
        } while (iscntrl(k)) ;
    for (i=0 ; i<length ;)
    {
      line[i++]=(toupper(k)) ;
      t=(clock()+(timeout*CLK_TCK)) ;
      while (rxstat()==0) if (t<clock()) return (0) ;
      k=rxchar() ;
      if (iscntrl(k)) break ;
    }
    return (i) ;
}
```

Listing 8.2: Fetching modem responses

first non-printable character received, or when string space is full), or false if a timeout occurred at any point. The timeout is reset whenever a printable character comes in, and the string may contain a partial line. If a line was received, import returns with the length of the null-terminated string and the string itself converted into upper case.

Our second function sends commands to a modem. It is called with a null-terminated string which it sends out of the serial port (or more accurately, it places it in whatever output buffer **txchar** might maintain). It returns false if the string could not be sent, otherwise it returns true, with the number of characters sent (Listing 8.3).

```
  int export (char *line)
{
unsigned int i ;
for (i=0 ;; i++)
    {
    if (line[i]==0) return (i)  ;
    else txchar(line[i]) ;
    }
}
```

Listing 8.3:   Sending modem commands

Our third function is at a slightly higher level, and uses **import** to wait (with a timeout) for a specific response. It returns false if the response did not arrive in the specified time, otherwise it returns true (Listing 8.4).

This completes our look at the **AT+FCLASS** command and other features of fax modems common to all the service classes. We turn now to the commands that are specific to each of the classes themselves.

```
int wait_for (char *match_string, int timeout)
{
char response_string[64] ;
while (import(response_string,64,timeout))
    {
    if (strstr(response_string,match_string)) return (1) ;
    }
return (0) ;
}
```

Listing 8.4:   Waiting for a specific response from a modem

9

Programming Class 1
Fax Modems

Introduction

If you want to know how to program a Class 1 fax modem, this is the chapter for you. In it, we deal with the way that Class 1 fax modems work in some detail. However, you'll find that understanding the T.30 session protocol is essential to making a Class 1 modem do its job. We also assume that anyone reading this understands the issues raised in Chapter 8, containing an introduction to EIA fax modems in general. The code fragments presented here are partial versions of the code in FAXCLAS1.C in the FAX subdirectory on the accompanying disk. The utility functions they include were shown in fragmentary form at the end of the last chapter, and full versions can be found in FAXUTILS.C.

Class 1 commands and parameters

The original version of the Class 1 specification should appeal to minimalists, which is precisely what it was designed to do. There are only six fax commands in the Class 1 set, only four of which actually do anything. There is only one extra response code.

The downside of this is that Class 1 commands do so little of the work that most of the logic involved in sending and receiving faxes becomes the responsibility of the applications software. Therefore, anyone writing code for Class 1 fax modems will find they are probably referring back to the ITU T.30 fax session protocol more frequently than to the EIA class 1 specification.

Table 9.1 shows all the possible Class 1 commands. The +FTH command sends an HDLC frame, while the +FRH command receives an HDLC frame. The +FTM

Table 9.1: Class 1 commands

	Transmission commands	Reception commands
HDLC frame data	AT+FTH=<mod>	AT+FRH=<mod>
Stream data	AT+FTM=<mod>	AT+FRM=<mod>
Silence	AT+FTS=<time>	AT+FRS=<time>

command sends T.4 data, while the **+FRH** command receives T.4 data. The **+FTS** command pauses with the transmitter switched off for specified period while the **+FRS** command waits until no more data is received and then pauses for the specified period of silence.

All Class 1 data commands must be the last commands on a line. This doesn't apply to the silence commands **+FTS** and **+FRS**, which can precede a data command provided that a semicolon is used as a separator.

The **<time>** parameter used for the two silence commands is a decimal number in the range 0–255 indicating a number of 10 ms intervals.

The **<mod>** parameter for the four data commands takes a numeric value indicating the fax modulation scheme to be used as shown in Table 9.2.

The Class 1 fax commands can (in theory) take any of the usual forms allowed by the **AT+** conventions, e.g. the test parameter form **AT+FRH=?** would give a list of possible parameters, the read parameter form **AT+FRH?** would give a list of the current parameter setting, and the set parameter form **AT+FRH=<mod>** forces the modem to do whatever the parameter in question demands.

In practice some of these are never used and should not be attempted. For instance, it makes no sense to use a command such as **AT+FTS?**, as this is a

Table 9.2: Class 1 modulation codes

<mod>	Bit Rate	Modulation
3	300 bps	V.21 channel 2
24	2400 bps	V.27 ter fallback
48	4800 bps	V.27 ter
72	7200 bps	V.29 fallback
96	9600 bps	V.29
73	7200 bps	V.17 long training
74	7200 bps	V.17 short training
97	9600 bps	V.17 long training
98	9600 bps	V.17 short training
121	12000 bps	V.17 long training
122	12000 bps	V.17 short training
145	14400 bps	V.17 long training
146	14400 bps	V.17 short training

command to transmit nothing for a specific interval and is not readable. Some modems will respond with an **OK** if this is tried, some with a 0 followed by an **OK**, while others will respond with **ERROR**.

Software should also assume that both the **+FTS** and **+FRS** commands have a range from 0–255 as specified rather than checking to make sure. While some modems respond correctly to **AT+FTS=?** with a range such as 0–255, others will respond with **ERROR**.

However, very little can be assumed about the other four commands. Not all modems can handle all the modulations and speeds listed. It isn't a deficiency in a modem if it can't handle a particular speed; different models have different specifications. As the commands take a range of numeric parameters, the modem should be interrogated to determine what it does support. A modem that supports every possible speed and modulation type would reply to a test parameter command with the line

```
3,24,48,72,73,74,96,97,98,121,122,145,146
```

which may or may not be followed by a final **OK**, depending on how the Class 1 specification was interpreted.

The only speeds and modulations that a modem has to support are the minimal ones specified in T.30. Since the philosophy behind the Class 1 command set is that the HDLC data commands should be used for negotiating while the stream data commands should be used for transfer of image data, it is safe to assume that both **+FRH** and **+FTH** will be able to take a value of 3, as the ability to negotiate at 300 bps using V.21 channel 2 is mandatory in T.30. Similarly, **+FRM** and **+FTM** must both accept the values 24 and 48 to provide support for sending and receiving faces at 2400 bps and 4800 bps using V.27 ter.

The following constraints on the choices that can be made from the available modulations and speeds make life for the implementor reasonably straight-forward:

- An ability to handle 2400 bps V.27 ter modulation is the minimum requirement for sending faxes. While T.30 permits a fax machine to have only 2400 bps capability, all current fax modems can handle 4800 bps. Some early model fax boards based on synchronous modems lacked this capability.

- All systems that can handle V.29 modulation can do so at both 9600 and 7200 bps.

- All fax modems that can handle V.17 modulations are assumed to be able to do so at all four speeds from 14400 to 7200 bps. If a system can handle the rarely found V.33 modulation, it is assumed it can do so at both 14400 and 12000 bps.

While the T.30 DIS/DTC frames allow for the possibility of a modem supporting V.29 without supporting V.27 ter and, vice versa, the possibility of a modem supporting V.17 without supporting both of the other modulation schemes is not allowed for.

It cannot be overstressed that software wishing to use selections other than mandatory values must interrogate the modem to check their acceptability prior to their use. Asymmetric capabilities are common in the modem world, and a capability to receive at a particular speed and modulation combination doesn't imply that a modem can use it for transmission (or vice versa). For example, consider the following captured dialogues:

```
AT+FRM=?                      at what speed can the modem receive T.4 data?
              24,48           V.27 ter modulation only
              OK              final OK
AT+FTM=?                      at what speed can the modem send T.4 data?
              24,48,72,96     V.27 ter and V.29 modulation
              OK              final OK
```

The modem in the above example can transmit faxes faster than it can receive. Technically speaking, it possesses a V.29 modulator but not a V.29 demodulator.

Class 1 data formatting

The **+FRH, +FTH, +FRM** and **+FTM** commands all have a number of things in common. Whenever any of these commands is issued successfully, the modem will initially respond with the **CONNECT** result code. Following this, if the command was **+FTH** or **+FTM**, the applications software must provide data to be sent. If it was **+FRH** or **+FRM**, the applications software must be prepared to receive data.

In both cases, the data has a common format, consisting of a stream of data terminated by a two byte sequence **<dle><etx>**, where **<dle>** is ASCII code 16 decimal (10h) and **<etx>** is ASCII code 3. If a **<dle>** code occurs in the data stream it is shielded by sending an extra **<dle>** with it. No other occurrences of **<dle>** are legal. This gives the following rules for sending data:

- All data is sent transparently until a **<dle>** is encountered.

- A **<dle>** is sent as **<dle><dle>**.

- The end of a data stream is indicated by sending **<dle><etx>**.

- All other sequences consisting of **<dle><any other character>** are illegal

- All data sent must obey whatever flow control options happen to be in force.

The corresponding rules for receiving data are as follows:

- All data is received transparently until a `<dle>` is encountered.

- A `<dle><dle>` is interpreted as a single `<dle>`.

- A `<dle><etx>` is interpreted as the end of the data stream.

- All other sequences consisting of `<dle><any other character>` are discarded.

- All data received must obey whatever flow control options happen to be in force.

Class 1 responses

Class 1 data commands that execute normally always have at least two responses: a primary response is returned by the modem to indicate that the command has been accepted; after subsequent data transfer has been accomplished, a secondary response is returned. While there are exceptions, the messages **CONNECT** and **OK** are positive responses while **ERROR** and **NO CARRIER** are negative responses.

Listing 9.1 shows a generic function for handling modem responses. It returns 1 for a positive response, 0 for a negative response, and –1 if a timeout occurred. (Timeout detection and subsequent retry is an essential part of the fax protocol.)

```
int framestat (unsigned int timeout)
{
    char result[16] ;
    while (import (result,16,timeout))
    {
      if (strstr(result,"OK")) return (1) ;
      if (strstr(result,"CONNECT")) return (1) ;
      if (strstr(result,"ERROR")) return (0) ;
      if (strstr(result,"NO CARRIER")) return (0) ;
    }
    return (-1) ;
}
```

Listing 9.1: Generic function to handle modem responses

While this function generalizes a number of common cases, understanding the primary and secondary responses to Class 1 commands is as important in knowing how to program these devices as understanding what the commands actually do. The following accounts of the four data commands provide all the necessary information.

AT+FTM=<mod> and its responses

The `AT+FTM=<mod>` command is used to send a raw data stream to a fax modem, which in turn sends it to the remote fax. There are two situations in which this is done; first, when the Training Check Frame (TCF) is sent before the start of phase C as a check on the line quality; and second, to send T.4 fax data in phase C.

The `CONNECT` response to this command shows it has been successfully completed, and the modem has started to transmit using the specific modulation requested. It also means the modem should have completed the proper training sequence. These sequences are specific to each modulation, and are needed for the two modems to synchronize their operations. Bear in mind that we are using synchronous communications here, and there is no start bit indicating when data is due to begin.

The point at which a fax modem should issue the `CONNECT` message in relation to its training sequence is not specified in Class 1. Some modems do this before they have completed their training sequence, while others wait until they have begun transmitting before notifying the computer they have a connection. The TR-29.2 committee issued a bulletin numbered TSB-43, which has been incorporated in the latest version of Class 1, recommending that the response should be issued at the start of training rather than the end. This is certainly more efficient, as all fax modems buffer data anyway, and don't have to send it until they are ready. However, the bulletin notes that existing implementations may not conform to this, so the precise state of the modulation process can't be inferred from the fact that a `CONNECT` result code is issued.

In all cases, once a carrier has been established and the modem has trained, it begins transmitting a continual stream of 1 bits while waiting for the applications software to begin sending raw data. If this doesn't occur within 5 s, the modem spontaneously generates an `ERROR` response, turns off its transmitter, and awaits further commands.

Since these 1 bits sent before the real data are a waste of telephone time, and money, the applications software should begin sending data to the modem as soon as possible after receiving a `CONNECT` response. However, the situation is complicated by the fact that some class 1 modems seem to expect the applications software to implement the recommended delay of 75 ± 20 ms between the reception of the confirmation frame and the start of phase C data. If you don't know whether your modem falls into this category (and there seems no way to find out short of trial and error) the best way of handling this is by inserting 95 ms worth of 0xff characters before the initial EOL code. The methods described on page 205 for sending timed 0 bits can be be adapted for this. You should also bear in mind the necessity for post-command delays described in our introduction to the EIA service classes.

All fax data should be structured according to the transmission rules we outlined earlier. The modem accepts the raw data according to the receive rules also outlined earlier, and retransmits them to the remote fax. Because the rate at which the modem receives the data is always greater than the rate at which it retransmits it, and because of the necessity to process the `<dle>` shielding characters, all data is buffered by the modem.

All data received by the modem is filtered to remove `<dle>` shielding, and is then retransmitted to the remote fax. When a final `<dle><etx>` pair is detected, the modem stops transmitting, turns off the carrier, and generates the secondary **OK** result code before waiting for another command.

One problem area is that while flow control constraints must be obeyed by the computer sending the data, none of the standards specify what should happen if the modem buffers become full as a result of flow control failure.

The situation regarding buffer underrun is also quite complicated. If buffer underrun occurs, and the last character sent was not a **NULL** (decimal or hexadecimal **0**), the modem should immediately turn off the transmit carrier, send an **OK**, and return to the command state. On the other hand, if the last character transmitted by the modem was a **NULL**, the modem is supposed to replicate the **NULL**s either until 5 s elapses, or until the computer sends more data. If more data arrives it should be sent, while if the 5 s timeout expires (representing the maximum scan line time in T.4 data) then the modem should send an **ERROR** result code and await another command.

The Class 1 specification originally stated that the replication of **NULL** octets is 'useful for generating TCF (1.5 s of 0s) and zero-fill within lines'. Unfortunately, the feature is fairly useless for either purpose. The generation of zero-fill within lines (required to pad lines out to the minimum scan line times as required by T.30) would only be possible if there were some way of telling whether the buffer was empty or not, as replication only occurs once the buffer is empty. As there is no Class 1 command that offers this information, automatic replication of **NULL**s can't be used for zero fill. Any attempt to time a delay without an empty buffer wouldn't result in any zero bits being filled until the buffer had emptied of its own accord, and there's no simple way of telling when that might be.

The replicating **NULL** feature can't be used for generating 1.5 s of 0s either. The theory is that application software could issue a single NULL and then wait for 1.5 s. However, the ambiguity noted earlier in the exact point at which the **CONNECT** message appears (it could be before or after the training sequence) makes it impossible to know whether the initial **NULL** would be immediately sent or not. This means that a 1.5 s timing delay is no guarantee of a proper TCF being sent.

The correct procedure for timing both fill bits for minimum scan-line times and TCF frames is to send a predetermined number of **NULL**s. The TR-29.2 committee Technical Services Bulletin TSB-43 offers the suggestions shown in Table 9.3 for the number of **NULL**s required to send a 1.5 s TCF at specific speeds.

Table 9.3: NULLs per TCF for various speeds

bps	number of NULLs
14400	2700
12000	2250
9600	1800
7200	1350
4800	900
2400	450

The figures are fairly obvious once we remember that the computer is sending data as 10-bit bytes, while the modem is sending the data on without start and stop bits as pure binary octets. The formula is

```
Number of NULLs = (bit rate/8) x 1.5
```

In the case of the TCF, the **NULLs** should be followed by a `<dle><etx>` sequence. The number of **NULLs** to send when zero-filling lines to minimum scan line times is more complicated and isn't mentioned in the Class 1 specifications. However, Table 9.4 shows the minimum number of bytes that should be sent per line for different bit rates and scan line times (some of the figures for 5 ms scan line times have been rounded up).

All Class 1 transmitting software does is count the bytes between consecutive EOL codes. If the number of bytes for a negotiated scan line time at a particular speed is below the figure in Table 9.4, then sufficient **NULL** bytes should be added before the second EOL is sent to the modem. This ensures the minimum scan line time is adhered to no matter what state the buffer might be in.

Table 9.4: Minimum scan line times translated into minimum bytes per lines for various speeds

	5ms	10ms	20ms	40ms
14400 bps	9	18	36	72
12000 bps	8	15	30	60
9600 bps	6	12	24	48
7200 bps	5	9	18	36
4800 bps	3	6	12	24
2400 bps	2	3	6	12

The position regarding termination of a **+FTM** command as a result of buffer underrun, where the last character sent was not a **NULL**, should not be taken to imply the redundancy of using a terminating `<dle><etx>` sequence, even though in both cases the modem is supposed to turn off its transmit carrier, send an **OK** and return to the command state. On a purely literal interpretation, there seems to be no point in terminating the data stream with a **<dle><etx>**. As T.4 data ends with a final RTC sequence, it can't finish with a **NULL**, so it could be argued that there's no need to terminate the data stream when just permitting the buffer to underrun should produce the same effect. Despite this, explicit ending of the data stream with the proper sequence is always the preferred option.

It is also important to realize that even where a **<dle><etx>** is inserted to end a data stream, the modem can't be relied on to issue an **OK** response until the buffer is empty. As a full 16K buffer would take well over 10 s to empty at 9600 bps, application software must be prepared for a wait of at least this for data to clear the buffer before being worried about the non-appearance of a result code.

The behaviour of a modem when data is backed up for transmission in its buffers is actually rather poorly handled in the Class 1 specification. Were the modem to issue the **OK** as soon as the **<dle><etx>** was detected, the 10 s wait for a response code would apply not to the final **OK** after the data had been sent, but to the response to the next **AT+F** command issued. It isn't clear from the Class 1 document exactly what should happen here.

Using AT+FTM to send a fax page

Listing 9.2 contains a code fragment showing a typical use of the **AT+FTM** command to send 1-D fax page at 9600 bps. It is called with a pointer to a file that we assume to be positioned at the start of an image, together with the length of the image. The **+FTM** command is issued, and once a suitable response is received, the image is sent out with **<dle>** shielding. After the RTC sequence, a **<dle><etx>** is sent to inform the modem that the data has ended, and we wait for an **OK** response.

The task of sending pages is complicated by the need to pad out each scan line to the minimum time required. The length required for each line is easily worked out from the bit rate; the number of bytes can be worked out from the bit rate using the formula

```
minimum_length=(((bitrate)/(1000/min_scan_line_time))/8)+1
```

All we have to do is count the bytes between successive EOL codes and add sufficient null bytes to bring the value up to the computed minscan as required.

The difficult part is detecting the EOL codes. If we could be sure the EOLs were byte-aligned the task would be simplified, as all we'd need to do is search for the bit pattern **xxxx0000** followed by **00000001**. Unfortunately, we can't be sure that EOLs are byte-aligned, so we have to count the number of successive 0 bits and find the EOL codes ourselves. We do this by keeping track of the number of trailing 0s in the last octet we sent out, using the algorithm

```
for (trailing0s=8 ; octet ; octet<<=1,trailing0s--) ;
```

One limiting case is where `octet==00000000`, in which case the loop immediately terminates with `trailing0s=8`. The other limiting case is `octet==00000001`, in which case we have to complete 8 left shifts before the loop terminates with `trailing0s=0`.

```c
int sendpage (FILE *faxfile,long int bytecount)
{
    unsigned char k ;
    char faxbuf[512] ;
    int x, i=0 ;

    export ("AT+FTM=96\r") ;
    if (framestat()!=1) return (0) ;

    do
    {
      if (bytecount > 512)
      {
        i=512 ;
        bytecount-=512 ;
      }
      else
      {
        i=(short)bytecount ;
        bytecount=0 ;
      }
      fread (faxbuf,1,i,faxfile) ;
      for (x=0 ; x<i ; x++)
      {
        k=faxbuf[x] ;
        txchar(k) ;
        if (k==dle) txchar(dle) ;
      }
    } while (bytecount) ;
    for (x=3 ; x ; x--)
      {
      txchar (0x0) ;
      txchar (0x08) ;
      txchar (0x80) ;
      }
    txchar (dle) ;
    txchar (etx) ;
    while (txstat()!=0) ;
    return (wait_for("OK",30)) ;
}
```

Listing 9.2: Sending a page with +FTM

The next step is to count to number of leading 0s in the next octet to be sent out, using this complementary algorithm

```
for (leading0s=8 ; octet ; octet>>=1,leading0s--) ;
```

We then add together **trailing0s** and **leading0s**; if they are over 10, we have found an EOL code and proceed to pad the line out as necessary.

AT+FTH=<mod> and its responses

The **AT+FTH=<mod>** command is used to send framed HDLC data to a remote fax. The applications software sends the data as a raw stream to the fax modem, which then handles all the necessary formatting. This includes adding the start flag, bit-stuffing the data according to HDLC rules to prevent flags occurring within the data, and adding the FCS frame check digit followed by a final flag at the end.

There are many parallels between the **AT+FTH** command and the **AT+FTM** command. For both commands, the modem issues an initial **CONNECT** response message once the selected modulation has been established. Just as the modem transmits a stream of idle 1 bits after an **AT+FTM** command if there is an interval between the establishment of a carrier and the start of the data stream from the applications software, so the **AT+FTH** command is followed by the fax modem sending a stream of flags while it is waiting for the applications software to begin sending data. Another item of identical behaviour is that if no data is received for 5 s after the **CONNECT** response, the modem generates an **ERROR** response, turns off the transmitter and awaits further commands.

The format of the data being sent in HDLC mode is identical to the format of the data being sent in raw mode, with **<dle>** characters marking the end of the data stream in the same way. The fax modem buffers data sent in exactly the same fashion, and flow control is used in the same manner to prevent the buffer overflowing.

However, there are three main differences between the behaviour of the modem (as it appears to the application software) when sending raw data and when sending HDLC framed data. The first is a minor one; there is no zero byte replication available if the buffer underruns. All that happens when the buffer underruns is that the modem should immediately add the FCS frame check sequence and the closing flag to the data to complete the frame.

The second (and main) difference between the behaviour of a fax modem in raw transmit and HDLC frame modes is in the secondary response, given by the modem after the data has terminated. This is always an **OK** after an **AT+FTM**. By contrast, ending the data stream after the **AT+FTH** command can give two possible secondary result codes. If the frame is either the final one in a transmit sequence or a null frame, the modem responds with an **OK**. If the frame is to be followed by another, the modem responds with another **CONNECT** message. The fax modem always inspects the P/F bit of each frame it sends. If this is a **1** then the frame is a final one, so the modem turns off its transmitter and carrier and responds with an **OK**. It then awaits further commands.

If the P/F bit is seen to be a **0**, the modem maintains its carrier, continues to transmit HDLC flags while its buffer is empty, and responds with a **CONNECT**. It then awaits more data, which must arrive within 5 s. Of course, if the receiving fax is expecting a frame within a shorter period of time, then the timeout specified in T.30 should be used. Length of this timeout is irrelevant.

The third difference between **+FTM** and **+FTH** is that the first negotiating frame in any sequence must be preceded by flags, which must last for at least 1 s, as required by the T.30 recommendation.

Using AT+FTH to send a frame

Listing 9.3 shows how straightforward sending a negotiating frame can be. The sendframe function is called with a character array containing the data to be sent, with the first byte of the array containing a count of the number of octets in the frame (which allows **NULL**s to be included in the data).

```
int sendframe (char *frame)
{
    int i ;

    export ("AT+FTH=3\r") ;
    if (framestat()!=1) return (0) ;
    for (i=1 ; i<=frame[0] ; i++)
    {
      txchar(frame[i]) ;
      if (frame[i]==dle) txchar(dle) ;
    }
    txchar(dle) ;
    txchar(etx) ;
    return (framestat()) ;
}
```

Listing 9.3: Sending a negotiating frame with **AT+FTH**

Receiving data

There are a number of similarities between how a Class 1 fax modem behaves when told to start receiving either streamed data with **+FRM** or framed data with

+FRH. For all the Class 1 data commands, the normal primary response is a **CONNECT** message, but we must remember that the only thing preventing a primary **CONNECT** response to either of the Class 1 transmission commands is an illegality somewhere in the command. A fax modem in transmit mode doesn't worry about carriers as all half duplex transmission is blind. As long as the modem is off-hook and the command has been properly issued, a fax modem having received such a command will send a carrier, training, send any preamble required, and then start transmitting either a stream of 1 bits (for **+FTM**) or a stream of HDLC flags (for **+FTH**), as described above.

This doesn't apply to a Class 1 reception command; receiving data is only possible if a carrier can be detected at the modulation selected. In this respect, a fax modem in receive mode is similar to a data modem, which also won't send a **CONNECT** response unless it finds a valid carrier.

This gives rise to two situations requiring primary responses to initial receive commands which don't arise when transmitting:

- The first is where the wrong carrier is detected, resulting in the **+FCERROR** message, which is a response unique to Class 1 fax modems in receive mode. When encountered, it means that, although the command was correct, the modulation encountered was different to that specified in the command. Remedial action can sometimes be taken by issuing the command again with a different parameter. Where this command is a result of incorrect interpretation of a fax negotiation or bad coding of the T.30 specification, there isn't much that can be done.

  However, the **+FCERROR** response could be encountered as a result of a timing problem or line noise. For instance, following the receipt of phase C data via an **AT+FRM** command, the usual step for a receiver would be to send an **AT+FRH=3** command to obtain the post-page message frame indicating an MPS, EOM or EOP at the negotiating speed. A response of **+FCERROR** to the **AT+FRH=3** command would indicate that though there was a carrier, it was not a V.21 signal. The transmitter wouldn't have decided to negotiate at a different speed, and couldn't just be waiting for a response to a frame we missed, as in that case we would have been unable to detect any sort of carrier. The obvious conclusion is that the transmitter is still sending T.4 data and the original termination of phase C was due to line noise. A suitable response would be either to return to phase C via **AT+FRM=<mod>** to pick up the rest of the data, or to use the **AT+FRS** command to wait until the transmitter had finished sending.

- The second situation requiring a primary response peculiar to Class 1 reception is when no carrier could be established at all. In other words, we need a response for a situation where the **+FRH** command times out. Given that a failure to receive an expected message could be considered to be the main T.30 mechanism for triggering error recovery, this is quite important. It may seem that the familiar **NO CARRIER** message would do this, but there is some doubt as to whether it is a valid primary response in the context of a Class 1 modem looking for a carrier.

The length of time the modem takes to respond spontaneously with a **NO CARRIER** command is not specified in the Class 1 documentation. There is nothing to stop the modem looking for a carrier forever. In practice, this timeout is often limited by the value set in register S7 (as with data calls), but this is not an official part of the EIA Class 1 specification. It is quite common for S7 to be set to anything upwards of 30 sec to cater for the initial phase A dialling or answering timeout. This interval is way beyond any of the T.30 timeouts, and is far too long to be useful within a fax session, so the **NO CARRIER** response cannot be considered a valid primary response when commanding a modem to establish a carrier.

Receive timeouts required in this situation include the 3 s timeout a transmitter uses when waiting for a response to a command before repeating it, and the 6 s timeout a receiver uses when waiting for a valid command. These have to be implemented by the application software and not by the modem. When a timeout expires, either of the received data commands can be aborted by sending any character to the modem. If the modem implements XON/XOFF flow control for received data that is transmitted back to the computer, these characters should be avoided. The best one to use for this is the ASCII CAN character (decimal 24 or hexadecimal 18h) as it is used in this way by the other EIA fax modem proposals.

The **AT+FRS** command used to wait for silence can also be aborted in this way if necessary. Like the receive data commands, there is no timeout set for the **+FRS** command, which could thus wait for silence indefinitely if no manual abort is available.

When either of the **+FRH** or **+FRM** receive commands, or the **+FRS** command, is terminated in this way, the modem will normally respond with an **OK**; but, as the ever-useful TSB-43 bulletin points out, the exact message to be used wasn't adequately defined in the original documentation. Depending on whether the modem was waiting for a carrier, was training or was waiting for data, different modems could reply with **NO CARRIER** or **ERROR** result codes as well as **OK**. Virtually the only thing you can be sure of is that there won't be a **CONNECT** response. TSB-43 comments

'It is advisable, for interworking with the greatest number of existing products, for DTEs to accommodate the possibility of receiving a result code other than **OK** upon terminating a receive operation.'

AT+FRH=<mod> and its responses

It may not be immediately apparent that there are two situations in which the **AT+FRH** command is issued. The first is that just described; the T.30 session has reached the stage where a command or response is due, and the **+FRH** command is used to tell the modem to detect a carrier as indicated by the **<mod>** parameter, and report back with the frame. Like all other HDLC frames, the second octet of the received frame is the control field; the fifth bit received in this octet is the P/F bit, set to a 1 if the frame is the final one in the received sequence, or reset to 0 if more frames are to follow.

The second type of situation in which the **AT+FRH** command is issued by an application is immediately after the delivery of a frame with a P/F bit of 0. In this context, the **+FRH** command isn't telling the modem to look for a carrier; the modem and application software both know one exists. This type of **AT+FRH** command is a request for the delivery of the next frame.

This immediately qualifies our observation that a **NO CARRIER** is not a valid primary response to an **AT+FRH**. This is only true when the command is used to tell the modem to look for a carrier and report back with a frame. If the **AT+FRH** is used to request delivery of a subsequent frame, the **NO CARRIER** response is valid, indicating that the carrier has been lost and no frame is available. Notice that the response in this case doesn't require the use of a timeout; the **NO CARRIER** refers not to a carrier that was never established, but to one that was lost.

Just as the modem adds opening and closing flags before sending the data provided to it by the computer onto a remote fax, so it also removes the opening and closing flags from the frame data it receives over the phone line before passing it onto the applications software. It also handles zero-bit deletion and error checking, with the result that the application software sees only the data that arrives between the opening and closing flags. The rules outlined above for formatting received data streams apply, so **<dle>** bytes must be filtered out of all raw data received from the modem, with the end of the data being marked by the **<dle><etx>** pair.

However, the 16 bit FCS frame check sequence is not removed from received data, and appears as two extra bytes tacked onto the end of the frame before the final **<dle><etx>**. If the application software needed to check the integrity of the frame, it could easily compute a CRC for all the data received apart from these last two bytes. However, there is no need to do so, as the fax modem does this checking automatically. If the CRC it calculates matches the CRC contained in the last two bytes of the frame, the modem responds **OK**.

If the CRC appears wrong, the modem responds with the secondary **ERROR** result code. This could occur as a result of a number of other things. For example, it would also be the response if the receive data buffers in the modem overflowed. This could either be a result of faulty flow control handling or it could be caused by a computer-modem bit rate that was too low. This is an unlikely occurrence under most circumstances, but could occur in error correction mode under certain (usually avoidable) circumstances.

The first octet received from a fax modem in any fax frame should always be **11111111**. If verbal result codes are turned on, and the application software starts receiving binary data once the first **CONNECT** error message is seen, the first byte received will actually be a line feed, as the **CONNECT** error message terminates with a `<cr><lf>` sequence. Software should therefore either make sure it doesn't start to receive data as soon as the **CONNECT** message is seen, but should wait for the final line feed. It is also a good idea, given the inconsistency of fax modem response codes under some circumstances, that a check is made to ensure the first byte of a frame really is the **11111111** fax address field.

Once the start of the frame data has been checked, it is vital that the applications software checks the P/F bit of each frame (the fifth bit received in the second octet of the frame). If this is a **0**, a further **AT+FRH** must be issued as another frame will follow. However, if the P/F bit is a **1**, the applications software must compose a suitable response and turn the line around with an **AT+FTH** to send it.

In the case of the P/F bit of **0**, the secondary **AT+FRH** command should be issued as quickly as possible (but not before the final line feed after an **OK** result code). The problem isn't that the modem might lose data from the next frame when receiving two HDLC frames in a row, as frames tend to be quite short and there should be no difficulty in the modem buffering all data received after the end of the frame. Responding within 1 ms is therefore not required, as there is no possibility of receive buffer overrun if the **AT+FRH** command is not issued quickly enough. However, there are situations in the T.30 specification where response times of 75 or even 50 ms may be called for, so the timing constraints are not trivial. Class 1 software cannot afford to hang around doing nothing in these circumstances, especially on multi-tasking computer systems where there may be unknown latencies causing delays in the receipt of modem data.

In the case of a P/F bit of 1, the application should continue with the session. This always involves some sort of transmission; alternating transmission and reception is the nature of a fax session. The TSB-43 note warns it is possible for the start of this transmission to be lost if it occurs before the remote fax has turned off its transmitter, with potentially disastrous results. The recommendation is that this situation be avoided by checking that the carrier has dropped, rather than assuming that P/F=0 means that it must have been.

This could be done by sending another **AT+FRH** command and waiting for a **NO CARRIER** response, but this method is slightly unsatisfactory as garbage generated while the carrier is dropped could result in an **ERROR** message instead. Of the other alternatives, issuing the **AT+FRS=1** command to wait for silence before sending is probably more efficient than using **AT+FTS** to introduce a fixed delay before transmitting, as there is no possibility of setting a delay that is too short.

I've never seen the software included in this book lose a fax session for this reason, even though it doesn't include any of the suggested checks for a dropped carrier before line turnaround. If I did come across the situation described in TSB-43, checking for a lost carrier is probably the first thing I'd try; but I prefer to work on the principle that if it isn't broken, then don't fix it.

```
int getframe (char *frame)
{
    clock_t t ;
    unsigned char k, lastchar=0 ;
    int i ;

    export ("AT+FRH=3\r") ;
    if (framestat()!=1) return (0) ;
    t=(clock()+(CLK_TCK*3)) ;
    for (i=0,lastchar=0 ; i<64 ; )
    {
      while (rxstat()==0) if (t<clock()) break ;
      k=rxchar() ;
      {
        if (k==etx) break ;
        lastchar=0 ;
        if (k!=dle) continue ;
      }
      else if (k==dle)
      {
        lastchar=dle ;
        continue ;
      }
        frame[i++]=k ;
    }
    return (framestat()) ;
}
```

Listing 9.4: Receiving a frame with **AT+FRH**

Using AT+FRH to receive a frame

Listing 9.4 shows the basics of receiving a frame with **AT+FRH**. The `getframe` function is called with a pointer to a frame buffer, and the function returns with 1 on success, with the frame stored in the buffer. If no frame arrived the function returns 0, and if the frame timed out it returns –1.

Bit ordering

We have pointed out before that the order in which bits of each byte are transmitted over a communications link are not from left to right as read, but are from right to left. The least significant bit of each byte is the first one sent or received over a communications link, so for all HDLC and streamed data, the octets are always reversed from the usual order. The 16-bit T.4 bit sequence

`000000000000001`, which is an EOL code with four fill bits, would always be sent or received as the sequence `000000010000000`.

Possibly more importantly, the T.30 control field of an HDLC frame is defined as the octet `11001000` for a final frame and `11000000` for an intermediate frame. These octets would be sent or received over a communications link as the hexadecimal values 13h and 03h respectively. All the facsimile information fields would likewise be inverted. The FIF translation table (Table 9.5) shows the mapping. If the x bit showing which station sent the last DIS is set to 1, the hexadecimal equivalent in the table must be ORed with 01h.

Table 9.5: Facsimile information field bit inversions

FRAME ID	BIT PATTERN	INVERTED CODE
Initial Identification		
DIS	00000001	80h
CSI	00000010	40h
NSF	00000100	20h
Command to send (polling)		
DTC	10000001	81h
CIG	10000010	41h
NSC	10000100	21h
Command to receive		
DCS	x1000001	82h
TSI	x1000010	42h
NSS	x1000100	22h
Phase B responses		
CFR	x0100001	84h
FTT	x0100010	44h
Post-message commands		
EOM	x1110001	8eh
MPS	x1110010	4eh
EOP	x1110100	2eh
PRI-EOM	x1111001	9eh
PRI-MPS	x1111010	5eh
PRI-EOP	x1111100	3eh
Post-message responses		
MCF	x0110001	8ch
RTP	x0110011	cch
PIP	x0110101	ach
RTN	x0110010	4ch
PIN	x0110100	2ch

Table 9.5: (continued)

FRAME ID	BIT PATTERN	INVERTED CODE
Other signals		
DCN	x1011111	fah
CRP	x1011000	1ah
ECM frames		
FCD	01100000	06h
RCP	01100001	86h
PPS	x1111101	beh
PPR	x0111101	bch
CTC	x1001000	12h
CTR	x0100011	c4h
EOR	x1110011	ceh
ERR	x0111000	1ch
RR	x1110110	6eh
RNR	x0110111	ech
FDM	x0111111	fch
NUL	00000000	00h

Allowances must be made for the reversal of the bits in each octet when composing or interpreting negotiating frames, e.g. bit 16 is the last bit of the second octet of the information field of a DIS frame, and indicates whether a receiver has 2-D decoding capability. A transmitter receives this as the most significant bit of the second octet, and so has to AND this byte with 80h to check whether 2-D data can be sent.

The reversal of bits received from or sent to the fax modem does not apply to control bytes. XON, XOFF, `<dle>` and `<etx>` are all sent and interpreted bytewise as normal; the sequence `08h  C0h` received from a fax modem is not a `<dle><etx>`.

AT+FRM=<mod> and its responses

The `AT+FRM=<mod>` command is used to receive a raw data stream from a fax modem. As with the `AT+FTM` command, there are two situations in which a raw data stream is transferred. The first is when the Training Check Frame (*TCF*) is received before the start of phase C as a check on the line quality, and the second is when T.4 fax data is received in phase C after a CFR confirmation response frame has been sent.

The **CONNECT** response to this command indicates that the carrier specified by the `<mod>` parameter has been detected. If a different carrier is detected, the **+FCERROR** response outlined earlier is returned instead. Many modems will return a **NO CARRIER** response to indicate that no carrier of any description could be detected within the timeout set by register S7, though as with the **+FRH** command, this is not specified in the EIA-578 Class 1 standard.

Once the **CONNECT** response (and any subsequent linefeed) has been received, anything else the fax modem sends the computer has to be data that has been received from the remote fax. As usual, the standard rules for processing any received data streams apply, so **<dle>** sequences must be filtered out of all raw data received from the modem, with **<dle><dle>** pairs being interpreted as a single **<dle>** and with the end of data being marked by the **<dle><etx>** pair.

The secondary response for the **AT+FRM** command is always **NO CARRIER**, which follows the end of the data and indicates that the remote transmitter has gone off. The line can now be turned around, and a message response frame sent.

Following execution of the **+FRM** command, any character sent to the modem before the secondary response (apart from flow control characters) is taken to be a signal to abort the command. The modem sends an **OK** result and waits for another command. This abort mechanism can be used in a number of ways:

- To enforce a timeout if **+FRM** is issued and no **CONNECT** message is returned after a given period of time.

- To cancel a session if the received data appears to be corrupted (if it doesn't conform to T.4 standards).

- To cancel a session in cases when a disk fills up and no more data can be received. As there is no way of telling how large a fax might be before it has completely arrived, this is an ever-present possibility. At 14400 bps, data can arrive as fast as 100K per minute.

It is a good idea to standardize on the ASCII CAN character (decimal 24 or hexadecimal 18h) to cancel receipt of phase C data. The fact that it is used like this in the other EIA fax modem standards, which implement similar data reception methods, makes it easier to convert software modules for different standards.

If an **AT+FRM=<mode>** command is cancelled, it should be followed by the **AT+FRS=<time>** wait for silence command; the cancellation doesn't stop the transmitter from sending data any quicker, but it does relieve the processor of the burden of handling received data that is simply going to be thrown away. In multitasking environments, this overhead is not trivial.

Using AT+FRM to receive a page

Listing 9.5 shows a typical use of the **AT+FRM** command to receive a fax page. The fax data is buffered by the **getpage** function and written to disk using the file pointer passed by the caller.

If we need to count the number of lines in the image for inclusion in a TIFF IFD, we would also perform EOL detection using the same **trailing0s** and **leading0s** algorithms described on page 207.

```
int getpage (FILE *faxfile)

{

    unsigned char k, lastchar=0 ;
    char faxbuf[512] ;
    int i=0 ;

    export ("AT+FRM=96\r") ;
    if (framestat()!=1) return (0) ;
    for (;;)
    {
      k=rxchar() ;
      if (lastchar==dle)
      {
        if (k==etx) break ;
        lastchar=0 ;
        if (k!=dle) continue ;
      }
      else if (k==dle)
      {
        lastchar=dle ;
        continue ;
      }
    faxbuf[i++]=k ;
    if (i==512)
      {
      fwrite (faxbuf,1,i,faxfile) ;
      i=0 ;
      }
    }
    if (i) fwrite (faxbuf,1,i,faxfile) ;
    return (wait_for("NO CARRIER",1)) ;
}
```

Listing 9.5: Receiving a page with **AT+FRM**

Flow control

All incoming and outgoing data (whether in frame or stream mode) is buffered by the modem, and flow control is mandatory in both directions. XON/XOFF flow control must be available independently in each direction, as flow control characters can appear either in received or transmitted data. CTS/RTS handshaking can be optionally implemented as an alternative to XON/XOFF flow control. For more background on flow control, see the lengthy discussion in Chapter 8.

ATD, ATA and ATH

The fact that these three universally available **AT** commands are not made part of the Class 1 command set but are relegated to an appendix and described as 'commonly used mechanisms' doesn't alter the fact that all Class 1 software assumes their availability. They are as essential to the fax Class 1 command set as any of the other commands.

The following slight differences applicable to their use in fax mode should be noted:

- Where the **ATD** dialling command is successful, and results in the **CONNECT** response, an implied **AT+FRH=3** is executed and the fax modem starts looking for the first HDLC frames from the answering fax.

- Where the **ATA** answering command is given, it also implies an **AT+FTH=3** command. The **CONNECT** response code should be followed by the data for the first HDLC frame (usually the station ID). This also applies if the answering mechanism is automatically triggered by a RING after an **ATS0=1** autoanswer command is given.

- The **ATH** command is always used after transmission or reception of a DCN disconnect frame to hang up. **ATH0** can be used if you don't trust the default. The **ATH** command used in Class 1 needs no escape sequence of the type **+++**, as a fax modem is almost always either in command mode, or is very quickly placed there. Escape sequences and guard times are not part of the Class 1 command set.

Sending a fax using a class 1 modem

The example in Table 9.6 shows a complete one-page fax being sent at 14400 bps. The fax is transmitted from a modem with the local ID of FAXOUT, and is received by a modem with the ID of FAXIN; note that these IDs do not conform to the numeric constraints of T.30. In Table 9.6, all the data sent to and received from the modem is shown, with the exception of streamed data sent (training and T.4). The data sent or received after **AT+FTH** or **AT+FRH** is shown in hexadecimal, and is reversed just as it appears when it arrives. The notes in the last column show the bits in their proper order.

Table 9.6: Sending a fax with a Class 1 modem

Pre-session modem setup		
AT+FCLASS=1		set class 1
	OK	
AT+FTM=?		find out capabilities
	3,24,48,72,73,74,96,97,98,121,122,145,146	
	OK	note : not all modems say OK
Phase A : Call establishment		
ATD<number>		dial: implied +FRH=3
	CONNECT	V.21 flags detected (preamble)
Phase B : Pre-message negotiations		
	ff	address octet
	3	control octet 11000000 = non-final frame
	40	FCF octet 00000010 = CSI
	20 20 20 20 20	FAXIN (backwards)
	20 20 20 20 20	
	20 20 20 20 20	
	4e 49 58 41 46	
	e 53	16-bit FCS
	<dle><etx>	end data
	OK	
AT+FRH=3		get next frame as control octet PF bit = 0
	CONNECT	carrier still present
	ff	address octet
	13	control octet 11001000 = final frame
	80	FCF octet 00000001 = DIS
	0 ee 78	00000000 01110111 00011110
	1 2	16-bit FCS
	<dle><etx>	end data
	OK	possible check for a dropped carrier here
AT+FTH=3		send our frame as control octet PF bit = 1
	CONNECT	V.21 flags (preamble) sent
ff		address octet
3		control octet 11000000 = non-final frame
43		FCF octet 11000010 = TSI
20 20 20 20 20		FAXOUT (backwards)
20 20 20 20 20		
20 20 20 20 54		
55 4f 58 41 46		
<dle><etx>		end of frame data

Table 9.6: (continued)

	CONNECT	still connected as control octet PF bit =0
ff		address octet
13		control octet 11001000 = final frame
83		FCF octet 11000001 = DCS
0 22 78		00000000 01000100 00011110
<dle><etx>		end of frame data
	OK	carrier dropped as control octet PF bit =1
AT+FTS=8		delay for 75 ms ± 20 ms
	OK	
AT+FTM=145		send using 14400 bps long training
	CONNECT	ready
<2700 x 0>		training sequence for 1.5 seconds
<dle><etx>		end data
	OK	
AT+FRH=3		now wait for response
	CONNECT	V.21 flags detected (preamble)
ff		address octet
13		control octet 11001000 = final frame
84		FCF octet 00100001 = CFR
ea 7d		16-bit FCS
<dle><etx>		end data
	OK	

Phase C : Message transmission

AT+FTM=146		setting 14400 bps short training
	CONNECT	ready
<T.4 data>		
<dle><etx>		end data
	OK	

Phase D : Post-page 1 exchanges

AT+FTS=8		delay for 75 ms ± 20 ms
	OK	
AT+FTH=3		we want to send post-page message
	CONNECT	V.21 flags (preamble) sent
ff		address octet
13		control octet 11001000 = final frame
2f		FCF octet 11110100 = EOP
<dle><etx>		end data
	OK	
AT+FRH=3		now wait for response
	CONNECT	V.21 flags detected (preamble)
ff		address octet
13		control octet 11001000 = final frame

Table 9.6: (continued)

	8c	FCF octet 00110001 = MCF
	a2 f1	16-bit FCS
	<dle><etx>	end data
	OK	

Phase E : Call hangup

AT+FTH=3		we want to disconnect
	CONNECT	V.21 flags (preamble) sent
ff		address octet
13		control octet 11000000 = final frame
fb		FCF octet 11011111 = DCN
<dle><etx>		end data
	OK	
ATH0		disconnect
	OK	

Receiving a fax using a Class 1 modem

The example in Table 9.7 is the exact reverse of the single page transmission given in Table 9.6.

Table 9.7: Receiving a fax with a Class 1 modem

Pre-session modem setup

AT+FCLASS=1		set class 1
	OK	
AT+FRM=?		find out capabilities
	3,24,48,72,73,74,96,97,98,121,122,145,146	
	OK	note : not all modems say OK

Phase A : Call establishment

	RING	wait for phone to ring
ATA		answer call : implied +FTH=3
	CONNECT	V.21 flags (preamble) sent

Phase B : Pre-message negotiations

ff		address octet
3		control octet 11000000 = non-final frame
40		FCF octet 00000010 = CSI
20 20 20 20 20		FAXIN (backwards)
20 20 20 20 20		
20 20 20 20 20 20		
4e 49 58 41 46		
<dle><etx>		end of frame data

Table 9.7: (continued)

	CONNECT	still connected as control octet PF bit =0
ff		address octet
13		control octet 11001000 = final frame
80		FCF octet 00000001 = DIS
0 ee 78		00000000 01110111 00011110
<dle><etx>		end of frame data
	OK	carrier dropped as control octet PF bit =1
AT+FRH=3		wait for response
	CONNECT	V.21 flags detected (preamble)
	ff	address octet
	3	control octet 11000000 = non-final frame
	43	FCF octet 11000010 = TSI
	20 20 20 20 20	FAXOUT (backwards)
	20 20 20 20 20	
	20 20 20 20 54	
	55 4f 58 41 46	
	21 67	16-bit FCS
	<dle><etx>	end data
	OK	
AT+FRH=3		get next frame as control octet PF bit = 0
	CONNECT	carrier still present
	ff	address octet
	13	control octet 11001000 = final frame
	83	FCF octet 00000001 = DCS
	0 22 78	00000000 01110111 00011110
	c6 44	16-bit FCS
	<dle><etx>	end data
	OK	
		note : maybe we ought to check for a dropped carrier here.
AT+FRM=145		receive using 14400 bps long training
	CONNECT	ready
training		
<dle><etx>		end of data
	OK	
AT+FTH=3		
	CONNECT	
ff		address octet
13		control octet 11001000 = final frame
84		FCF octet 00100001 = CFR
<dle><etx>		end data
	OK	

Table 9.7: (continued)

Phase C : Message transmission		
AT+FRM=146		setting 14400 bps short training
	CONNECT	
	<T.4 data>	
	<dle><etx>	end data
	NO CARRIER	
Phase D : Post-page 1 exchanges		
AT+FRH=3		we wait for the post-page message
	CONNECT	V.21 flags (preamble) detected
	ff	address octet
	13	control octet 11001000 = final frame
	2f	FCF octet 11110100 = EOP
	33 66	16-bit FCS
	<dle><etx>	end data
	OK	
AT+FTH=3		now send for response
	CONNECT	V.21 flags sent (preamble)
ff		address octet
13		control octet 11001000 = final frame
8c		FCF octet 00110001 = MCF
<dle><etx>		end data
	OK	
Phase E : Call hangup		
AT+FRH=3		we wait for a disconnect command
	CONNECT	V.21 flags (preamble) detected
	ff	address octet
	13	control octet 11000000 = final frame
	fb	FCF octet 11011111 = DCN
	<dle><etx>	end data
	OK	
ATH0		
OK		

ECM with Class 1 modems

Sending an error-corrected fax with a Class 1 modem should be simply a question of implementing the rather complicated T.30 error correcting protocol by using the **+FTH** command to transmit data, or the **+FRH** command to receive data. On the hardware side, the modem needs to be able to handle fast HDLC transfers. As for the application software, the fax data has to be segmented into proper facsimile data frames of the correct size and with the correct headers, and the quite extensive set of ECM frames need to be supported, but this is simply a question of getting the details right. The work of generating the frame check sequences when transmitting, and using them to identify any bad frames when receiving, should all be taken care of by the firmware in the modem, just as when HDLC frames are exchanged when negotiating.

The basic Class 1 command set, with its six commands, is sufficiently primitive (or flexible, if you prefer) to handle ECM. The following examples of ECM transmission and reception show a one-block fax being sent and received. The special ECM error correction and flow control frames are both used in this session.

Table 9.8: Sending a fax with ECM using a Class 1 modem

Pre-session modem setup		
AT+FCLASS=1		set class 1
	OK	
AT+FTM=?		find out transmit stream data capabilities
	3,24,48,72,96	V.21 V.27 ter and V.29
	OK	
AT+FTH=?		find out transmit frame data capabilities
	3,24,48,72,96	V.21 V.27 ter and V.29 also
	OK	
Phase A : Call establishment		
ATD<number>		dial: implied +FRH=3
	CONNECT	V.21 flags detected (preamble)
Phase B : Pre-message negotiations		
	ff	address octet
	3	control octet 11000000 = non-final frame
	40	FCF octet 00000010 = CSI
	20 20 20 20 20	FAXIN (backwards)
	20 20 20 20 20	
	20 20 20 20 20	
	4e 49 58 41 46	
	<fcs>	16-bit FCS
	<dle><etx>	end data
	OK	good frame
AT+FRH=3		get next frame as control octet PF bit = 0
	CONNECT	carrier still present

Table 9.8: (continued)

	ff	address octet
	13	control octet 11001000 = final frame
	80	FCF octet 00000001 = DIS
	0 ee f8 44	00000000 01110111 00011111 00100010
	<fcs>	16-bit FCS
	<dle><etx>	end data
	OK	good frame
		note : maybe we ought to check for a dropped carrier here.
AT+FTH=3		send our frame as control octet PF bit = 1
	CONNECT	V.21 flags (preamble) sent
ff		address octet
3		control octet 11000000 = non-final frame
43		FCF octet 11000010 = TSI
20 20 20 20 20		FAXOUT (backwards)
20 20 20 20 20		
20 20 20 20 54		
55 4f 58 41 46		
<dle><etx>		end of frame data
	CONNECT	still connected as control octet PF bit =0
ff		address octet
13		control octet 11001000 = final frame
83		FCF octet 11000001 = DCS
0 6 f8 4		00000000 01100000 00011111 00100000
<dle><etx>		end of frame data
	OK	carrier dropped as control octet PF bit =1
AT+FTS=8		delay for 75 ms ± 20 ms
	OK	
AT+FTM=96		send using 9600 bps
	CONNECT	ready
<1800 x 0>		training sequence for 1.5 seconds
<dle><etx>		end data
	OK	
AT+FRH=3		now wait for response
	CONNECT	V.21 flags detected (preamble)
	ff	address octet
	13	control octet 11001000 = final frame
	84	FCF octet 00100001 = CFR
	<fcs>	16-bit FCS
	<dle><etx>	end data
	OK	good frame

Table 9.8:   (continued)

Phase C : Message transmission

AT+FTH=96		set HDLC frame transmission at 9600 bps
	CONNECT	ready for first frame
ff		address octet
3		control octet 11000000 = non-final frame
6		FCF octet 01100000 = FDF
0		frame 1 data follows
<phase C data>		256 octet frame as bit 28=0
<dle><etx>		end data
	CONNECT	ready for next frame 2
ff		address octet
3		control octet 11000000 = non-final frame
6		FCF octet 01100000 = FDF
1		frame 2 data follows
<phase C data>		256 octet frame as bit 28=0
<dle><etx>		end data
	CONNECT	ready for next frame 3

(this carries on for 63 more frames)

ff		address octet
3		control octet 11000000 = non-final frame
6		FCF octet 01100000 = FDF
42		frame 67 data follows
<phase C data>		256 octet frame as bit 28=0
<dle><etx>		end data
	CONNECT	ready for next frame
ff		address octet
3		control octet 11000000 = non-final frame
86		FCF octet 01100001 = first RCP frame
<dle><etx>		end data
	CONNECT	ready for next frame
ff		address octet
3		control octet 11000000 = non-final frame
86		FCF octet 01100001 = second RCP frame
<dle><etx>		end data
	CONNECT	ready for next frame
ff		address octet
3		control octet 11000000 = non-final frame

Table 9.8: (continued)

86		FCF octet 01100001 = third RCP frame
<dle><etx>		end data
	CONNECT	ready for next frame
<dle><etx>		null frame tells modem to drop carrier
	OK	done that, ready for command

Phase D : Post-page exchanges

AT+FTS=8		delay for 75 ms ± 20 ms
	OK	
AT+FTH=3		we want to send post-page message
	CONNECT	V.21 flags (preamble) sent
ff		address octet
13		control octet 11001000 = final frame
bf		FCF 1 octet 11111101 = PPS
2f		FCF 2 octet 11110100 = EOP
0 0 42		page 1 block 1 frame = 67
<dle><etx>		end data
	OK	
AT+FRH=3		now wait for response
	CONNECT	V.21 flags detected (preamble)
	ff	address octet
	13	control octet 11001000 = final frame
	bc	FCF octet 00111101 = PPR
	00 03 00 00 00	bitmap of 256 possible frames shows
	00 00 00 f8 ff	that frames 9 and 10 were corrupted,
	ff ff ff ff ff	but that frames 1-8 and 11-67 arrived
	ff ff ff ff ff	intact ; though frames 68-256 are
	ff ff ff ff ff	shown as bad, we didn't sent these so
	ff ff ff ff ff	they are irrelevant
	ff ff	
	<fcs>	16-bit FCS
	<dle><etx>	end data
	OK	good frame

Phase C : Message transmission

AT+FTH=96		set HDLC frame transmission at 9600 bps
	CONNECT	ready for first frame
ff		address octet
3		control octet 11000000 = non-final frame
6		FCF octet 01100000 = FDF
8		frame 9 data follows
<phase C data>		256 octet frame as bit 28=0
<dle><etx>		end data
	CONNECT	ready for next frame 2
ff		address octet

Table 9.8: (continued)

3		control octet 11000000 = non-final frame
6		FCF octet 01100000 = FDF
9		frame 10 data follows
<phase C data>		256 octet frame as bit 28=0
<dle><etx>		end data
	CONNECT	ready for next frame 3
ff		address octet
3		control octet 11000000 = non-final frame
86		FCF octet 01100001 = first RCP frame
<dle><etx>		end data
	CONNECT	ready for next frame
ff		address octet
3		control octet 11000000 = non-final frame
86		FCF octet 01100001 = second RCP frame
<dle><etx>		end data
	CONNECT	ready for next frame
ff		address octet
3		control octet 11000000 = non-final frame
86		FCF octet 01100001 = third RCP frame
<dle><etx>		end data
	CONNECT	ready for next frame
<dle><etx>		null frame tells modem to drop carrier
	OK	done that, ready for command

Phase D : Post-page exchanges

AT+FTS=8		delay for 75 ms ± 20 ms
	OK	
AT+FTH=3		we want to send post-page message
	CONNECT	V.21 flags (preamble) sent
ff		address octet
13		control octet 11001000 = final frame
bf		FCF 1 octet 11111101 = PPS
2f		FCF 2 octet 11110100 = EOP
0 0 42		page 1 block 1 frame = 67
<dle><etx>		end data
	OK	
AT+FRH=3		now wait for response
	CONNECT	V.21 flags detected (preamble)
	ff	address octet
	13	control octet 11001000 = final frame
	ec	FCF octet 00110111 = RNR

Table 9.8: (continued)

	`<fcs>`	16-bit FCS
	`<dle><etx>`	end data
	OK	good frame
`AT+FTH=3`		ask if receiver ready yet
	CONNECT	V.21 flags (preamble) sent
`ff`		address octet
`13`		control octet 11001000 = final frame
`6f`		FCF octet 11110110 = RR
`<dle><etx>`		
	OK	
`AT+FRH=3`		
	CONNECT	V.21 flags detected (preamble)
	`ff`	address octet
	`13`	control octet 11001000 = final frame
	`8c`	FCF octet 00110001 = MCF = all
		frames OK
	`<fcs>`	16-bit FCS
	`<dle><etx>`	end data
	OK	good frame
Phase E : Call hangup		
`AT+FTH=3`		we want to disconnect
	CONNECT	V.21 flags (preamble) sent
`ff`		address octet
`13`		control octet 11000000 = final frame
`fb`		FCF octet 11011111 = DCN
`<dle><etx>`		end data
	OK	
`ATH0`		disconnect
	OK	

Sending a fax with ECM using a Class 1 modem

Given the complexities of the ECM protocol, the sample session above isn't terribly difficult to implement. Note that both **+FTM** and **+FTH** have to be tested before dialling, as it isn't known at that stage whether an ECM transfer will be negotiated. In fact, the receiver sends back a DIS with both bits 27 and 31 set, indicating that it can also handle T.6 coding. The receiver in this session is more capable than the transmitter, as it also shows in bits 11–14 (set to 1101) that it can handle all possible modulation schemes up to V.17, whereas the transmitter can only manage 9600 bps using V.29.

The transmission itself is quite straightforward, though it's worth emphasizing that as it is half duplex, the transmitter has no idea whether the data is being received properly. The only peculiarity is the null frame used to switch off the transmitter at the end of phase C. The usual method of doing this, sending a frame with the control octet P/F bit set, cannot be used in this context as the line isn't yet being turned around.

Receiving a fax with ECM using a Class 1 modem

The next sample session (Table 9.9) is the reverse of the single page transmission given above.

Table 9.9:   Receiving a fax with ECM using a Class 1 modem

Pre-session modem setup		
AT+FCLASS=1		set class 1
	OK	
AT+FRM=?		find out receive stream data capabilities
	3,24,48,72,73,74, 96,97,98,121,122, 145,146	V.21 V.27 ter V.29 and V.17
	OK	
AT+FRH=?		find out receive frame data capabilities
	3,24,48,72,73,74, 96,97,98,121,122, 145,146	V.21 V.27 ter V.29 and V.17
	OK	
Phase A : Call establishment		
	RING	wait for phone to ring
ATA		answer call : implied +FTH=3
	CONNECT	V.21 flags (preamble) sent
Phase B : Pre-message negotiations		
ff		address octet
3		control octet 11000000 = non-final frame
40		FCF octet 00000010 = CSI
20 20 20 20 20		FAXIN (backwards)
20 20 20 20 20		
20 20 20 20 20		
4e 49 58 41 46		

Table 9.9: (continued)

<dle><etx>		end of frame data
	CONNECT	still connected as control octet PF bit =0
ff		address octet
13		control octet 11001000 = final frame
80		FCF octet 00000001 = DIS
0 ee f8 44		00000000 01110111 00011111 00100010
<dle><etx>		end of frame data
	OK	carrier dropped as control octet PF bit =1
AT+FRH=3		wait for response
	CONNECT	V.21 flags detected (preamble)
	ff	address octet
	3	control octet 11000000 = non-final frame
	43	FCF octet 11000010 = TSI
	20 20 20 20 20	FAXOUT (backwards)
	20 20 20 20 20	
	20 20 20 20 54	
	55 4f 58 41 46	
	<fcs>	16-bit FCS
	<dle><etx>	end data
	OK	good frame
AT+FRH=3		get next frame as control octet PF bit = 0
	CONNECT	carrier still present
	ff	address octet
	13	control octet 11001000 = final frame
	83	FCF octet 00000001 = DCS
	0 6 f8 4	00000000 01100000 00011111 00100000
	<fcs>	16-bit FCS
	<dle><etx>	end data
	OK	good frame
AT+FRM=96		receive using V.29 9600 bps
	CONNECT	ready
training		
<dle><etx>		end of data
	OK	
AT+FTH=3		
	CONNECT	V.21 flags (preamble) sent
ff		address octet
13		control octet 11001000 = final frame
84		FCF octet 00100001 = CFR
<dle><etx>		end data
	OK	

Table 9.9: (continued)

Phase C : Message reception
AT+FRH=96

	CONNECT	
	ff	address octet
	3	control octet 11000000 = non-final frame
	6	FCF octet 01100000 = FDF
	0	frame 1 data follows
	<phase C>	256 octet frame as bit 28=0
	<fcs>	16-bit FCS
	<dle><etx>	end data
	OK	good frame

AT+FRH=96

	CONNECT	
	ff	address octet
	3	control octet 11000000 = non-final frame
	6	FCF octet 01100000 = FDF
	1	frame 2 data follows
	<phase C>	256 octet frame as bit 28=0
	<fcs>	16-bit FCS
	<dle><etx>	end data
	OK	good frame
		(this carries on for 6 good frames)

AT+FRH=96

	CONNECT	
	ff	address octet
	3	control octet 11000000 = non-final frame
	6	FCF octet 01100000 = FDF
	8	frame 9 data follows
	<phase C>	indeterminate number of octets arrive
	<dle><etx>	end data
	ERROR	bad frame (discard data)

AT+FRH=96

	CONNECT	
	ff	address octet
	3	control octet 11000000 = non-final frame
	6	FCF octet 01100000 = FDF
	a	frame 11 data follows - missing frame?
	<phase C>	256 octet frame as bit 28=0
	<fcs>	16-bit FCS
	<dle><etx>	end data
	OK	good frame 11 - line noise hit frame 9+10

(this carries on for 55 more frames)

Table 9.9: (continued)

AT+FRH=96		
	CONNECT	
	ff	address octet
	3	control octet 11000000 = non-final frame
	6	FCF octet 01100000 = FDF
	42	frame 67 data follows
	\<phase C>	256 octet frame as bit 28=0
	\<fcs>	16-bit FCS
	\<dle>\<etx>	end data
	OK	good frame
AT+FRH=96		
	CONNECT	
	ff	address octet
	3	control octet 11000000 = non-final frame
	86	FCF octet 01100001 = one RCP frame
	\<fcs>	16-bit FCS
	\<dle>\<etx>	end data
	OK	good frame

Phase D : Post page 1 exchanges

AT+FRH=3		we only need to detect one RCP frame
	CONNECT	look for post-page message
	NO CARRIER	woops - transmitter not ready
AT+FRH=3		try again for the post-page message
	CONNECT	V.21 flags (preamble) detected
	ff	address octet
	13	control octet 11001000 = final frame
	bf	FCF 1 octet 11111101 = PPS
	2f	FCF 2 octet 11110100 = EOP
	0 0 42	page = 0 block = 0 frame = 66
	dc 24	16-bit FCS
	\<dle>\<etx>	end data
	OK	good frame
AT+FTH=3		now send our report on the bad frames
	CONNECT	V.21 flags (preamble) sent
ff		address octet
13		control octet 11001000 = final frame
bc		FCF octet 00111101 = PPR
00 03 00 00		bitmap of 256 possible frames shows
00 00 00 00		that frames 9 and 10 were corrupted,
f8 ff ff ff		but that frames 1-8 and 11-67
ff ff ff ff		arrived intact ; though frames 68-256
ff ff ff ff		are shown as bad, we didn't sent these
ff ff ff ff		so they are irrelevant.
ff ff ff ff		

Table 9.9: (continued)

```
ff ff ff ff
<fcs>                              16-bit FCS
<dle><etx>                         end data
                  OK               good frame


Phase C : Message transmission
AT+FRH=96
                  CONNECT
                  ff               address octet
                  3                control octet 11000000 = non-final
                                   frame
                  6                FCF octet 01100000 = FDF
                  8                frame 9 data follows
                  <phase C>        256 octet frame as bit 28=0
                  <fcs>            16-bit FCS
                  <dle><etx>       end data
                  OK               good frame
AT+FRH=96
                  CONNECT
                  ff               address octet
                  3                control octet 11000000 = non-final
                                   frame
                  6                FCF octet 01100000 = FDF
                  9                frame 10 data follows
                  <phase C>        256 octet frame as bit 28=0
                  <fcs>            16-bit FCS
                  <dle><etx>       end data
                  OK               good frame
AT+FRH=96
                  CONNECT
                  ff               address octet
                  3                control octet 11000000 = non-final
                                   frame
                  86               FCF octet 01100001 = one RCP frame
                  <fcs>            16-bit FCS
                  <dle><etx>       end data
                  OK               good frame


Phase D : Post page exchanges
AT+FRH=3                           get the post-page message
                  CONNECT          V.21 flags (preamble) detected
                  ff               address octet
                  13               control octet 11001000 = final frame
                  bf               FCF 1 octet 11111101  = PPS
                  2f               FCF 2 octet 11110100 = EOP
                  0 0 42           page = 0 block = 0 frame = 66
```

Table 9.9: (continued)

	dc 24	16-bit FCS
	<dle><etx>	end data
	OK	good frame
AT+FTH=3		send receiver not ready to make some time
	CONNECT	V.21 flags (preamble) sent
ff		address octet
13		control octet 11001000 = final frame
ec		FCF octet 00110111 = RNR
<dle><etx>		end data
	OK	
AT+FRH=3		get the RNR response
	CONNECT	V.21 flags detected (preamble)
	ff	address octet
	13	control octet 11001000 = final frame
	6f	FCF octet 11110110 = RR
	<fcs>	16-bit FCS
	<dle><etx>	end data
	OK	good frame
AT+FTH=3		now send our response
	CONNECT	V.21 flags (preamble) sent
ff		address octet
13		control octet 11001000 = final frame
8c		FCF octet 00110001 = MCF
<dle><etx>		end data
	OK	
Phase E : Call hangup		
AT+FRH=3		we wait for a disconnect command
	CONNECT	V.21 flags (preamble) detected
	ff	address octet
	13	control octet 11000000 = final frame
	fb	FCF octet 11011111 = DCN
	<fcs>	16-bit FCS
	<dle><etx>	end data
	OK	good frame
ATH0		
	OK	

Possibly the only notable thing about this ECM session is that the first entry into phase D required two attempts: this sometimes happens when phase C is terminated on the first RCP frame received, as the transmitter is not only sending two more frames but is also going to delay for around 75 ms. It is possible to use the **+FRS** command to wait for the transmitter to go off, but it is not unknown on some modems and operating systems for the response to this command to be sufficiently delayed to cause loss of the first post-message command. Repeating the **+FRH** command should be done as a matter of course as long as the 3 s response timeout hasn't expired, so the method used to recover from the first failure to receive the post-message command is not exceptional.

There are some other ECM features worth noting. For instance, if the transmitter had received four PPR frames rather than one, an immediate re-entry to phase C would not be possible. Instead, an exchange of EOR and ERR frames could have enabled the session to continue with the errors uncorrected; alternatively, the transmitter could have sent a CTC frame telling the receiver to continue to correct. This frame is accompanied by the first two octets of the DIS information field, containing the speed to be used. From the viewpoint of the transmitter, the CTC exchange comes after receipt of the PPR at the end of phase D, just before re-entry into phase C (see Table 9.10).

Table 9.10: Handling PPRs with a Class 1 modem

AT+FTH=3		4th PPR received means we must respond
		with either an EOR or a CTC command
	CONNECT	V.21 flags (preamble) sent
ff		address octet
13		control octet 11001000 = final frame
13		FCF octet 11001000 = CTC
0 c		00000000 00110000 = 7200 bps V.29
<dle><etx>		end data
	OK	
13		
AT+FRH=3		wait for response
	CONNECT	
	ff	address octet
	13	control octet 11001000 = final frame
	c4	FCF octet 00100011 = CTR
	<fcs>	16-bit FCS
	<dle><etx>	end data
	OK	good frame
Phase C : Message transmission		
AT+FTH=72		set HDLC frame transmission at 7200 bps
		continue at new speed

The **AT+FTS=8** command, to delay for 75 ± 20 ms, is only used by the transmitter when it changes modulation, e.g. when switching from V.21 negotiating speed to phase C modulation. No delay is needed when switching from reception at one modulation to transmission at another, as this is a line turnaround, and there is already a mandatory preamble of 1 s at this point.

Problems with Class 1 ECM

Regrettably, we don't live in an ideal world, and sending error corrected faxes with some Class 1 modems turns out not to be as straightforward as our examples show. In their ability to handle ECM, not all modems are created equal. We can distinguish three different levels of support for this feature.

The most desirable level is when a modem behaves in the manner shown in the sample sessions. Although a modem conforming to the Class 1 specification ought to be able to handle ECM in this way, I've found most fall into the other two groups: the second group can't handle high speed HDLC, but at least they don't lie about it; the third group consists of modems that can't handle ECM, but think they can. We'll look at these in turn.

Modems that don't claim to support ECM

There are a number of reasons why modems are unable to support the fast HDLC options needed for ECM. Some use the least expensive chipsets and often lack V.29 modulators, so they can't handle V.29 reception either. While it isn't surprising to find this type of modem unable to manage ECM, there are some top-of-the-range models that also don't support this. This is not because of a lack of processing power, but a straightforward decision by the manufacturer not to implement the relevant commands. Here's an example dialogue from such a modem:

```
AT+FTM=?
            24,48,72,73,74,96,97,98,121,122,145,146
            OK
AT+FRM=?    24,48,72,73,74,96,97,98,121,122,145,146
            OK
AT+FRH=?
            3
            OK
AT+FTH=?
            3
            OK
```

The modem can handle all possible modulation schemes in stream mode for both transmission and reception. It's worth noting that this modem doesn't reproduce the common error we've seen earlier, of claiming to be able to transfer faxes at 300 bps (which isn't an option supported by T.30). But while it handles 300 bps V.21 negotiations, it has no facilities for sending and receiving HDLC frames at faster speeds.

Strictly speaking, this doesn't mean the modem can't handle ECM. It is possible to send and receive HDLC frames at high speeds using the **+FTM** and **+FRM** commands, but all the work has to be handled by the software. This isn't simply a question of adding opening and closing flags, and generating and validating CRCs for the frame checking sequence. The difficult part of implementing HDLC lies in implementing the transmitter idle mode, the receiver hunt mode and zero-bit insertion and deletion. All are complicated by the fact that HDLC is a bit-oriented protocol; most modern CPUs and UARTs are well suited to manipulating, sending and receiving data in discrete 8-bit chucks, but are very inefficient at handling large quantities of individual bits.

Most of the HDLC control fields fall naturally into octets; the flag sequence is **01111110**, the FCS is a 16-bit CRC, the addressing and control fields are both octets, as are all the T.30 frame IDs. Unfortunately, there is no guarantee that the octets will fall neatly into line with byte boundaries.

In the case of a fax frame, it is clear that as soon as the transmitter sees the fax address octet **11111111**, it will need to stuff a 0 bit in the sequence after the first five consecutive 1 bits. The octet would be sent as the nine-bit sequence **11111011**, and even if the contents of the whole of the rest of the frame originally consisted of octets neatly aligned on byte boundaries, it would immediately become shifted right by one bit, and the original octet boundaries would vanish.

In fact, every byte the CPU sends to serial port needs to be examined in conjunction with the previous octet transmitted to see if any bit-stuffing is needed, and most bytes will need to be shifted around, with some portions combined with leftover bits from the previous byte and the remaining portions becoming leftover bits, which in turn need to be combined with the succeeding byte to be sent.

Some parts of the HDLC protocol would need to be handled at the interrupt level. For instance, the transmission of flags when the transmitter is idling could not be handled on a polled basis. Even here, though, the exact bytes used for sending flags would vary from frame to frame. Although the flag sequence consists of bitstrings with the value **01111110**, if the previous frame ended on the fifth bit of a byte, its closing flag would not have begun on a byte boundary. The flag would have begun with the sequence **xxxxx011**, and the idling flags would then consist of pairs of octets with the values **11110011 11110011**. In a similar way, a receiver would need to examine every bit of each byte received to check for both opening and closing flags, and also to handle any zero-bit deletion.

The burden of handling HDLC is not trivial. Virtually all serious synchronous communications software products use dedicated on-chip circuitry. Devices such as the Z8530 Serial Communications Controller include automatic zero bit insertion and deletion, address field recognition, idle flag insertion and CRC generation and detection. While it is possible to program all this in software using the Class 1 stream modes, doing so simply turns a fax modem into a rather limited and expensive synchronous-asynchronous converter. The additional CPU burden imposed by the multiple levels of bit shifting also serves to make this theoretical option impractical on many systems which are either multi-tasking or simply not fast enough.

Modems that claim to support ECM but don't

The fact that a modem claims to handle fast HDLC speeds doesn't always make it true. The following is a typical dump from a modem that leads you to believe that it can handle ECM with no problems:

```
AT+FRH=?
            3,24,48,72,73,74,96,97,98,121,122,145,146
            OK
AT+FTH=?
            3,24,48,72,73,74,96,97,98,121,122,145,146
            OK
```

Some modems with this type of response handle ECM beautifully on good lines but fail dismally under adverse conditions. The reasons for failure are not always apparent. I have used some modems that always produce **NO CARRIER** responses to any **+FRH** command apart from **+FRH=3**, which leads me to think they are always looking for a V.21 preamble. Others always respond to **+FTH** commands with an **ERROR** code.

Other modems start transmitting or receiving perfectly well, but then fail, with errors occurring after a few frames of the session. I noticed that the number of successful frames varied with the modulation and speed used, leading me to suspect the problem was one of flow control. Investigations with debugging scripts and breakout boxes has convinced me there are some modems that implement perfectly good flow control when sent data with **+FTM**, but which show no evidence of any flow control events at all when sending data with **+FTH**. This type of modem inevitably fails in the middle of a block and is unable to recover.

Some modems give the impression that they lack the processing power to receive handle HDLC frames at fast speeds. This is more processor-intensive than sending, but the symptoms are very similar to the lack of flow control on transmission; the modem begins receiving properly but then starts to receive frames with errors. This symptom can also occur when an otherwise ECM-capable modem is connected to a computer at too slow a speed. It is quite easy to do this by mistake, as the arithmetic involved doesn't make the mistake immediately obvious.

The usual speed for a computer to fax modem link is 19200 bps asynchronous. If a fax modem is receiving at 14400 bps synchronously, the addition of start and stop bits means it has to transfer this data to the computer at a speed of 18000 bps. When receiving in stream mode using **+FRM=146** there is no problem, as the speed of the connection is adequate for the data that needs to be transferred. This isn't true when receiving ECM data using **+FRH=146** in frame mode, because there is a great deal of additional control information that needs to be included with the data; unlike stream mode, phase C data is not delivered raw. A 19200 bps link to a computer gives a margin of only 1200 bits, or 120 characters, for each second.

One HLDC frame contains 256 octets plus at least 5 flag, control, address and FCS octets. On average, 7 frames will arrive each second over a 14400 bps link. Each frame is accompanied by text messages; the initial **CONNECT** and final **OK** with their associated **<cr><lf>** sequences total 17 bytes. The final **<dle><etx>** sequence takes an extra two bytes, and an indeterminate number of extra **<dle>** characters can also appear in the data. In addition, the **AT+FRH=146** command, with its **<cr>** needs to be issued for each frame; this takes up an extra 11 bytes per frame. The total overhead of 30 bytes for each of 7 frames gives us at least an additional 210 characters to be transferred over the link to the computer each second, and we have already worked out that there is only room for an extra 120. The position with 64 octet frames would be even worse, as these generate four times the overhead, or 840 characters.

The conclusion is that a fax modem capable of 14400 bps error corrected reception needs to be linked to a computer much faster than 19200 bps. The minimum speed will vary with the type of data and frame size, but any speed less than 23000 bps must be considered unreliable. For most users, a rate of 38400 bps, the next step up from 19200 bps, will provide plenty of leeway.

Enhancements to class 1

There are currently a number of enhancements to Class 1 being planned. However, none of the proposals should stop any existing Class 1 software from functioning, and downward compatibility ought to be assured.

Adaptive answering

Many Rockwell Class 1 command sets includes the **AT+FAE** command for adaptive answering, which enables a modem to answer the telephone when it rings and discriminate between fax and data calls. Adaptive answering is turned on with the **AT+FAE=1** command, but the default is **AT+FAE=0**, which turns adaptive answering off. Though this is not included in the official EIA-578 Class 1 specification, the similar **AT+FAA** command for adaptive answering is an optional part of the EIA-592 recommendations in both the draft used for most Class 2 modems and the official Class 2.0 version.

When adaptive answering is on, a call from a data modem provokes the response **DATA**, with the modem automatically switching to **+FCLASS=0**. An application seeing this should issue the **ATO** command to go on line. When the call terminates, the **AT+FCLASS=1** command must be issued to re-enable adaptive answering, as it only functions in fax modes. A call that comes in from a fax when adaptive answering is on is treated in the same way as it would be without this feature, and the T.30 preamble and handshake sequence begins as usual.

While the ability to distinguish between fax and data calls is included in many fax modems, I am not aware of any standard for telling voice from fax calls. This is typically the function of 'fax switches', of which there are a number of models available. Some simply switch between fax and voice, while others also control answering machines. Some depend upon the optional CNG tone for their proper functioning, and switch calls through to voice if no CNG is detected. Others, including so-called passive fax switches, function in the opposite way, and switch calls through to a fax if no voice is detected.

It is becoming quite common for home office fax machines to include some kind of fax switch capability, especially on models that include answering machines. Fax modems capable of distinguishing between voice and fax calls are few and far between, and there isn't universal agreement on their usefulness or reliability. While there is certainly no standard way of handling voice/data discrimination using software commands, Annex D to ITU Recommendation T.30 describes how fax switches should function using the CNG tone in conjunction with an outgoing voice message (OGM).

The best long-term solutions to sharing voice and fax on one line depend upon the availability of advances in telephone technology. Distinctive ring signals usually require an extra payment to be made to the telephone company providing the service. With distinctive ringing, the same telephone line can have two or more numbers, with each number having a separate and distinctive ring signal. One phone number can be distributed as a fax number and another as a voice number. Modems capable of detecting these distinctive rings are beginning to become available, and can easily be programmed to answer a call only if the distinctive ring pattern corresponding to the fax number is detected. It is also possible that the provision of calling line identification signals (CLI) will result in an improved ability to differentiate types of caller.

Class 1.0

A version of the latest draft of Class 1 has recently been submitted to the ITU Study Group 8 as a proposed international standard provisionally known as T.class1. There is one significant difference between the international T.class1 document and the latest EIA-578 U.S. domestic Class 1 draft. It is that T.class1 proposes that a modem with T.class1 capability should respond to **AT+FCLASS=?** with a 1.0 instead of a simple 1, and that the command **AT+FCLASS=1.0** should be used to configure the modem for T.class1 operation.

The rationale behind this proposal is that the ANSI/EIA-592 Class 2.0 recommendation has been submitted to the ITU Study Group 8 for consideration as standard T.class2 at the same time. As this document requires that **AT+FCLASS=2.0** is used to configure the modem (with **AT+FCLASS=2** being *de facto* reserved for the unapproved 1990 version of SP-2388), standardizing on the format of a digit with a decimal point for all service classes does make some sort of sense.

While there are different implications for modems that claim conformance with Class 2 or 2.0, the differences between **+FCLASS=1** and **+FCLASS=1.0** are insignificant. Software that works with Class 1 modems will probably work with Class 1.0 modems too, as the command set appears to be identical. I would be surprised if approval of T.class1 resulted in two different kinds of service Class 1 in the same way as we have two different versions of service Class 2.

Nevertheless, it may well be a good idea for application software to future-proof its operation by checking with the **+FCLASS=?** command for both Class 1 and 1.0 compliance. Subsequent setting of the service class using **AT+FCLASS=1.0** will allow software to support modems that conform the proposed ITU standard without also conforming to the ANSI/EIA version.

10

Programming Class 2 Fax Modems

Introduction

If you want to learn how to program a Class 2 fax modem, this is the chapter you'll need, as here we deal with how they work in some detail. Unlike Class 1 devices, these modems don't require an intimate knowledge of the T.30 session protocol, but an understanding of the five phases of a fax session (as described in Chapter 6) would be useful. As with all fax modem programming, we also assume that anyone reading this chapter also understands the issues raised in Chapter 8.

Class 2 and Class 2.0

This book adopts current industry practice of referring to EIA fax modems as being either Class 1, 2 or 2.0 devices. A user (and any software) can tell the difference between the classes by looking at the response to the **AT+FCLASS=?** command, described in detail in Chapter 8. To summarize, a modem whose response includes a **1** has Class 1 capability, a response including a bare **2** has Class 2 capability, while a response including a **2.0** has Class 2.0 capability.

This makes perfect sense as well as being common practice, but the EIA see things differently. While they agree about identifying Class 1 modems, the official position is that modems complying with service Class 2 are those which include a **2.0** in their response to the **+FCLASS=?** enquiry. Modems responding with a bare **2** are not EIA compliant modems at all.

The situation is like something out of Alice in Wonderland. The EIA call Class 2 what everyone else calls Class 2.0. What everyone else uses, and refers to as Class 2, is a standard that the EIA now say is nothing to do with them: though they originated the specification and maintain the copyright, they don't officially sell it to the public, and classify all modems that use it as a basis for their commands as manufacturer-specific devices. It's one thing to decide not to ratify a standards proposal, but another to attempt to de-legitimize an existing *de facto* standard by retrospectively using its name for something entirely different.

This is highly unsatisfactory for a number of other reasons. The EIA deliberately give the impression that Class 2.0, being 'official', is somehow better and more dependable than Class 2. In fact, for software developers the reverse is probably true. Both the approved and unapproved proposals contain at least as many optional features as mandatory ones, but the older proposal is more settled and the range of features used by various models is a known factor for software designers. It is unclear what features in Class 2.0 will become standard offerings.

The ambiguities in the specification leading to differing implementations of Class 2, and the most common Class 2 modem bugs and quirks, are all fairly well known: there are many software packages for such modems that have written around the problems. However, Class 2.0 modems are an unknown quantity in this respect. There is currently a handful of manufacturers marketing Class 2.0 modems. Preliminary reports indicate there is already some controversy between two of these vendors over how to implement at least one of the compulsory commands. This implies that, even if the approved proposal is tighter than the draft, it is still not specified in quite enough detail to prevent manufacturers from interpreting it differently.

Preliminaries

Using a Class 2 fax modem can be remarkably easy. Though the first two examples in the following sections don't quite explain what the modem commands do, and also assume no network or operator errors; they are designed to show how easy it can be to send and to receive faxes. We go into detail later.

Setting up a Class 2 modem is quite straightforward, and consists usually of sending the **AT+FCLASS=2** command and using a 19200 bps serial port setting. More details on the **AT+FCLASS** commands and other setup requirements such as flow control were given in Chapter 8, while we discuss advanced Class 2 setup strings later in this chapter.

The format of the fax data sent and received using a Class 2 fax modem is identical to that used with a Class 1 modem. This was discussed in Chapter 9, to which readers are referred for the details (see page 202).

How to send a fax with a Class 2 modem

Sending a one-page fax pre-encoded in 1-D T.4 format requires the following simple steps on virtually any Class 2 fax modem:

1. Issue an ordinary **ATD** dial command and ignore everything until you see an **OK**.

2. Send an **AT+FDT** (data transmit) command and wait for a **CONNECT**. Ignore everything before the **CONNECT** message.

3. Wait for an XON; then send the T.4 encoded fax in a data stream, ending with a **<dle><etx>** pair. Wait for the fax modem to respond with an **OK**. Your transmission software must be able to support XON/XOFF flow control, and must always keep the modem supplied with data to send out. Apart from those requested for flow control, no pauses in the data are allowed.

4. Send out the end transmission **AT+FET=2** command. The modem hangs up the line and responds **OK**.

How to receive a fax with a Class 2 modem

Receiving a T.4 format fax via a properly set Class 2 fax modem is possibly easier than sending, as it is a passive procedure:

1. Tell the modem you want to receive a fax with the **AT+FCR=1** command. The modem will respond with **OK**.

2. Wait for the modem to detect an incoming call and respond with **RING**, then answer the call by sending out **ATA**. Ignore everything until you see another **OK**.

3. Send out an **AT+FDR** (data receive) command and wait for a **CONNECT**. Ignore everything before the **CONNECT** message.

4. Send a DC2 (ASCII 18), and then receive T.4 encoded fax in a data stream. Look for the final **<dle><etx>**. Ignore everything that comes in until the fax modem comes back with an end of transmission **FET:** report. Remember the **FET** parameter and wait for an **OK**.

5. If the **FET:** parameter was **0**, loop back to step 3 and receive another page. If the **FET** parameter was **2**, send a final **AT+FDR** (data receive) command and ignore everything until a final **OK**.

More detail on sending

The way in which the T.30 protocol divides a session into five phases is basic to understanding the philosophy behind the commands used in Class 2, because Class 2 fax modem commands essentially tell the modem to move from one phase to another.

Table 10.1 should make the relationship between the session phases and fax commands clearer for the single page send outlined above. This time we have included a call setup section, all the modem responses (not simply those we act upon), and have made it a two page fax. A short indication of what each response means is also given. Let's look at this in more detail.

Table 10.1: Example of a Class 2 fax transmission

Pre-session modem setup		
AT+FCLASS=2		
	OK	
AT+FLID=îID stringî	set local ID	
	OK	
AT+FDCC=VR,BR,WD,LN,DF,EC,BF,ST		set any modem parameters (the contents of this command are not letters ; the letters are the names of 8 single-digit numeric fields).
	OK	
Phase A : Call establishment		
ATD<number>		
	+FCON	Fax preamble detected
	+FNSF:<nsf>	NSF detected
	+FCSI:<remote id>	CSI detected
	+FDIS:<codes>	DIS detected
	+FPOLL	if the answerer is pollable
	OK	phase A complete
Phase B : Pre-message negotiations		
AT+FDT		
	+FDCS:<codes>	TSI and DCS sent TCF, training OK phase B complete
Phase C : Message transmission		
	CONNECT	high-speed carrier sent
	<xon>	ready to start sending data
<T.4 data>		page 1 sent
<dle><etx>		end of page 1
	OK	phase C complete
Phase D : Post-page 1 exchanges		
AT+FET=0		send RTC, another page to come, send MPS
	+FPTS:1	MCF detected
	OK	phase D complete
AT+FDT		now back to phase C for page 2

Table 10.1: (continued)

Phase C : Message transmission		
	CONNECT	high-speed carrier detected
	<xon>	ready to start sending data
<T.4 data>		page 2 is sent
<dle><etx>		end of page 2
	OK	phase C complete
Phase D : Post-page 2 exchanges		
AT+FET=2		send RTC, no more pages, send EOP
+FPTS:1		MCF detected
		phase D complete
Phase E : Call hangup		
	+FHNG:0	DCN sent
	OK	back on hook
		phase E complete

Pre-session modem setup

The **AT+FCLASS=2** tells the modem to operate as a Class 2 device. The second setup command tells the modem what local ID is to be used. Readers of Chapter 6 will remember that the local ID is included in TSI frames on sending and in CSI frames when receiving. Anyone who has bought a fax machine will be familiar with the chore of poring over the manual and pressing obscure and meaningless sequences of little buttons to program the local ID into a fax. This is one of the few areas where the ease of use of fax modems is an improvement on that of fax machines, as you can program the ID number into a fax modem with a simple command such as:

AT+FLID="+12 345 6 7890"

The LID in **AT+FLID** stands for Local ID. This is easy to remember, which is just as well as one of the weak points about fax modems is that (unlike fax machines) they keep forgetting who they are. In theory, a few bytes of non-volatile memory (*NVRAM*) in a modem should allow it to remember the ID between sessions, but most modems don't have this. Consequently, the local ID needs to be reset before each fax session.

The T.30 recommendation states that the FIF (Facsimile Information File) in the ID frames must contain 20 characters, which must be either spaces, the + sign or digits in the range 0–9 (ASCII characters 32, 43 or 48–57). However, most fax modems allow this range to be extended. The test form of the command allows users to ascertain the range of possible values, with a typical dialogue being:

```
AT+FLID=?
        (20)(32-127)
        OK
```

The initial (20) indicates that up to 20 characters can be entered, with the (32–127) showing that any character between space and delete can be part of the string. This is inaccurate, as the fact that all ID strings should be enclosed in "quotes" means that the " character (ASCII 34) cannot be entered as part of the ID. Most modems that accept alphabetic IDs convert all letters to upper case. If there are fewer than 20 characters, an ID is padded with spaces. Some modems pad on the left, some on the right.

The behaviour of a numeric-only fax machine when it receives an alphanumeric ID is unspecified, so the usual format of an ID being a telephone number should be adhered to for safety, especially whenever faxes are being sent to or received from unknown fax machines.

If a null string **""** is entered as the local ID, the modem will not send any TSI or CSI frames.

Programmers should note that the **+FLID** command is sometimes quite badly implemented and should be treated with care. **+FLID** bugs tend to get noticed more than others because in most fax software, there is a user option to set the content of the ID, so it is one of the few commands that is directly accessible.

Five common bugs are as follows:

- The modem powers up with the **+FLID** space full of garbage.

- The modem won't accept the full 20 characters, when terminated with a " (though 20 characters and no final " is accepted).

- The modem returns an **ERROR** code when the ID string format is violated (too many characters or no trailing "), but it still stores the string quite happily.

- The modem includes the " characters in its ID frames.

- The modem doesn't support the **AT+FLID?** form of the command.

Class 2 session parameters

The final setup command we have to cover is the rather cryptic

```
AT+FDDC=VR,BR,WD,LN,DF,EC,BF,ST
```

which sets up session parameters. Familiarity with these parameters is needed for many Class 2 commands and responses, so we discuss them in some detail here. If you are tempted to jump straight into the innards of fax sending, it should be stressed that the Class 2 parameter table (Table 10.2) is central to how Class 2 modems work. The **+FDCC** setup command simply provides a convenient starting point for discussing this topic. We'll get back to phase A of the sending process later.

Table 10.2: Class 2 fax parameters (values marked * are optional and need not be implemented)

Label	Function	DIS/DCS bits	Values	Description	
VR	Vertical	bit 15	0	Normal, 98 lpi	
	resolution	"	1	Fine, 196 lpi	
BR	Bit Rate	bits 11-14	0	2400 bps V.27 ter	
		"	1	4800 bps V.27 ter	
		"	2 *	7200 bps V.29 or V.17	
		"	3 *	9600 bps V.29 or V.17	
		"	4 *	12000 bps V.17	
		"	5 *	14400 bps V.17	
WD	Page Width	bits 17-18	0	1728 dots in 215 mm	
		"	1 *	2048 dots in 255 mm	
		"	2 *	2432 dots in 303 mm	
		bit 34	3 *	1216 dots in 151 mm	
		bit 35	4 *	864 dots in 107 mm	
LN	Page Length	bits 19-20	0	A4, 297 mm	
		"	1 *	B4, 364 mm	
		"	2 *	unlimited length	
DF	Data format		0	1-D modified Huffman	
		bit 16	1 *	2-D modified Read	
		bit 26	2 *	2-D uncompressed mode	
		bit 31	3 *	2-D MMR (T.6)	
EC	Error checking	bit 27	0	Disable ECM	
		bits 27-28	1 *	ECM 64 bytes/frame	
		bits 27-28	2 *	ECM 256 bytes/frame	
BF	Binary File	bits 51, 53	0	Disable BFT	
	Transfer	"	1 *	Enable BFT	
ST	Scan Time/Line	bits 21-23		VR=normal	VR=fine
		"	0	0 ms	0 ms
		"	1	5 ms	5 ms
		"	2	10 ms	5 ms
		"	3	10 ms	10 ms
		"	4	20 ms	10 ms
		"	5	20 ms	20 ms
		"	6	40 ms	20 ms
		"	7	40 ms	40 ms

AT+FDCC is just one instance of a family of commands, reports and responses, including the closely related **+FDIS** and **+FDCS** commands, as well as **+FDCC**. Together, they provide the Class 2 fax modem programmer with something approaching the same sort of control over the T.30 negotiating process as a Class 1 programmer may have by directly accessing the negotiating frames themselves.

Of course, setting up and reporting on the contents of T.30 DIS/DCS frames by a Class 2 fax modem couldn't be handled by manipulating the octets that make it up, as this would force Class 2 application software to handle T.30 bit-fields to determine the session parameters, with all the attendant problems of bit ordering and extend fields. This is contrary to the philosophy behind the design of Class 2, which aims to present a more understandable and intuitive programming interface. Instead, nearly all the essential T.30 parameters are reduced to a string of eight readily understandable named numeric fields separated by commas, as in our example above. The valid values and meanings of the eight individual ordered fields, with the T.30 bitfields to which they correspond, are shown in Table 10.2.

+FDCC

The **+FDCC** command gives a user the same sort of functionality as the option buttons on many top-of-the-range fax machines. Just as you can press a button labelled FINE to toggle between sending a fine resolution (which takes longer) and a normal resolution fax (which is quicker), so you can use **+FDCC** to alter the VR parameter to precisely the same effect. If you want faxes to be sent or received really reliably, you may opt to force the BR bit rate parameter to 0 or 1 instead of allowing the modem to use higher speeds that might be available. A list of possible parameters can be obtained with the test parameter form of **+FDCC** as in the following example:

```
AT+FDCC=?
            (0,1),(0-5),(0-2),(0-2),(0,1),0,0,(0-7)
            OK
```

Referring back to the previous table, this response shows we have a modem that can handle normal and fine resolution, any speed from 2400 to 14400 bps, horizontal page dimensions of 2048 dots in 255 mm and 2432 dots in 303 mm, as well as the standard 1728 dots in 215 mm, either A4 or B4 or unlimited length faxes, either 1-D or 2-D T.4 formats, and all possible scan line times. However, it can't do ECM or BFT, handle small size faxes, handle uncompressed mode or T.6 encoding.

Just as the existence of a FINE button on a fax doesn't mean every fax will be sent in fine resolution, so the capabilities of the modem are not the same as the parameters that will actually be negotiated for any given session. For some parameters, such as speed, the setting cannot be forced through the **+FDCC** setting, as the T.30 protocol is used to ensure that a mutually compatible speed will be negotiated with a remote fax once a connection is made.

For other parameters, such as the resolution and page dimensions, the relevant portion of the T.30 frame sent by the caller is taken directly from how the modem is set up, and the caller isn't negotiating with the other fax as much as it is informing it what to expect; while virtually all faxes and fax modems can handle both normal and fine resolutions, the key point is that with a pre-encoded T.4 fax image, the resolution isn't open to negotiation. If it has been encoded in standard resolution, sending an image as if it were fine would squash it in half.

The read parameter form of the **+FDCC** command can be used to determine if such problems will occur. It allows a user to ask the modem what parameters it intends to use:

```
AT+FDCC?
            0,5,0,2,0,0,0,0
            OK
```

In this example, the modem tells us it will attempt to negotiate to send a normal resolution fax at 14400 bps, with the dimensions 1728 dots across a 215 mm indefinitely length page, using 1-D Huffman coding and 0 ms scan time per line.

In case you hadn't yet worked it out, the final form of the **+FDCC** command is used in our pre-session stage, and is the one that sets the parameters. While a complete range of parameters can be specified, only those needing to be set need have to be included. The rest can be missed out or replace with commas. For example, we could give the following command to the same modem whose parameters we just read with **AT+FDDC?**:

```
AT+FDCC=1,,,,1
            OK
```

The effect would be to change parameters VR and DF from 0 to 1. Instead of a normal resolution 1-D fax, the modem will now negotiate on the basis of sending a fine resolution 2-D fax. We can check this has had the desired effect by asking the modem (using the read parameter form again) what it intends to negotiate this session:

```
AT+FDCC?
            1,5,0,2,1,0,0,0
            OK
```

Note that the Class 2 specification does not specify what defaults a fax modem should use for **+FDCC** parameters. Modems fresh out of the box all seem to default, sensibly enough, to normal resolution, 1-D, 1728 dots across a 215 mm page, with the highest speed and lowest scan line times possible. However, the best approach to programming parameters on Class 2 modems is to use the test parameter form of the command so that we can determine what the possible options are before setting them.

+FDIS and +FDCC

The family of Class 2 commands and responses which includes **+FDCC** all either take or return parameters with this same general form. A proper understanding of their use and meaning is as central to your success in writing or customizing Class 2 based software as the proper use of the basic negotiating frames is to success in understanding Class 1 software. Though there are some similarities, Class 2 fax parameter commands are quite different to the T.30 DIS, DCS and DTC frames, so it's worth continuing with this topic in some detail.

If you look back to the setup portion if our Class 2 fax sending table (Table 10.1), you'll find instances of some of the other commands and responses that reference the Class 2 parameters in a slightly different way. We find an **+FDIS:** report in phase A, and you'll notice in phase B that there is an instance of the related **+FDCS:** report. Both of these are slightly easier to understand than the example of **+FDCC** cited earlier. The **+FDIS:** is followed by a spontaneously generated report of the contents of the DIS frame received by the calling fax as part of phase A, while **+FDCS:** reports on the DCS frame that is subsequently sent out by the caller just prior to training.

+FDCS

The **+FDCS** command is by far the easiest; it simply holds the parameters being used during the current session. If the modem is transmitting, **+FDCS** holds parameters based on the last DCS frame the modem sent, while if the modem is receiving, the parameters are taken from the DCS frame that was last received. At the start of a session, **+FDCS** is set to zero, so the command **AT+FDCS?**, when issued if the modem is on-hook, will always return 0,0,0,0,0,0,0,0.

Once a session has begun, **AT+FDCS?** returns the same values given in the **+FDCS:** report which is spontaneously generated during phase B. **+FDCS** contains no writable parameters, and is read-only.

+FDIS

The **+FDIS** command is far more complex than **+FDCS**. Though its use in phase A to report on a real DIS frame received is quite intuitive, it can also be a little deceptive, as the main use of **+FDIS** is not simply to report on frames received. It also duplicates the function of **+FDCC**, in that is serves to hold, maintain and report on a list of preferred session parameters for the current session.

The naming convention here can actually be quite confusing. The parameters set via **+FDIS** are used by an answering fax, which generates a DIS frame based on its contents. However, the contents of its own **+FDIS** are also used by a transmitting station, which uses them in conjunction with a received DIS frame to generate a DCS frame.

So we have four possible uses of **+FDIS**, all with identical parameter formats. One reports on the DIS frame of an answering fax, while the other three enable software to control the fax parameters for both sending and receiving sessions in the same way as

+FDCC. In fact, in all our examples so far, all the occurrences of **AT+FDCC** could be replaced with **AT+FDIS** with exactly the same responses and effects.

The two commands are not quite identical, though the distinction between them is subtle. There are two user-defined parameter lists held in a fax modem's memory. One is maintained by **+FDIS**, and is used for control of the current session; the other is a master list maintained by **+FDCC**, used to initialize **+FDIS** on power-on and after each fax session.

Whenever a change is made to the **+FDCC** parameter list, the list maintained by **+FDIS** is also re-initialized to match. The reverse is not the case. A change made to the **+FDIS** parameter list does not affect the **+FDCC** list, and has only a temporary effect until the end of the next session, when **+FDIS** is once again re-initialized from **+FDCC**.

Parameter changes made using **+FDIS** are only valid for the current session, while changes made using **+FDCC** are valid for all subsequent sessions from the point at which they are made. It might reasonably be asked what the point of having two such similar commands might be. Certainly, it is better programming practice not to assume any default conditions and to reinitialize the modem explicitly to any state required. Arguably, there is an increased possibility of memory corruption in any device attached to an outside electrical system like a telephone network, and therefore neither **+FDIS** nor **+FDCC** should be assumed to be inviolate.

The Class 2 specifications give no advice on when to use **+FDIS** and **+FDCC**. Though we can assume the **+FDCC** command will take a little longer, as it has more to do, the difference is unlikely to be measurable. If a modem has NVRAM, this could be used this to keep a permanent record of any changes to **+FDCC**. Since NVRAM can be written to only a limited number of times, using **+FDCC** where **+FDIS** would do is probably undesirable. However, most Class 2 modems have no NVRAM for this purpose, so the question is academic.

The only occasion where **+FDIS** is specified as being the correct command to use is when changing parameters within a session. In other words, **AT+FDIS** can be used within a session at any point where phase B is being re-entered. This is something that the receiver can request (by rejecting a page), in which case the sender may decide to drop down a speed. Alternatively, the sender can decide to re-enter phase B at the end of a page, most commonly to change the resolution being used in cases where normal and fine resolution pages need to be sent in the same session.

We return to this use of **AT+FDIS** when on-hook later in this chapter. However, while all Class 2 fax modems support the use of **AT+FDIS** to set parameters before a fax session, some do not support the use of the command in mid-session to change parameters. There is no technical reason for this, merely a deficiency in the content of the modem firmware. Most users don't know and can't tell whether their Class 2 modems can use **AT+FDIS** when off-hook, as their fax software usually takes care of that for them.

This does, however, bring us on to the differences between fax modems.

Five types of modem-dependent quirks

Not all modems behave the same or should be treated the same as far as Class 2 parameter settings are concerned. There are at least five types of these differences. We'll look at this in relation to the fax parameter commands just discussed, but in fact you could write a reasonable length quirks section on virtually any Class 2 command.

Three commands can be used to find out what type of modem is being used: the **AT+FMFR?** enquiry returns an ASCII string that identifies the manufacturer (or OEM); the **AT+FMDL?** enquiry returns a string to identify the model (or chipset manufacturer); while **AT+FREV?** identifies the version or revision level. These commands are fairly well supported, as most manufacturers have more to gain than they have to lose from giving their modems a definite identity, and the programming of the response strings is really pretty trivial to manage. Unlike the various data mode ATIn commands, which are often not comparable between manufacturers and models, the three identification commands **+FMFR**, **+FMDL** and **+FREV** can be interpreted in the same way on all Class 2 modems.

The fact the Class 2 modems are so easy to identify make it possible to write software that takes account of their inconsistencies and incompatibilities, and enable software designers to work around any bugs and handle any differences.

Optional features

The first type of difference is simply because many of the parameters are optional. These differences between modems can usually be catered for by judicious inspection of test parameter responses such as **AT+FDIS=?** and subsequent handling of the different cases.

The original Class 2 draft stated that as "T.30 does not provide for the answering station to specify all speeds exactly using a DIS frame ... implementation of some BR codes by an answering DCE is manufacturer specific". I'm not exactly sure what that means, but I suspect it refers to ambiguities in the way that Class 2 parameter set was designed. This brings us to our second set of differences.

Manufacturers' discretion

Let's examine the **AT+FDCC=,3** command to set the speed to 9600 bps a little more closely. On a modem that handles both V.17 and V.29 modulations (any modem that works at 14400 bps) there is ambiguity in the attempt to restrict the fax speed to 9600 bps. There are currently two fax modulation schemes which support 9600 bps: V.29, which offers 9600 bps with fallback to 7200 bps; and V.17, which offers 14400 bps with fallback to 12000, 9600 and 7200 bps. We pointed out earlier that V.17 is going to be more reliable than V.29 for a 9600 bps connection.

You might want to send long faxes using a 14400 bps fax modem to a destination that also has 14400 bps fax capability, but over a phone line known to give intermittent problems. It may be that most V.17 14400 bps connections you make eventually result in corrupted faxes, but that all the V.17 9600 bps connections are perfect. In this situation, you might consider trying to force the modem to use the slower speed of V.17 9600 bps, as it will result in better reception.

On the other hand, an alternative situation may be that you want to send to a modem that has V.17 capability, but don't want to use V.17 as it is in some way broken. There were a number of early model fax modems and machines where the implementation of V.17 was less than satisfactory, and slight incompatibilities resulted in worse connections than the more established V.29 modulation. In this situation, you would want to force the modem to send a fax at V.29 9600 bps.

However, there's no way of telling a Class 2 fax modem if you have a preference for V.17 rather than V.29. The command set allows you to restrict the speed to 9600 bps, but you can't do anything to specify the modulation to be used.

This is primarily a problem for the transmitter. If you re-read Chapter 6, you'll see that one of the side-effects of the way modems negotiate is that, with the sole exception of 2400 bps, a receiver can't specify its capabilities in terms of speed, only in terms of modulation scheme. It can inform a transmitter that it can handle V.27 ter only, or V.29 only, or both, or V.27 ter and V.29 and V.17 also. So, like a transmitting Class 2 fax modem, a receiver cannot explicitly tell a transmitter to use V.17 but only at 9600. In view of this, it makes sense to regard a restriction to 9600 bps only as equivalent to a restriction to V.29 when used by a sending Class 2 fax modem.

The important point is that there are areas where the Class 2 specification is deliberately loose, so the same command (such as **AT+FDCC=,3**) could result in different types of connection on different modems, with neither of them being wrong.

Other areas where modems behave differently because manufacturers are free to decide what to do include the algorithm used to determine whether a train is successful, and the strategy the modem adopts if it fails to train. I suspect that some modems report training as being successful under pretty dubious conditions, while others seem to drop down in speed almost immediately.

Different interpretations

The third type of difference is due to the same sort of imprecision seen when we discussed the **AT+FCLASS** command. This again is something the Class 2 specification permits. The ordered list of possible parameter values returned by the **AT+FDCC=?** command could be separated either by spaces or commas. A continuous range of values could be presented by showing the lower and upper limits separated by a hyphen. Subparameter ranges of values are additionally supposed to be enclosed in brackets.

The following examples of parameter reports are given by four modems with chipsets from different manufacturers, all of which are sitting on my desk:

```
(0,1),(0-5),(0-2),(0-2),(0,1),0,0,(0-7)          (Rockwell)
(0,1),(0-3),(0,1),(0-2),(0-2),(0),(0),(0-7)      (Sierra)
(0-1),(0-3),(0-3),(0-2),(0-2),(0),(0),(0-7)      (Exar)
(0,1),(0-5),0,(0,2),(0-1),0,0,(0-7)              (AT&T)
```

All responses were preceded and followed by <cr><lf> sequences (not shown). The responses were all followed by OK, around which the Rockwell, Sierra and AT&T modems placed additional <cr><lf> sequences, leading to blank lines before the OK. The Exar modems missed out the <cr><lf> before the OK, so no blank lines were evident.

The Rockwell and AT&T-based modems regards a bare 0 as a single value, while the Sierra and Exar-based modem regards it as a range of one. Values of 0 and 1 are interpreted by the Exar-based modem as a range of two, but by the Rockwell and Sierra-based modems as individual possibilities. Meanwhile, the AT&T-based modem can't decide whether possible values of 0 and 1 are a range or not.

All these responses as well as other possibilities are quite legally within the Class 2 specification. The commas could be replaced by spaces, and ranges such as (0-2) could be replaced by (0,1,2) quite legally.

Partial implementations

The fourth type of variation that has to be taken into account can be illustrated by looking at the read parameter forms of various commands. Many manufacturers simply don't implement all of these, and you'll see a command sequence like the following:

```
AT+FDIS=?
            (0-1),(0-3),(0-3),(0-2),(0-2),(0),(0),(0-7)
            OK
AT+FDIS?
            ERROR
```

This was the Exar-based modem, of course. The lack of a blank line before the OK and the reporting of possible VR values of 0 and 1 as a range (0-1) are the giveaways. This isn't a serious omission: the read parameter form of the various commands isn't necessary given the fairly copious reporting that goes on in a Class 2 session.

On a more general note, everyone should be aware that partial implementation of the Class 2 specification is the rule. There isn't a single modem that implements every feature, or even one that gives every report.

Some Class 2 features are listed as optional. Other genuinely optional features include error correction, adaptive answering, quality checking, data conversion and transition to Class 1. However, it would be untrue to say that most omissions from the command set of the average Class 2 modem are because the relevant features are optional. The reality is that manufacturers have always felt free to pick and choose which commands to implement, and if a feature isn't essential then no matter what the specification says, it will be regarded as optional. The criterion seems to be whether the modem will talk to other fax machines, not whether it conforms to the draft Class 2 specification. It is difficult to say that manufacturers have their priorities wrong. The fact that Class 2 has such a rich set of commands makes it easy to leave bits out and still retain a high level of functionality, which is usually superior to most low-end fax machines.

As well as contributing to the general impression of anarchy that pervades the Class 2 market, this presents something of a problem to anyone writing a book. I've taken the pragmatic view that where a feature is one that nobody implements, it gets mentioned in the main body of the text if it's relevant to something that is discussed anyway. Otherwise, it gets a summary mention in the NOTES directory on the accompanying disk, where it is included on the basis that the fact that nobody has implemented a command in the past doesn't mean that it isn't going to be implemented in the future.

Bad implementations

It's difficult to be charitable about bugs, but they do exist. Often, it's a puzzle to know what to make of them. Consider the following sequence captured from the Exar fax modem used as an example earlier in the chapter:

```
AT+FDCC=?
            (0-1),(0-3),(0-3),(0-2),(0-2),(0),(0),(0-7)
            OK
AT+FDCC?
            0,1,0,2,0,0,0,5
            OK
AT+FDCC=,3
            OK
AT+FDCC?
            ,,3,0,2,0,0,0,5
            OK
```

The initial **AT+FDCC=,3** command should have set the modem to default to its maximum speed of 9600 bps for the next session, as it had previously been set to 4800 bps. On checking to see what the parameters are via **AT+FDCC?**, it looks as if something has gone wrong, as the initial VR parameter has vanished from the list.

I tried again using the full initialization sequence of

`AT+FDCC=0,1,0,2,0,0,0,5`

and everything was back to normal. I first thought the modem simply didn't understand a set parameter command sequence which included initial defaults, but on looking more closely, I noticed that the VR parameter hadn't actually vanished; it had been replaced by a comma.

The two following test sequences revealed what was happening:

```
AT+FDIS=RUBBISH!
                OK
AT+FDCC?
                R,U,B,B,I,S,H,!
                OK
AT+FDCC=GARBAGE!
                OK
AT+FDCC?
                G,A,R,B,A,G,E,!
                OK
```

The modem was not actually parsing the parameter list, but was storing the first eight text characters it came across in an eight-character memory store. As values were not checked, any characters were acceptable, including silly ones. The modem was able to accept and act on correct values, so presumably it checked the validity of parameters while composing a DCS frame once phase A was complete. This was fortunate, as the 8-byte scratch area used by `+FDCC` wasn't reserved for this use exclusively. The same area was used by both `+FDIS` and the unrelated and supposedly read-only command `AT+FAXERR=?` (which should produce a report of valid hangup codes). Admittedly, this code is unsupported by Exar chips, and isn't often used by software even where it is available, but the corruption is still worrying.

The inability of the modem to recognize legal settings of something as important as the Class 2 parameter set where the command contains defaults is a potentially serious bug. It makes universal initialization strings like this one we suggested earlier unusable:

`AT+FDIS=0,,0,2,0,0,0,0`

Of course, using `AT+FDCC=?` to determine the capabilities avoids the need to use defaults, so the bug is neither fatal nor difficult to write around; but a bug it most certainly is.

Lest anyone think I'm picking unfairly on Exar, it's worth mentioning that my AT&T modem gives an **ERROR** response to commands such as `AT+FDCC=,3` which have missing parameters, while some Sierra bugs can be found in the NOTES directory on the accompanying disk. Rockwell only escape because their success in marketing their modem chipsets has led to bugs in their firmware being duplicated as a *de facto* standard by other manufacturers.

Code for setting up fax modems

Listing 10.1 contains a code fragment showing how a Class 2 fax modem may be set up. The procedure is simply to issue the **AT+FDCC=?** command to determine the modem's capabilities, and then select those offering the fastest performance. The problem is how to parse the parameter list correctly given the problems outlined above.

```
process_parameters ()
{
        char result[64] ;
        unsigned char resolution, bps, compression, field, inbracket ;
        export ("AT+FDCC=?\r") ;
        import (result,64,35) ;
        wait_for ("OK",2) ;
        for (x=0,field=0,inbracket=0 ;; x++)
        {
          if (isdigit(result[x]))
          {
              if (field<1) resolution=result[x] ;
              else if (field<2) bps=result[x] ;
              else if (field<5) compression=result[x] ;
          }
          if (result[x]==0) break ;
          if (result[x]==0x28) inbracket=1 ;
          if (result[x]==0x29) inbracket=0 ;
          if ((result[x]==0x2c)&&(inbracket==0)) field++ ;
          if (field==5) break ;
        }
        strcpy (result,"AT+FDCC=0,0,0,2,0,0,0,0\r") ;
        result[8]=resolution ;
        result[10]=bps ;
        result[16]=compression ;
        export (result) ;
        wait_for ("OK",1) ;

}
```

Listing 10.1: Setting up a Class 2 fax modem

The solution used in the fragment is to assume that the rightmost digit of each field contains the parameter with the highest performance, and that all fields are separated by commas. Commas inside brackets are ignored for the purpose of deciding when a field has ended. This approach can quite happily deal with lists of the form (0,1),(0-5),(0-2),(0-2),(0,1),0,0,(0-7) which we saw earlier. The code here is only interested in resolution, speed and compression method, but the method could be amended for other parameters if necessary.

After the code in Listing 10.1 has been executed, we can start sending or receiving data. This seems like a good cue to return from our discussion of Class 2 session parameters back to our analysis of fax sending with Class 2 modems.

Phase A

This is quite straightforward once you remember what is actually happening in terms of the T.30 session protocol, though there are a number of interesting points typical of the way in which Class 2 modem firmware is implemented.

The modem dials the remote fax, which answers the call. At this point, all Class 2 modems generate a CNG tone. A CED from the answerer is neither reported nor looked for. The modem does look for is a standard T.30 preamble, and when it detects this it reports it with the **+FCON** response. If it cannot detect a remote fax, a Class 2 modem will hang up the line.

The time allowed for this to happen usually depends upon the setting in register S7, but this isn't part of the Class 2 specification. The T.30 recommendation allows about 35 sec from the time the call is answered before a caller gives up. Most modem manufacturers normally start timing the S7 register from the point at which the call is dialled, not the point at which it is answered. However, one unfortunate effect of the lack of detail that modem manufacturers offer users on the precise implementation of any of the fax specifications is that it isn't usually possible to be sure what normal behaviour is supposed to be.

In any event, once the preamble is detected, the initial negotiating frames from the answering fax are then received and reported on. There are three possible types of frame, and while there is a superficial similarity in the three possible reports, the information contained in the frame is actually treated very differently in each case:

- NSF frames are usually ignored, but can be reported as a series of hexadecimal digits in the order in which they were received. Each value should be prefixed with a space, and the whole line should start with **+FNSF:** as a leading identifier.

- CSI frames are reported using **+FCSI:** followed by the ID in normal character form. Applications software doesn't have to do anything with the contents of the CSI frames, but usually the identification string is stored and used as part of a fax log kept on disk.

- DIS frames are the only preliminary ones that have to be received. The correspondence between the DIS frame and the **+FDIS:** report is very loose.

One bit of the received DIS frame isn't included in **+FDIS:** but has its own special reporting method. This is bit 9, which is set to show that the answering fax machine has documents available for polling. The modem gives a separate **+FPOLL** response when this is detected, and the caller has the option either to send or receive if the **+FPOLL:** report is given. Most fax modems neither produce the **+FPOLL:** response nor have any polling capability.

The Class 2 specification states that the DIS frame should be followed by the final **OK** result code. Once this arrives, phase A is over and the fax modem can move on to phase B.

Phase B

Recall that phase A began with the issuing of the ATD dial command and proceeded without intervention until the fax modem had received the initial negotiating frames from the answering machine, ending with the final **+FDIS** report. At this point, the sending fax modem stops automatic operation and awaits further instructions. There are a number of reasons why the caller can't decide what action to take until after phase A. Some examples of phase A responses that affect the subsequent progress of the session are as follows:

- If a confidential fax is being sent, the caller may need to verify the ID returned in the **+FCSI:** report before proceeding.

- If the **+FPOLL** response is seen, the caller may decide to poll and receive a fax rather than send one (provided the modem supports polling).

- If the caller wants to send a precoded fine resolution fax (and had previously set **AT+FDIS=1** or **AT+FDCC=1** so that the fax modem knows about this), it is essential to check the **+FDIS:** report in case the receiver can only handle normal resolution. If this is the case, the transmitter may want to abort the session instead of sending a fax that will be received in a s-t-r-e-t-c-h-e-d form. Similarly, if the receiver can only handle A4 pages, and the sender has a longer fax to send, there is a mismatch which may be something the software cannot handle.

- It is possible on some modems for the session parameter previously set up by **AT+FDIS** or **AT+FDCC** to be altered at the last minute. For instance, if the caller has both fine and normal resolution versions of the same image, it is not unreasonable to delay a decision as to which will be sent until after the receiver's DIS frame has been received. If the receiver can only handle 4800 bps (there are fax modems and machines that can't receive except using V.27 ter), then it makes sense to send the normal resolution image, and it is still possible to issue **AT+FDIS=0** at this point as no negotiations have yet taken place. On the other hand, it makes sense to send the fine resolution version to a receiver capable of 14400 bps. There are models which don't like **AT+FDIS** commands issued once the modem is on line, though, so this last minute adjustment isn't always possible.

The only case where the information supplied by the modem during phase A can safely be ignored is when all the caller wants to do is to send a 1-D coded A4 page to whatever machine answers the phone. The 1-D coded A4 page is the lowest common denominator of the fax universe which every working fax can handle.

The AT+FDT command

The basic fax sending command is **AT+FDT**, instructing the modem we want to send a fax, and that it should proceed with the necessary pre-message negotiations. The first use of this command is at the start of phase B, causing the modem to go to the start of phase C. It handles all the fax negotiations and training automatically, ending with entry into whatever high-speed transfer rate is decided upon.

The fax modem begins phase B by sending a V.21 channel 2 preamble. It continues by optionally sending the ID of the transmitting station in a TSI frame. The ID sent is the one that has previously been programmed using the **AT+FLID=** command. Remember that this is an optional frame: if the ID string is empty then no ID will be sent at all.

Following any TSI frame (or following the sending of the V.21 carrier if no TSI is sent), the fax modem continues by sending a DCS frame. The content of the DCS frame is arrived at through the standard negotiation procedure of comparing the received DIS frame with the parameters contained in the local FDIS store; as described above, these can be either set directly using the **AT+FDIS=** command, or are copied from the default modem parameters as reported by **AT+FDCC**. The DCS frame is arrived at using the best fit for each of the available parameters as follows:

- Fine resolution will be used only if both FDIS:VR=1 and DIS bit 15=1. Otherwise normal resolution will be used.

- The highest bit rate common to both FDIS:BR and DIS bits 11–14 will be used. As noted earlier, it is not possible to force a particular modulation scheme explicitly.

- Where FDIS:WD and DIS bits 17–18 and 33–38 specify different widths, the one closest to the default of 1728 pixels on a 215mm line will be used.

- Where FDIS:LN and DIS bits 19–20 specify different lengths, the shorter will be used. This is quite clear, as unlimited length implies that B4 is possible and B4 implies A4 is possible. A4 capability is mandatory.

- The availability of all settings for scan line times is compulsory, but users should note that setting a high value for the scan line time can negate the benefit of having a high speed modem. If a scan line time of 40 ms is used, then no matter what speed a fax is sent at or what its contents are, it must take at least 45 s to send a normal resolution A4 page with 1143 scan lines in 297 mm, while a fine resolution page will take over 90 s and a superfine page over 3 min. Wherever possible, scan line times of 0 ms should be used, as this is most efficient.

Following transmission of the DCS frame, the fax modem automatically sends the TCF (training check frame) consisting of 1.5 s of 0 bits sent at the speed specified in the DCS frame, and it then returns to the 300 bps negotiating speed and waits for a response. There are two possible responses:

- If the receiver decides that the training sequence indicates the connection is suitable for use at the negotiated speed, it responds with a CFR (confirmation) frame. The transmitting modem switches over to the higher speed negotiated for transmission and reports this with a **CONNECT** message.

- If the training sequence was not received properly, then a FTT (failure to train) frame is received instead. The transmitter then has to decide whether to drop to a lower speed, try again at the same speed, or abort the session. As described earlier, both the algorithm a receiver uses to determined the adequacy of the training sequence and the strategy adopted by the transmitter when an FTT is received are manufacturer-dependent.

Once the **CONNECT** message appears, phase C has begun, and the modem establishes the high-speed carrier used for transferring the fax data itself.

Phase C

Phase C consists of sending the encoded data to the fax modem. The rules for doing this were described earlier in the chapter. While the Class 2 specification states that all transmitted data must conform to the ITU T.4 recommendation, it does not state what action (if any) should be taken if the data fails to adhere to this. Many fax modems don't seem to do any checking of phase C data at all, and send everything on transparently until a **<dle><etx>** occurs.

One point to remember is that, even when the **<dle>** shielding characters are taken out, the data stream is still not sent on to the remote fax transparently. The fax modem has to detect the end of each scan line by looking for EOL sequences, and then ensure than the minimum scan line time that has been negotiated is observed by padding out each line as necessary with 0 bits.

The data should not be sent out as soon as the **CONNECT** message is seen. Instead, the fax modem indicates its readiness to accept T.4 data by returning an XON character (ASCII DC1, decimal 17, hexadecimal 11).

The separation of the end of phase B from the start of phase C is actually of less consequence than the fact that it is the XON character in particular which is used as a phase C prompt to start sending data. This is a problem because XON is more frequently used as a flow control character than as a prompt. The XON/XOFF flow control protocol for transmitted data is a required feature of all fax modems. Hardware flow control using CTS and RTS lines is also optionally provided, but this is as an additional alternative to XON/XOFF rather than as a replacement.

Many implementations of XON/XOFF flow control at operating system level are not transparent, so XON characters are not passed through to the application but are swallowed up by the communications handler drivers. The XON/XOFF handler included with Microsoft Windows is an example of this type of communications handler. Users wanting to use Class 2 fax modems on such systems have to find a way around this problem; if the initial XON is swallowed up somewhere and the application never sees it, it may never be able to start phase C.

There are at least five possible ways of resolving the conflict between the requirement of the fax application software to see the initial XON, and the requirement for adequate flow control during phase C:

- Use hardware flow control instead of XON/XOFF. The usual method is to activate CTS/RTS handshaking, though strictly speaking only the CTS line is needed for output flow control. This is not always a trivial exercise. You must have a modem that supports CTS handshaking, and you must also ensure that this option is activated after entry to fax mode via **+FCLASS=2**. It also requires an operating system handler that supports CTS handshaking, and that this option is activated in the communications handler. The hardware must be capable of generating the appropriate modem status interrupts reliably, and finally, if an external modem is being used, a reliable and correctly wired cable from the serial port to the modem is needed.

- The next method is to switch off XON/XOFF flow control at the operating system level and avoid provoking flow control by means of the **AT+FBUF?** command. We explored this method in Chapter 8, so we shan't repeat it here. However, it doesn't work on all modems: some don't implement **+FBUF** at all, while others have such small buffers that using this method is impractical.

- The third method also involves switching off XON/XOFF at the operating system level, and also doesn't work on all modems. It is simply to handle XON/XOFF flow control at the application level rather than hand it over to the operating system. If you are using a modem with a large buffer and generous thresholds, this may well be a feasible alternative. For example, we saw in Chapter 8 that modems are available with XOFF and XON thresholds set 13107 and 3276 bytes from buffer full and buffer empty levels, respectively, and that the XON threshold is the more critical. Where a fax is being sent at the top speed of 14400 bps, the applications software must be able to resume sending within 1.82 s of an XON being sent. It would be most unusual for an application to be denied access to the CPU for that length of time, and to be frank, an operating system that behaved in such a way is probably unsuitable for real-time communications.

● The fourth method is by far the simplest, but is a definite kludge. It is simply to ignore the XON message altogether. A timeout is set on receipt of the **CONNECT** message, and T.4 data is sent once this timeout expires. The optimum length of the timeout varies from modem to modem: it must not be long enough to provoke an error, but must be sufficient to ensure that the XON has been sent. I've found that a time of around 1.5 s works for most modems, but don't be surprised if your particular model needs a different delay.

● The last method is the best one for situations where hardware handshaking is not possible. It consists of leaving XON/XOFF output flow control disabled until the initial XON is received, and then switching it on just before phase C data is sent. Once the final **<dle><etx>** has been sent and acknowledged, XON/XOFF can be disabled once again.

Bit ordering in phase C

We've discussed the ordering of the bits within bytes many times before. However, the most complicated bit ordering problem in the entire fax universe is reserved for Class 2 fax modems.

Recall that the bit ordering of bytes as read is from left to right (MSB→LSB). In contrast, the order in which the bits in a byte are transmitted over an asynchronous connection is from right to left (MSB←LSB). The TIFF file format makes special provision for this difference in the way fax images can be stored by means of the FillOrder tag, with FillOrder=1 being used for the case where the first bit (MSB) in the first byte is also the first bit in the image. In contrast, FillOrder=2 refers to the case where the last bit (LSB) in the first byte is the first bit in the image.

We also saw that when using Class 1 modems, if the bits comprising a fax image are stored as normal MSB->LSB bytes using FillOrder=1, all bytes have to be inverted before transmission and after reception (see page 215).

The TR-29.2 committee, who originated the Class 2 fax modem standard, had the good idea of relieving the applications software of the burden of having to handle the inversion by having a simple command, **AT+FBOR**, which has the sole function of telling the modem what sort of bit ordering the application software intends to use. While the bit ordering of the control characters used such as `<dle>` and `<etx>` is not affected by +FBOR, the same setting should apply to both sending and receiving data. Unfortunately, this apparently simple concept has turned out disastrously wrong.

The first problem lies in the way the two options are defined. The Class 2 specification states

'There are two choices:

direct: the first bit transferred of each byte on the DTE-DCE link is the first bit transferred on the PSTN data carrier.

reversed: the last bit transferred of each byte on the DTE-DCE link is the first bit transferred on the PSTN data carrier.'

The command **AT+FBOR=0** selects direct bit ordering for phase C data, while the command **AT+FBOR=1** selects reverse bit ordering for phase C data.

It isn't immediately obvious that these definitions result in a TIFF FillOrder=1 image needing to be sent using FBOR=1 reversed bit order, while a TIFF FillOrder =2 image should be sent using FBOR=0 direct bit order. Arguably the usage is counter-intuitive. The situation is summarized in Table 10.3.

Most would agree with the TIFF Class F document, which states that 'facsimile data appears on the phone line in a bit-reversed order relative to its description in CCITT Recommendation T.4.' However, what TIFF class F calls a bit-reversed order is what the Class 2 document calls a direct bit order. There is clearly a lot of scope for getting the settings muddled up.

This brings us to what is probably the most notorious bug in the fax modem universe. Because the **+FBOR** command is not universally implemented, it makes sense to write Class 2 software to cater for the default. This is supposed to be **+FBOR=0**. In other words, direct bit ordering is used for all data received and transmitted. However, most Class 2 fax modems default to *direct* bit ordering when *transmitting* faxes but default to *reversed* bit ordering when *receiving* faxes.

Table 10.3: Bit orderings in Class 2

First	7	6	5	4	3	2	1	0	Last	First	0	1	2	3	4	5	6	7	Last
MSB→LSB										MSB←LSB									
Order as read										Order as transmitted									
TIFF FillOrder=1										TIFF FillOrder=2									
Class 2 reversed bit order										Class 2 direct bit order									
FBOR=1										FBOR=0									
Default for class 2 reception										Default for class 2 transmission									

The Class 2 fax modem default is simply not consistent, so no matter what bit ordering the application software decided to use for storage of fax images, the bits in each byte would have to be reversed for either sending or receiving data.

Whatever the origin of this bug, which is usually blamed on Rockwell, the reason for its continued existence is that virtually all Class 2 fax software is written around it. The alternative (using **+FBOR**) is not reliable, as **+FBOR** is not universally implemented.

Ending phase C

The T.4 recommendation states that the end of phase C should be marked by sending six consecutive EOL codes, known as the RTC (return to control) sequence. When transmitting with a Class 2 modem, the RTC should not be added to the end of the data stream. Indeed, a modem is supposed to strip out repeated EOL codes, on the basis that some fax machines treat two EOLs as an RTC.

Instead of sending an RTC, a transmitter that has completed sending T.4 data indicates this by appending a **<dle><etx>** sequence, as described earlier. Once the **<dle><etx>** is detected by the fax modem it responds with an OK message.

The fax modem ought not to end phase C by sending the RTC as soon as it detects the **<dle><etx>** sequence. Instead, it should prepare and hold onto the RTC without transmitting it. The modem then waits for the next command (which the OK response should have triggered) and, meanwhile, transmits 0 fill bits as padding if it has no data left to send.

Starting phase D

Once the **<dle><etx>** has been sent to the fax modem after the end of the T.4 data and the OK prompt is returned in response, the fax modem has to be told how to proceed with phase D. Although the fax modem itself is still clinging stubbornly to the end of phase C by padding out the last line, and still has to send the RTC sequence and return to negotiating speed, we have to all intents and purposes entered phase D of the T.30 session protocol.

In practice, phase D is usually straightforward. In our Class 2 fax sending table above, the transmitter simply sends an **AT+FET=0** command to the modem if it wants to send another page; this causes the modem to send the RTC, return to negotiating speed and send the T.30 MPS (multi-page signal) phase D command. Alternatively, if there are no more pages in the document, the transmitter sends an **AT+FET=2** command to the modem which triggers the same sequence of events, with the difference that the T.30 EOP (end of procedure) command is sent instead. In both cases, the modem then turns the line around and waits for a response from the receiving fax.

Table 10.4: Type of Class 2 post-message commands

	T.30 frame	Meaning
AT+FET=0	MPS	another page follows
AT+FET=1	EOM	another document follows
AT+FET=2	EOP	no more pages

When the response arrives, it is conveyed in the form of an FPTS: (page transfer status) report, which contains a single numeric parameter. The most common report is FPTS:1, indicating that the receiver has sent an MCF (message confirmation) frame as a notification that the page has been successfully received.

For some reason, the Class 2 specification refers to the phase D post-message commands as 'page punctuation'. The presentation of the **AT+FET** command in the Class 2 documentation is far more complex as it can take up to 16 different parameters. However, ten of these apply only to error-correction mode, which is not widely supported. As most fax modems only support non-ECM operations, the responses specific to error-correction mode are omitted from Table 10.4. We also omit the three PRI-Q versions of **+FET**. (I haven't seen any Class 2 software that generates any of the PRI-Q (procedure interrupt request) messages, and I've never found occasion to use them myself.) Once we ignore these we are left with only three phase D commands, for another page, another document, and no more pages. We show only the normal forms of the **+FET** command, together with the relevant T.30 post-message commands that they trigger. The first and last of these commands are very common, and we used them in Table 10.1.

The **AT+FET=1** command, which the Class 2 specification describes as being used to send another document, needs a little more examination. The implication of the description is that if a sender wants to send a number of documents, the first one of which is four pages long, then the first three pages should end with the **AT+FET=0** command and the last should end with the **AT+FET=1** command, following which the first page of the next document should be sent. A further implication is that the receiver can tell from the fact that the **AT+FET=1** command is used before page 5 that the first four pages belong together and page 5 does not.

However, a closer look reveals that the real difference is that **AT+FET=0** sends an MPS frame while **AT+FET=1** sends an EOM frame. The only difference between the two frames is that an MPS implies a return to the beginning of phase C on receiving a response, while an EOM implies a return to the beginning of phase B and the renegotiation of the session. All a receiving fax knows is that a session consists of a series of pages, and the concept of a session being divided into documents is simply not part of the T.30 conceptual framework.

The only point of returning to phase B rather than phase C in the middle of a transmission is when a renegotiation of the session parameters really is needed.

As we have intimated already, this can be accomplished in mid-session by sending a new **+FDIS** command when on-line. For example, if a document which mostly contains handwriting has one page with newsprint on, it makes sense to send the page with newsprint in fine resolution and the rest of the fax in normal resolution. The procedure for doing this would be as shown in Table 10.5.

Table 10.5: Use of +FET=1 to change parameters between pages

AT+FET=1		different resolution to come, send EOM
	+FPTS:1	MCF detected
	OK	phase D complete
AT+FDIS=1		change VR parameter to fine
	OK	modem acknowledges change
AT+FDT		now back to phase B for next page
	+FDCS:<codes>	new DCS has been sent
		training OK
		phase B complete
	CONNECT	high-speed carrier detected
	<xon>	ready to start sending data

Note that the **+FET=1** command has nothing to do with whether a page belongs in one document or another, and its only meaning is that there will be a renegotiation of parameters before the next page. In the above example, the fine resolution page was part of the same document as the other pages.

Anyone misled by the wording of the Class 2 recommendation who decides to issue the **AT+FET=1** command between successive documents in a session where those documents have the same resolution, coding method and page dimensions is simply wasting their time in renegotiating parameters that haven't changed.

Phase D responses

We have already seen that phase D responses are conveyed in the form of FPTS: reports. While eight possible values are returned, some of these only apply to error-corrected mode, and so aren't dealt with here. The remaining FPTS: reports are shown in Table 10.6.

This is the set that most Class 2 modems claim to support. By far the most common of these is the **+FPTS:1** page OK report.

Most faxes don't do much in the way of quality checking, and are happy to report all received pages as being good. However, those systems that do check received data and evaluate it for T.4 formatting errors can use the RTN or RTP frames to force the session to resume with phase B rather than phase C, giving

Table 10.6: Class 2 phase D FPTS reports

Report	T.30 frame	Meaning
+FPTS:1	MCF	Page received OK
+FPTS:2	RTN	Page bad : renegotiate/retrain
+FPTS:3	RTP	Page good : renegotiate/retrain
+FPTS:4	PIN	Page bad : interrupt requested
+FPTS:5	PIP	Page good : interrupt requested

the transmitter the chance to drop to a lower speed. The **AT+FDIS=** command with an altered BR parameter can be used to change speed after a retrain request. The syntax is almost identical to that used earlier for changing the resolution after sending the **AT+FET=1** command. However, it is wise to use the command **AT+FDCS?** to check on current session parameters, and reset the VR resolution and BR bit rate parameters. The alternative is to use the form with a leading comma, indicating the resolution remains unchanged. However, we noted earlier that some modems are a little peculiar about **AT+FDIS=,BR** sequences.

Whatever the transmitter decides, a new TCF training check frame is part of phase B, so the receiver too has the opportunity to demand a lower speed. Unless the transmitter wants to issue a revised set of **+FDIS** parameters, the transition back to either phase B or phase C is handled automatically by the modem. The controlling software simply has to remember that FPTS codes of 2 or 4 require the same page to be resent, while codes of 0, 3 and 5 allow the session to proceed normally.

If a report of **+FPTS:4** and **+FPTS:5** requesting a procedure interrupt is received in response to one of the PRI-Q **AT+FET** commands (4, 5 or 6), the modem will assume that an interrupt really is required, and will suspend the session. It signals this by generating the **+FVOICE** message, indicating that the telephone handset should now be picked up. If the session is to be resumed, it has to be restarted with a set of basic **ATD** and **ATA** commands sent to the transmitter and receiver, respectively. We've already pointed out that this type of manual intervention is the last thing most fax modem users want. Many modems don't support this type of operation, and lack the capability to generate procedure interrupts or **+FVOICE** messages.

The PIP and PIN frames that cause the **+FPTS:4** and **+FPTS:5** reports can also be generated spontaneously by a receiver. In this case, a transmitter is free to ignore the interrupt request, but should resume the session at the start of phase B. This is handled automatically on issuing another **AT+FDT** command. In fact, there seems no other way to handle PIN or PIP frames, as there is no Class 2 command that enables a return to voice communications.

The fax modem will exit phase D and enter phase E automatically if it has been sent the **AT+FET=2** command (indicating that the last page has just been sent) and the receiver has issued a page confirmation. In these circumstances, phase E is entered directly after the modem issues its **AT+FPTS:1** report.

In all other circumstances, the **AT+FDT** command is a pretty ubiquitous method of exiting phase D. A fax modem always keeps track of where it is in a session and what valid actions can be taken, and can be relied on to know that the **AT+FDT** command can imply different courses of action in different circumstances.

After an **AT+FET=0** command and a subsequent **+FPTS:1** report, the modem will always re-enter phase C directly. However, after an **AT+FET=1** command, or after any **+FPTS** report apart from **+FPTS:1**, the **+FDT** command causes the modem to behave in much the same way as it did when the same **+FDT** command was issued in phase B. The modem in fact returns to exactly that point; it generates and sends a new DCS frame and **+FDCS:** report, it sends a new TCF training check frame, and then waits for a confirmation before moving to phase C.

Phase E and other disconnections

A transmitter ends a session in phase E by first sending a DCN disconnect frame, and then by hanging up. As we've just seen, the most usual instance of this is when a transmitter sends the **AT+FET=2** command indicating that the last page has just been sent, and the modem responds with an **AT+FPTS=1** report to say that the receiver has issued a confirmation. You don't need a command to enter phase E in this context, as if you are using a Class 2 fax modem, it's an automatic consequence of receiving a confirmation for the last page of a session.

However, according to T.30, any type of call release constitutes phase E, and there are many other possible causes for a disconnection, all of which are to some extent the result of an error. Even if there is no error, a transmitter may not want to proceed with a session. For instance, if a receiver cannot handle a 2-D coded fax, but the transmitter doesn't have either a version coded one-dimensionally or any facilities for on-line conversion, it would be quite reasonable to decide not to send the fax at all.

The command to proceed unconditionally to phase E is **AT+FK**. It is available both to Class 2 transmitters and receivers, and causes a fax modem to send a DCN frame and disconnect the line as soon as possible. There is no point in sending a DCN frame if it won't be recognized by the remote machine; the fax modem waits until a suitable point in the session before terminating. When a modem has generated its **+FCDS:** report and entered phase C, it is a good idea for software that wants to abort a session because it doesn't like the **+FDCS** parameters to offer a little help in getting out of phase C, by sending an immediate **<dle><etx>** and waiting for an **OK**, before sending the **AT+FK** command.

While most fax modems will respond to an ATH command when **+FCLASS=2** and will hang up immediately, this procedure is not recommended, as the DCN frame will not be sent. This can cause problems for the fax at the other end of the line; it wouldn't be entirely fair to expect every fax to operate in a completely bug-free manner. Unexpected disorderly disconnections are basically bad manners.

Whether a disconnection is the result of a session ending normally or not, a fax modem always tells you what it thinks the cause of the hangup is via the **+FHNG** report. A normal disconnection generates either a **+FHNG:0** report or a **+FHNG:00** report; leading zeros are optional. The **+FHNG:** report is always followed by a final OK.

In much the same way as the contents of the **+FDCS:** report are available for later inspection via the **AT+FDCS?** query, so also the code returned by **+FHNG:** is kept available for later inspection. However, for some reason the command to obtain the code isn't **AT+FHNG?**, which isn't a valid command. The **AT+FAXERR** command is used instead; the code held by **+FAXERR** is retained until the next dial attempt begins.

Since the complete list of **+FHNG** hangup status codes is relevant to both transmission and reception of faxes using Class 2 modems, we postpone discussion of them until we've dissected receiving faxes in the same way as we've dissected sending faxes.

Sample code for sending a single page with a Class 2 modem

The fragment in Listing 10.2 shows how a single page fax can be sent with a Class 2 modem. The process parameters function referred to is left as an exercise, but should be based on the code in Listing 10.1. The actual phase C code is similar to that used when sending with a Class 1 modem, and assumes a file pointer **faxfile** positioned at an image whose size is held in **bytecount**.

```
int sendpage (FILE *faxfile,long int bytecount)
{
            unsigned char k ;
            char faxbuf[512] ;
            char result[64] ;
            int x, i=0 ;
            export ("ATD") ;
            export (phone_number) ;
            export ("\r") ;
            for (;;)
            {
                if (!(import (result,64,100))) return (0) ;
                if (strstr(result,"NO DIALTONE")) return (0) ;
```

Listing 10.2: Sending a single page fax using a Class 2 modem

```
        if (strstr(result,"BUSY")) return (0) ;
        if (strstr(result,"NO ANSWER")) return (0) ;
        if (strstr(result,"NO CARRIER")) return (0) ;
        if (strstr(result,"FHNG")) return (0) ;
        if (strstr(result,"FDCS")) process_parameters ;
        if (strstr(result,"OK")) break ;
    }
    export ("AT+FDT\r") ;
    wait_for ("CONNECT",1) ;
    while ((rxchar())!=xon) ;
    do
    {
      if (bytecount > 512)
      {
          i=512 ;
          bytecount-=512 ;
      }
      else
      {
          i=(short)bytecount ;
          bytecount=0 ;
      }
      fread (faxbuf,1,i,faxfile) ;
      for (x=0 ; x<i ; x++)
      {
          k=faxbuf[x] ;
          txchar(k) ;
          if (k==dle) txchar(dle) ;
      }
    } while (bytecount) ;
    for (x=3 ; x ; x--)
      {
      txchar (0x0) ;
      txchar (0x08) ;
      txchar (0x80) ;
      }
    txchar (dle) ;
    txchar (etx) ;
    wait_for("OK",30) ;
    export ("AT+FET=2\r") ;
    wait_for("OK",2) ;
    export ("ATH0\r") ;
    wait_for ("OK",2) ;
    return (0) ;
    }
```

Listing 10.2: (continued)

Table 10.7: Example of a Class 2 fax reception

```
Pre-session modem setup
AT+FCLASS=2
                        OK
AT+FLID="ID string"                     set local ID
                        OK
AT+FDCC=VR,BR,WD,LN,DF,EC,BF,ST         set any modem parameters
                        OK
AT+FCR=1                                enable reception
                        OK
                                        now wait for phone to ring
Phase A : Call establishment
                        RING            modem reports a ring
ATA                                     answer the phone
                                        send V21 flags preamble
                                        send CSI and DIS
                        +FCON           detect V21 flags preamble
Phase B : Pre-message negotiations
                        +FNSS:<nsf>     NSS detected
                        +FTSI:<id>      TSI detected
                        +FDCS:<codes>   DCS detected
                                        TCF
                                        training OK
                        OK              phase B complete
AT+FDR
                        +FCFR           send CFR confirmation
                        +FDCS:<codes>   final DCS report
Phase C : Message reception
                        CONNECT         high-speed carrier detected
<dc2>                                   ready to start receiving data
                        <T.4 data>      page data
                        <dle><etx>      end of page, RTC detected
                                        phase C complete
Phase D : Post-page 1 exchanges
                        +FPTS:1,<lines> page status report
                        +FET:0          MPS detected, another page to
                                        come
                        OK
AT+FDR                                  send MCF confirmation
Phase C : Message transmission
                        CONNECT         high-speed carrier detected
<dc2>                                   ready to start receiving data
                        <T.4 data>      page data
                        <dle><etx>      end of page, RTC detected
                                        phase C complete
Phase D : Post-page 2 exchanges
                        +FPTS:1,<lines> page status report
```

Table 10.7: (continued)

	+FET:2		EOP detected, last page
	OK		
AT+FDR			send MCF confirmation
Phase E : Call hangup			
	+FHNG:0		DCN sent
	OK		back on hook
			phase E complete

More detail on receiving

Table 10.7 does for receiving faxes what Table 10.1 did for sending them. It expands on the very basic description outlined at the start of the chapter, adding a setup section and including all the modem responses, not simply those we act on. The received fax shown here is two pages long, and the table is also annotated with a short indication of what each response means.

Let's look at this in a little more detail. This section is considerably shorter than the analysis of fax transmission, as many of the comments made earlier regarding a number of Class 2 commands apply with only minor modifications to reception.

Pre-session modem setup

There are three main differences between the pre-session setup sequence for receiving and that for sending a fax. The first two of these are trivial, but the third is important:

- An ID set through the **+FLID** command is used to generate a CSI frame when receiving, whereas the corresponding frame when sending was the TSI.

- Parameters set with the **+FDCC** command are used directly to generate a DIS frame when receiving, in contrast to the position if transmitting, where the parameters are used in conjunction with a received DIS frame to work out what DCS frame should be sent. In most other respects, the account of **+FDCC** given on page 252 applies to fax reception as well as transmission. In particular, note that **AT+FDIS** is just as acceptable an alternative to **+FDCC** as it is when transmitting.

- The last difference is the most important. A special command, **AT+FCR=1**, is used to set a Class 2 fax modem in receive mode. What this command does is set bit 10 of the DIS frame to 1, indicating to a caller that fax reception is possible. The default for this bit in a Class 2 fax modem is always 0, and setting **+FCR** to 1 is therefore an essential step in preparing to receive a fax.

Phase A: Call establishment

The most obvious difference between the transmit and receive versions of phase A is that instead of being initiating by dialling, a receive phase A is initiated by the fax modem reporting that it detects a ring on the line. This is typical of a fax session, which is generally transmitter-driven throughout, with the receiver restricted to a passive role; it can respond (and occasionally request) but it cannot command.

We discussed the general issues involved in answering a call in Chapter 8. Readers should refer to page 194 for details of the relative merits of using the **RING** report with the **ATA** command to answer a call, as opposed to using **ATS0=1** to set autoanswer on.

Answering data calls in fax mode

One optional but reasonably widely implemented Class 2 feature that can affect the messages returned after issuing an **ATA** is adaptive answering, controlled by the **+FAA** parameter. This enables a fax modem to detect automatically whether an incoming call is a data or fax call. Note that adaptive answering does not distinguish between fax and voice calls; there is no Class 2 command for this. A number of manufacturers have attempted proprietary solutions to this problem, most of which are notable for their lack of universal reliability.

To turn adaptive answering on, issue the **AT+FAA=1** command. It can be switched off with the **AT+FAA=0** command, and you can see if your modem supports adaptive answering by issuing an **AT+FAA=?** query in the usual way. If adaptive answering is available, this will return a list of valid values including a 1.

When a data call comes in with **+FCLASS=2** and adaptive answering is turned on, the modem automatically switches to **+FCLASS=0** (data mode) and issues a normal data mode **CONNECT** message. An application seeking to support both data and fax modes should activate its data communications option at this point, but if the call is from a fax, the corresponding message is **+FCON**. The exact sequence that leads to this message is described next.

Auto-answering a fax

Once the call is answered, the modem follows the standard T.30 procedure for an autoanswer fax. It waits for between 1.8 and 2.5 sec and then cycles through the following steps:

1. The modem sends the 2100 Hz CED tone, maintained continuously for between 2.6 and 4.0 s.

2. The modem then delays for 75 ± 20 ms and sends out the fax preamble of 1 s of V.21 HDLC flags (continuous 1 bits).

3. This is followed by the DIS frame (optionally preceded by a CSI frame if one is available).

4. If there is no answering preamble for 3 s, the cycle start again with step 1.

If timer T1 expires, the modem goes back on-hook and reports the correct hangup status code for the situation where a non-fax call comes in, which is **+FHNG:1**. Note that while T1 is supposed to be 35 ± 5 s, it is often set to the value of modem register S7. We suggested in Chapter 6 that a calling fax could reasonably set this to 135 s as part of the setup stage.

If the answering preamble from a calling fax is detected, the modem reports this with the **+FCON** message and then goes to phase B automatically.

Phase B : Pre-message negotiations

When using a Class 2 fax modem, phase B falls into two distinct parts. The first portion of phase B proceeds automatically. It consists of the reporting of all received frames in the same way that parallel frames are reported during transmission phase A. The format of the reports is identical to the corresponding frames described on page 261:

- NSS frames are reported via **+FNSS:** in an identical format to that of the NSF report.

- TSI frames are reported as ASCII strings via **+FTSI:** in the same way as the CSI frame.

- DCS frames are the only preliminary frames that have to be received. The contents are translated into class fax parameters as shown in Table 10.2 and reported via **+FDCS:**.

Once the DCS frame has arrived, the fax modem begins to receive the training check. As already mentioned, the algorithm used to decide whether this is satisfactory is something manufacturer-dependent. If the modem decides it was unsatisfactory, it sends an FTT frame automatically and returns to the start of phase B. However, if the modem decides the TCF was adequately received, it reports back with an **OK** and awaits further instructions.

Phase B: Confirming receipt

The second portion of phase B now begins. From the point at which the modem was told to answer the telephone, no intervention has been needed, and both phase A and the first part of phase B have proceeded automatically. However, now that training has been successful, the sending of a CFR confirmation frame would irrevocably commit the receiver to accept a page of data.

The reason why the session pauses at this point is not made clear in the Class 2 documentation. However, it is a good idea for there to be some point in the session where the receiver is given both the time to do the necessary housekeeping for the imminent reception of the fax, and also the option to disconnect. Without such a delay, a transmitting fax might have to begin sending data within 1/10 s of receiving a CFR frame, which is an example of the sort of real-time response constraint that Class 2 fax modems are designed to avoid.

A receiver has around 3 s to respond to the TCF, which makes the point just before the CFR is sent a good place to pause. There should be no problem with ensuring that everything ready at the receiver given that sort of time scale.

Housekeeping chores in preparation for the incoming fax

The housekeeping needed for receiving a fax is almost entirely concerned with the naming and creating of a disk file in which to store the incoming fax. On a single-user single-tasking system such as DOS, this is typically an extremely quick operation. However, this is not true for other operating systems; as a general rule, the more complex the operating system, the longer the time that should be allowed for disk housekeeping. Multi-tasking systems such as Windows and multi-user multi-tasking systems such as UNIX can both be slower to create files and also at allocating the necessary system resources to the fax receive process. Distributed systems and workstations using remote drives over networks can also slow down the process.

One other housekeeping chore that is often needed is to ensure that any non-transparent XON/XOFF flow control is switched off, as received fax data is a binary bitstream which can easily contain these characters as part of the data. It is essential that they are passed through to the applications software intact.

Even where a single-user system is being used, it wouldn't be a good idea to proceed with a fax reception without first opening the file to be used for storage of the incoming fax, as it is always possible for the file creation attempt to fail. To issue a CFR under these circumstances would be the equivalent of trying to receive on a stand-alone machine that had run out of paper.

Possible reasons for refusing to receive a fax

The second reason why a fax modem pauses for a confirmation to receive at this point is to allow a receiver the option not to accept the fax after having inspected the **+FTSI:** and **+FDCS:** reports. For instance, the **+FTSI:** report might reveal that the incoming fax was the latest junk fax transmission from your friendly neighbourhood life insurance salesperson, which may well be something you don't want to waste disk space or paper on.

Alternatively, the **+FDCS:** report may reveal that the fax was something you may not necessarily want to process. For instance, you might have a time-critical fax application which has occasional periods of intense activity when faxes come in very frequently. At these times you would not be too happy about receiving a fax at 2400 bps, which would be liable to tie up your phone line for six times longer than a 14400 bps fax.

If a decision not to proceed with fax reception is made, the only option is to issue the **AT+FK** response to abort the session. There is no command available to force the modem to renegotiate with a different set of Class 2 parameters. This isn't a deficiency in the Class 2 command set as much as a function of the way that the T.30 session protocol works.

Once the software controlling the Class 2 modem has performed all its house-keeping, and has a file ready to the receive that data, it can proceed with the final stage of phase B. It issues the **AT+FDR** command, which allows the fax modem to release the pending CFR frame and move into phase C.

AT+FDR

The **AT+FDR** command is the basic command used at all stages during Class 2 fax reception. We saw in our account of fax transmission how the modem keeps track of the state it has reached, and adjusts the action it takes when receiving an **AT+FDT** command to take account of the context. This is just as true of the **AT+FDR** command. In fact, it is normally the only Class 2 command that is ever issued when receiving.

The acronyms used to manufacture the Class 2 command set are often a matter for conjecture, but the identification of Fax Data Receive as being the source of the **+FDR** command is a particularly misleading interpretation. The description of the command in the index to the draft Class 2 standards as 'Receive Phase C data' is something of an oversimplification. What the command actually does is instruct the modem to continue with the session. The action that is subsequently taken varies with the context, but one action that never immediately occurs is for the modem to start receiving fax data.

From phase B to phase C

Once the **AT+FDR** command has been issued, transition from phase B to phase C proceeds normally. The fax modem releases the confirmation frame, and then moves to the negotiated high-speed transfer rate to await the fax data.

Suitable reports are usually issued along the way. The fax modem generates a **+FCFR** report as soon as it has confirmed that the training sequence was satisfactory, and follows this with a **CONNECT** as soon as the high-speed carrier is detected. However, between the **+FCFR** and the **CONNECT** message, other reports are also possible. The most common report at this stage is another **+FDCS:** showing the parameters that have been set by the final DCS before the last successful training check.

A typical Class 2 dialogue from initialization to the start of phase C runs as shown in Table 10.8.

The final **CONNECT** message, indicating that the high-speed carrier has been detected, is an indication that phase C data is about to be received.

Table 10.8: Class 2 prompts and commands up to the start of phase C

Command	Response
AT	
	OK
AT+FDCC=?	
	(0,1),(0-5),(0-2),(0-2),(0,1),0,0,(0-7)
	OK
AT+FDCC=1,5,0,2,1,0,0,0	
	OK
AT+FCR=1	
	OK
	RING
ATA	
	+FCON
	+FTSI: "+44 81 570 0758 "
	+FDCS: 0,5,0,2,0,0,0,0
	OK
AT+FDR	
	+FCFR
	+FDCS: 0,5,0,2,0,0,0,0
	CONNECT

Phase C data

The fax modem always pauses for permission before sending phase C data to the computer. This permission should be granted by sending a DC2 code (ASCII 18, hexadecimal 12). If fax data comes in before the DC2 is sent, the modem will attempt to buffer it, but as a fax modem buffer is often only 100 bytes, the period of grace this offers is seldom long enough to allow anything substantial to be done. This is why all housekeeping in preparation for the expected fax should already have been handled during the pause before the training confirmation has been sent.

The DC2 prompt also differentiates the fax data from the modem dialogue. If the data followed immediately after the **CONNECT** message, the various differences in the way in which responses are presented could make distinguishing the exact start of the data difficult. We discussed these differences at some length in Chapter 8; some of them are legal, such as the difference between verbose and numeric result codes, while others are more dubious, such as the varying number of new line codes that get issued.

The rules for receiving data, described earlier, are substantially the same rules as are used in Class 1 modems, described in Chapter 9. All **<dle>** characters in the fax data are sent to the computer as **<dle><dle>** pairs, and the end of the fax data is marked by a **<dle><etx>** sequence.

Cancelling a fax

One extra control character allowed when receiving faxes is the **<can>** character (decimal 24, hexadecimal 18), used to cancel receipt of a fax. It is effectively the same as the **AT+FK** abort session command. The provision to abort receipt of a fax by sending a **<can>** is most useful in cases when a disk fills up and no more data can be received. It isn't possible to anticipate this eventuality as there is no way of telling how large a fax might be before or during transmission. All a receiver can do is watch it grow and hope there is enough disk space to hold it all. A fax session conducted at the maximum speed of 14400 bps will fill up a disk at the rate of over 100K per minute in phase C. It doesn't take that long for a disk to become full.

The flow control headaches that plague Class 2 fax transmissions are largely irrelevant to fax reception. While the design of Class 2 modems means that all transmitted faxes must be sent to the modem faster than the modem can retransmit them down the line, there is no such constraint for received faxes. There is generally no problem in an application receiving data as fast as the modem will send it.

Flow control

Flow control, though mandatory for transmitted data, is optional for data received using Class 2 fax modems. This is different to Class 1 devices, which have to support flow control for both received and transmitted data. Even where received flow control is available, it should only be used in exceptional circumstances (such as while doing a disk save). The limited size of buffers in most fax modems makes things difficult for applications that depend upon flow control. The lack of flow control between faxes means that data will inevitably back up in the modem. Developers should try and write applications that don't require any flow control when receiving faxes. In any event, where XON/XOFF flow control is used, it must be available unidirectionally if fax data (which can contain flow control characters) is to be received properly.

Bit ordering

Most Class 2 fax modems default to reversed bit ordering when receiving fax data, which is the opposite of the default used for sending faxes. If data received during phase C is stored directly in a TIFF file, then the IFD will need to set FillOrder=1.

The +FPTS report

The `<dle><etx>` marking the end of the fax data is followed by a `+FPTS:` report of the received page transfer status. While there is some similarity with the `+FPTS:` response generated when sending a fax, this is in many ways a very different beast. The most immediately obvious difference is that the page transfer status report for a received fax page does not look the same. It takes more than one parameter, with multiple parameters being separated by commas. Though a total or five parameters are defined in the specification, only the first two of these are mandatory.

- The first of these compulsory `+FPTS:` values is the most important, and contains a code indicating the type of post-page response (PPR) the modem intends to give.

- The second compulsory value is always the next in sequence, and consists of the line count (lc) that was received in the page just finished. Needless to say, the count applies to scan lines rather than lines of text from an original source document.

- The optional third value is a bad line count (blc), which consists of the total number of all scan lines that violate the T.4 coding scheme.

- The optional fourth value holds the largest consecutive bad line count (cblc).

- The final optional parameter holds a lost byte count (lbc) with the total number of bytes lost due to buffer overflow. Typically, these bytes will have been lost at the start of a page.

For a number of quite technical reasons, the `+FPTS:` report is pretty useless on most fax modems. For instance, the sort of status report that is typically generated by modems with one of the common Rockwell chipsets is invariably of the form:

```
+FPTS: 1,2219,0,0
```

This response was actually received for a page which contained 365 scan lines. In fact, the same response was obtained no matter what the size of the transmitted page, or what the quality of the received fax. Neither the line count nor the other parameters are to be relied on, and this report is best ignored.

Counting the lines in a received fax

The unreliability of the **+FPTS:** report presents a problem for applications software that needs to know how many scan lines are contained in an image, for inclusion in a TIFF IFD, for instance. The only option here is to treat phase C data received with a Class 2 modem in much the same way as Class 1 data has to be treated, and perform its own searching for EOL codes.

The task of searching received T.4 data for EOL codes is, in theory at least, considerably simplified if the EOL codes are all byte-aligned. Unfortunately, byte-alignment of EOL codes is yet another of those Class 2 parameters that cannot be relied on. Most of the popular chipsets don't support it.

This means that applications aiming for some degree of universality are best advised to do their own Class 2 fax EOL checking bit by bit. One advantage of this is that it is relatively simple (at least for 1-D coding) to do quality checking at the same time. Regrettably, it is possible that the effort will be nullified by an inability of the modem to issue a PRI-Q post-page response to return to phase B. The facility for modifying the post-message response by writing to **+FPTS** should always be treated cautiously.

Phase D

As with transition to phases B and C, a fax modem moves into phase D automatically. The **+FTPS** report, which follows the **<dle><etx>** marking the end of the phase C data, is itself followed by a **+FET** report, showing the type of post-message command the transmitter has issued on its own return to phase D. The single parameter that this takes matches exactly the parameter taken by the **AT+FET=** command used to send a post-message command.

Table 10.9: Possible values included in the **FET:** report

code	T.30 frame	Meaning
FET:0	MPS	another page follows
FET:1	EOM	another document follows
FET:2	EOP	no more pages
FET:4	PRI-MPS	procedure interrupt - another page follows
FET:5	PRI-EOM	procedure interrupt - return to phase B
FET:6	PRI-EOP	procedure interrupt - no more pages

All Class 2 fax modems can be relied upon to support the MPS and EOP frames through the **FET:0** and **FET:2** reports, just as they can all be relied upon to support **AT+FET=0** and **AT+FET=2**. A number of Class 2 fax modems cannot handle EOM codes, which causes problems with some documents. It must be stressed that well over 99% of all faxes are sent using only the MPS and EOP frames. Users have to go out of their way to sent other frames, and many top-of-the range expensive fax machines don't readily support faxing using EOM and PRI-Q frames.

After the **FET:** report, the modem pauses yet again for another **AT+FDR** command.

Post-message housekeeping

As with the **+FDR** command before the first page, application software needs to do some disk housekeeping between pages. If fax pages are being kept in separate files, the file for the last page needs to be closed and the file for the next page (assuming **FET:0**) needs to be opened. In the case of multiple pages being saved in a single file (such as TIFF file containing multiple pages), there are IFDs that need writing and internal file pointers that need updating.

Whenever a file is being closed or a new file is opened, it is essential that any post-page confirmation messages be sent after the necessary disk housekeeping, not before. It would be premature to confirm fax receipt before all the T.4 data had been successfully saved to disk. If closing a file results in an error, the application should issue an **AT+FK** abort session command. Both the receiver and transmitter would then be able to recognize that the fax had not been successfully sent, and would write the error to their fax logs.

Once an application has finished all its housekeeping, the **AT+FDR** command can be issued. The post-page response frame the fax modem is holding onto is immediately released. In the normal case where **FPTS:<ppr>** is 1, this frame will be a standard MCF message confirmation frame. If the last page resulted in an **FET:0** report (the transmitter sent an MPS frame), the session resumes at the start of phase C. If the last page resulted in an **FET:2** report, then the session proceeds to phase E and a disconnection. Both these cases are shown in Table 10.9.

Phase E

When the **AT+FDR** command is issued after the **FET:2** report, a normal phase E disconnection occurs. The transmitter sends a DCN disconnect frame, the fax modem receives it and generates the **+FHNG:0** report showing a normal end to the session. A final **OK** indicates that the phone has disconnected and the session is finally over.

We have already mentioned that there are many possible **+FHNG** codes, which can be retrieved by means of the **AT+FAXERR?** query as well as being generated as part of the **+FHNG:** report. The complete list appears in Table 10.10.

Table 10.10: +FHNG status codes

Codes 0 - 9 are reserved for call placement and termination errors. Only four of these are defined.

0	Normal end of session
1	Call answered without successful fax handshake
2	Call aborted by user (via either +FK or CAN)
3	No Loop Current

Codes 10 - 19 are reserved for Transmit Phase A errors. Only two of these are defined.

10	Unspecified Phase A error
11	No answer or no DIS detected (T.30 T1 timeout)

Codes 20 - 39 are reserved for Transmit Phase B errors. Nine of these are defined.

20	Unspecified error
21	Remote cannot receive or send (bits 9 and 10 in DIS both 0)
22	Command frame error
23	Invalid command received
24	Response frame error
25	DCS sent three times without response
26	DIS/DTC received 3 times; DCS not recognised
27	Failure to train at either 2400 bps or +FMINSP value
28	Invalid response received

Codes 40 - 49 are reserved for Transmit Phase C errors. Only two are defined.

40	Unspecified error
43	DTE to DCE data underflow (data not sent fast enough)

Codes 50 - 69 are reserved for Transmit Phase D errors. Nine of these are defined.

50	Unspecified error
51	Response frame error
52	No response to MPS repeated 3 times
53	Invalid response to MPS
54	No response to EOP repeated 3 times
55	Invalid response to EOP
56	No response to EOM repeated 3 times
57	Invalid response to EOM
58	Unable to continue after PIN or PIP

Codes 70 - 89 are reserved for Receive Phase B errors. Only five errors are defined.

70	Unspecified error
71	Response frame error
72	Command frame error
73	Expected frame not received (T.30 T2 timeout)
74	Failure to resume phase B after EOM (T.30 T1 timeout)

Table 10.10: (continued)

---

*Codes 90 - 99 are reserved for Receive Phase C errors. Four of these are
defined.*

90	Unspecified error
91	Missing EOL after 5 seconds (line too long)
93	Modem receive buffer overflowed
94	CRC or frame error in ECM or BFT modes

*Codes 100 - 119 are reserved for Receive Phase D errors. Four codes are
defined.*

100	Unspecified error
101	Response frame error
102	Command frame error
103	Unable to continue after PIN or PIP

Sample code for receiving a single page with a Class 2 modem

Listing 10.3 is a quite straightforward translation of the basic Class 2 receive sequence into program code. Once again, we assume that a file pointer is passed to **getfax**, which again contains a **process_parameter** function that works out the metrics of the received images (which can then be stored in a TIFF IFD or a similar structure). Phase C reception is again almost identical to that for Class 1 modems, and the suggestions in the last chapter for code to detect EOLs in the received data could equally well be incorporated here.

Optional Class 2 features

There are a number of optional Class 2 features which I've never seen working. For completeness, details on their functioning are included in the NOTES directory on the accompanying disk. The features covered there are as follows:

- Error Correction.

- Data Conversion.

- Block Mode.

- Transition to Class 1 command set.

- Polling.

```
int getfax (FILE *faxfile)

{
        unsigned char k, lastchar=0 ;
        char faxbuf[512] ;
        char result[64] ;
        int i=0 ;

        export ("ATA\r") ;
        for (;;)
        {
          if (!(import (result,64,35))) return (0) ;
          if (strstr(result,"FHNG")) return (0) ;
          if (strstr(result,"FDCS")) process_parameters ;
          if (strstr(result,"OK")) break ;
        }
        for (;;)
        {
          export ("AT+FDR\r") ;
          for (;;)
          {
              if (!(import (result,64,35))) return (0) ;
              if (strstr(result,"FHNG")) return (0) ;
              if (strstr(result,"FDCS")) process_parameters ;
              if (strstr(result,"CONNECT")) break ;
          }
          txchar(dc2) ;
          for (;;)
          {
              k=rxchar() ;
              if (lastchar==dle)
              {
                  if (k==etx) break ;
                  lastchar=0 ;
                  if (k!=dle) continue ;
              }
              else if (k==dle)
              {
                  lastchar=dle ;
                  continue ;
              }
          faxbuf[i++]=k ;
          if (i==512)
              {
              fwrite (faxbuf,1,i,faxfile) ;
              i=0 ;
```

Listing 10.3: Receiving a fax with a Class 2 modem

```
                        }
                }
        if (i) fwrite (faxbuf,1,i,faxfile) ;
        for (;;)
                {
                if (import (result,64,20)==0) return (0) ;
                if (strstr(result,"FET")) break ;
                }
        if (wait_for ("OK",5)==0) return (0) ;
        if (strstr(result,"0")) continue ;
        if (strstr(result,"1")) continue ;
        if (strstr(result,"2")) break ;
        }
export ("AT+FDR\r") ;
if (wait_for ("FHNG:0",5)==0) return (0) ;
if (wait_for ("OK",5)==0) return (0) ;
export ("") ;
export ("ATH0\r") ;
return (wait_for ("OK",2)) ;
}
```

Listing 10.3: (continued)

One thing these features show is that the fax modem market has developed into a no-frills business. All these are powerful features which would add utility to any computer faxing application. So few are implemented because users of fax modems have traditionally been starved of programming information and have been forced to restrict themselves to using the facilities provided with the fax software that comes with the modem. To reduce costs, modem manufacturers tend to bundle the cheapest software, which in turn tends to be written for the lowest common denominator. As there is no point in adding features to a fax modem which can't be used by the software it comes with, manufacturers don't bother to implement optional features. To do so would just add to the research and development and manufacturing costs.

Fax on demand

A number of commercial fax on demand services are now available. They are superficially similar to fax polling, in that the direction of transmission is from the answering fax to the calling fax rather than vice versa. However, fax on demand is not the same as polling, which was designed as a facility for automatic document collection. The technological ancestry of fax on demand lies in the T.30 procedures for manual establishment of fax session.

Typical fax on demand systems answer the phone automatically. They play a series of messages and ask for responses in the form of DTMF codes. The typical voice dialogue is shown in Table 10.11.

Table 10.11: Typical fax-on-demand dialogue

Voice Prompt	Action
"Thank you for calling the Acme Shoeshop fax-you-a-catalogue service. Please press 1 for ladies' footwear, 2 for men's footwear, and 3 for children's footwear "	
	(Press 1)
"Please press 1 for boots 2 for shoes and 3 for slippers"	(Press 1)
"Please key in your size followed by *"	
	(Press 5*)
"Thank you for calling Acme Shoeshop fax-you-a-catalogue service. Please press the receive button on your fax machine NOW"	

At this point you hear a CNG calling tone, identical to the tone heard when a fax calls you, rather than the CED tone you hear when you call a fax. Pressing the manual receive button on your fax machine causes it to receive a fax as if it had just answered the phone. Your fax sends a DIS, the Acme shoeshop sends a DCS, and after a successful training check sequence and the fax arrives as if they had called you.

The way this is implemented in software is for an application to follow the table for normal fax reception given above, with one difference: the **ATA** command to go off-hook is triggered not by the telephone ringing, but in response to a keypress by an operator. Such facilities for manual answering are fairly universal in fax software, as the situation of an operator picking up a ringing telephone only to discover a CNG tone is very common.

Though fax-on-demand is not the same as fax polling, it isn't unusual to find that some fax machines aren't too fussy and will allow a document to be retrieved either on receipt of a DIS frame or a DTC frame. Indeed, a DIS frame from a caller in response to the DIS frame from the answering fax is implicitly recognized in the T.30 recommendation as a method of turning the line around and reversing the roles of caller and answerer. What polling does is leave the identity of the caller and answerer intact, and instead reverse the not-quite-identical roles of sender and receiver.

Common subsets of Class 2

As far as I am aware, no modem attempts a full implementation of the Class 2 specification. The lowest common denominator for Class 2 modems is the following command set.

1. Conventional commands and responses

Table 10.12: Standard AT commands and responses used in fax modem dialogues

ATD	dial a number
ATA	pick up the phone
RING	telephone is ringing
CONNECT	carrier sent/detected
OK	finished one command, ready for another

Various other commands are universally accepted as related in Chapter 8. However, the standard AT commands shown in Table 10.12 are the most essential. Note that some modems may end some on-hook reports without an OK response.

2. On-hook commands

Table 10.13: Minimal set of on-hook (set-up) Class 2 commands for fax modems

AT+FCLASS=?	to find out what the command set the modem supports
AT+FCLASS=2	to enter class 2
AT+FCLASS?	to find out what class is active
AT+FLID=<local ID>	to set the local ID
AT+FDCC=?	to find out the modem capabilities (+FDIS also OK)
AT+FDCC=VR,BR,WD,LN,DF,EC,BF,ST	set any modem parameters (+FDIS also OK)

As pointed out in Chapter 6, the only data rates guaranteed to be supported are 2400 bps when **+FCLASS=0** and 19200 bps when **+FCLASS=1**. While only the **+FDCC** command is listed, **+FDIS** is also universally supported and can be substituted for all on-hook commands. This is the only redundancy in our minimal command set (Table 10.13).

3. Off-hook commands

Note that stream mode only is supported for both receive and transmit, and that **+FDT** takes no conversion parameters. Unidirectional XON/XOFF flow control must be supported to stop the modem's buffer overflowing when receiving. Transmitted data sent to the modem will have the least significant bit of each byte transmitted first: but for data received from the modem, the most significant bit of each byte will have been received first (Table 10.14).

Table 10.14: Minimal set of off-hook (session control) Class 2 commands for fax modems

`AT+FCR=1`	`set receive mode`
`AT+FDR`	`resume receive session`
`<dc2>`	`Begin sending received fax data`
`<can>`	`Stop sending received fax data and abort`
`AT+FDT`	`transmit fax data`
`AT+FET=0`	`end a page, another to follow`
`AT+FET=2`	`end the last page`
`AT+FK`	`abort the session (send DCN)`

4. Modem reports and prompts

Table 10.15: Minimal set of Class 2 reports and prompts for fax modems

`+FCON`	`Fax preamble detected`
`+FCSI:<id>`	`Called subscriber identification report`
`+FDIS:VR,BR,WD,LN,DF,EC,BF,ST`	`Report of capabilities sent by answerer`
`+FTSI:<id>`	`Transmitting subscriber identification report`
`+FDCS:VR,BR,WD,LN,DF,EC,BF,ST`	`Report on session parameters decided by transmitter`
`+FCFR:`	`Confirmation of training check successful`
`<xon>`	`Begin sending fax data for transmission`
`+FPTS:<code>`	`Page transfer status (0=successful)`
`+FET:<code>`	`Post page command (0=another page 2=end of session)`
`+FHNG:<code>`	`Hangup status code (0=success)`

While it is often said that Class 2 modems relieve a computer of the need to conform to the stringent timing requirements of the ITU T.30 requirements, this is not strictly true. Fax remains a real-time activity for both Class 1 and Class 2 modems. In particular, the constraints in the transmission and reception of fax data are the same for both types, and modem buffers must not be allowed to fill up on reception or empty on transmission. Though Class 2 modems do take care of the more stringent timings, the requirement for a receiver to respond to all commands within 3 s ±15% still remains.

The command set listed above is a pragmatic minimum. Despite the fact that the Class 2 specification was never approved, and that manufacturers have felt free to 'play fast and loose' with the requirement, there must be serious doubts cast over the Class 2 credentials of any fax modem that doesn't implement the above commands. Most modems actually implement far more of the basic command set without adding much to the functionality. It is also common for modems to claim that they implement portions of the specification that are of no use whatsoever. Typical examples of this are to be found among modems that use one of the Rockwell fax chipsets.

The Rockwell subset

While most fax modem manufacturers are very miserly with technical information, Rockwell produce fairly extensive manuals for the components they manufacture, and it is probably for this reason that the Rockwell implementation of Class 2 has become best known. It is even used by programmers who have modems without Rockwell chipsets, as most other manufacturers don't publish or distribute a fraction of the information that Rockwell give away freely.

While Rockwell produce low-end modem chipsets that only handle Class 1 commands, their flagship products supporting fast data speeds such as 14400 and 28800 bps support both Class 1 and Class 2 commands. These implement all the commands included in the lowest common denominator subset, accept the partial parameter settings for **+FDCC** and **+FDIS** (such as **AT+DIS=,3** to set the bit rate only), and implement the following commands as specified in the Class 2 documentation (see Table 10.16).

Table 10.16: Rockwell additions to the minimal command set

`AT+FMFR?`	`Return manufacturer ID`
`AT+FMDL?`	`Return model ID`
`AT+FREV?`	`Return revision ID`
`AT+FAA=?`	`Return adaptive answer capabilities`
`AT+FAA=<code>`	`Set adaptive answer (0=fax only 1=data and fax)`
`AT+FAA?`	`Return adaptive answer setting (default 0)`
`AT+FBUF?`	`Return buffer information <bs>,<xoft>,<xont>,<bc>`
`AT+FPHCTO=?`	`Return phase C time out capabilities`
`AT+FPHCTO=<ms>`	`Set phase C time out (100 ms units, 0-255)`
`AT+FPHCTO?`	`Return phase C timeout (100 ms units, default 30)`
`AT+FAXERR?`	`Return fax error (last FHNG: code)`
`AT+FBOR=?`	`Return phase C bit order capabilities`
`AT+FBOR=<code>`	`Set phase C bit order (0=direct 1=reversed)`
`AT+FBOR?`	`Return phase C bit order (0=direct 1=reversed)`

The most recent RC288AC VFC modem chipset adds support for polling operations as follows:

`AT+FCIG=<ID>`	`Set polling ID`
`AT+FLPL=<code>`	`Document available for polling (0=off 1=on)`
`AT+FSPL=<code>`	`Enable polling (0=off 1=on)`

Setting **AT+FSPL=1** also enables the **+FPOLL** message if a caller detects that DIS bit 9 is set.

Rockwell modems are also better than many at telling you about commands they don't implement. For example, **AT+FCQ=?** and **AT+FBUG=?** will both return with a 0, indicating that neither copy checking nor HDLC frame reporting is supported. There are even some correct responses that are undocumented, such as **AT+FECM=?** returning 0 indicating that error correction mode is not available.

In general, users programming Rockwell-based modems would probably be well advised to check carefully that features described in the manuals behave as advertised. This applies to features described as unavailable as well as those described as available, since it is well known that in the computer business the state of the manuals generally lags behind the state of whatever they purport to describe. This maxim also applies to books. It is quite likely that some portions of this book will be out of date by the time you read it. Regrettably, this paragraph is unlikely to be one of them.

Other subsets

The Rockwell subset given above isn't one that is necessarily followed by all modems using Rockwell chips. Apart from the fact that a number of firmware features are customizable by OEMs using a Rockwell supplied utility called ConfigurACE, manufacturers are able to use the Rockwell modem data pump (*MDP*) without having to use the Rockwell micro controller unit (*MCU*); and as the Rockwell chipset is supplied with optional object code, using both the MDP and MCU can still leave modem builders free to develop their own firmware, in more or less the same way that users can buy a fax modem and decide to write their own software. Apart from Rockwell, other chipset manufacturers include AT&T, Exar, Cirrus, Sierra and UMC.

Many manufacturers who write their own firmware only implement EIA Class 1, and don't manufacture any Class 2 compliant modems at all. Other manufacturers have not released details of their Class 2 implementations into the public domain, which can make purchasing one of their modems for development something of a gamble. They may claim a better performance than models based on standard Rockwell components, but without programming information, users are at the mercy of those few major software developers who have a direct line to the manufacturers' R&D department, who choose whether or not to make use of additional features.

The future of Class 2

Since its abandonment by the TR-29.2 committee, the Class 2 specification has effectively been frozen. While manufacturers seeking to implement new fax features (such as the expected 28800 V.34 half duplex standard) are now free to extend Class 2 in any way they want, to do so would quite possibly invite software incompatibility problems which are best avoided.

It is likely that if there is to be a move, it will be from Class 2 to Class 2.0, driven by the need to use a command interface with a standard method of accessing features that the unapproved standard does not cover. Anyone who needs to implement the state-of-the-art in fax modem standards will then have to look at the Class 2.0 standard. This is covered in the next chapter.

11

Programming Class 2.0 Fax Modems

Introduction

If you have a Class 2.0 modem, you possess a device which is probably at the leading edge of fax modem technology. For reasons we go into shortly, the standard is not yet completely mature, so this chapter doesn't present a complete guide to programming Class 2.0 devices, but presents the standard as a development of Class 2 fax modems. You should therefore understand enough of the previous chapter to follow the transmission and reception flowcharts.

Why bother with Class 2.0?

Despite its favour with official regulatory bodies, the 2.0 standard is still something of an oddity. More than two years after it was approved, there are many more modems conforming to the incompatible 1990 draft of SP-2388 than there are modems conforming to the official specification. Manufacturers are rapidly bringing out state-of-the-art V.34 28800 bps data modems incorporating V.17 fax with the *de facto* unofficial Class 2 software interface rather than the approved Class 2.0 version.

There is nothing wrong with this. We have already seen that although compliance with the approved 2.0 standard dictates that the `AT+FCLASS=?` query produces a response including a 2.0, a plain 2 is an equally valid response. It simply indicates compliance with an unapproved standard. This is why modems adhering to the older standard are called Class 2 modems while those adhering to the newer standard are called Class 2.0 modems. Class 2.0 models may also support Class 2 or offer Class 1 operation either in addition to Class 2 or instead of it. I know of no manufacturers marketing modems that handle only Class 2.0.

Arguments against Class 2.0

In the last chapter we noted that the Class 2 command set is very rich, though no modems implement all of it; every one is a partial implementation. There is a good argument for this: the more features you implement, the more complex the code is, the longer it takes to develop, the more it costs and the greater the chance of there being bugs in it. On this reasoning, as it is possible to provide a reasonably functional Class 2 fax modem without implementing all the commands, there is no reason to add features that are not needed simply because they are in the specification.

If you're a manufacturer that is going to implement one of the EIA specifications only in part, you may as well do a partial Class 2. You are legally quite secure in describing your modem as Class 2, as this is a manufacturer-defined command set. It means what you want it to mean, and in any case, nobody will notice your partial implementation in the crowd of other partial implementations. Another good reason for using Class 2 is that there is a lot of compatible software around, and it tends to be written to tolerate minor deviations from the abandoned draft standard.

Contrast this with the position if you manufactured a Class 2.0 device. Not only are you committed to implementing all the non-optional portions of the recommendation, but you are also giving up compatibility with all the available Class 2 software in favour of the much smaller set of Class 2.0 software. The truth is that Class 2 will do just as well as Class 2.0 for most common fax operations.

Arguments in favour of Class 2.0

Despite the apparent redundancy of Class 2.0, the number of modems supporting the official standard is slowly increasing, partly for marketing reasons and partly for technical reasons. There is a marketing bonus to having a modem that conforms to the latest approved standard rather than one that could be categorized as both superseded and officially abandoned. From the technical viewpoint, support for the most recent changes in the ITU T-series recommendation is best handled by adopting a standard supported by a body committed to keeping it as up-to-date as possible. Any risks inherent in the strategy of going for 2.0 can be minimized by offering additional support for one or more of the other classes. Class 1 is especially suitable for this.

None of this means that Class 2 is doomed. Anyone who thinks that technical advances will inevitably drag unofficial industry standard practices along with them displays woeful ignorance of the inertia that a base of installed software can bring. It is also noteworthy that most Class 2.0 modems currently available implement virtually none of the optional features of the standard, and therefore offer few obvious tangible performance benefits over the current generation of Class 2 modems. There is still no immediately obvious reason for a purchaser to specify Class 2.0 modem compatability.

Translating from Class 2 to Class 2.0

Though the two classes use different commands, it is relatively easy to translate between them, especially if you restrict operations to the universal command set outlined in the last chapter. Thus, our account of Class 2.0 doesn't repeat the same pattern as the earlier exposition of Classes 1 and 2. It would be pointless to present the same information with different command names. Instead, we adopt the realistic position that as Class 2.0 is derived from Class 2, we can present the approved standard in terms of its differences from the unapproved draft.

Anyone disliking this can always consult the official specification directly. Unlike the earlier draft Class 2 document, the approved standard is readily available, but I think this way of approaching the standard would be a mistake. Given the current state of the market, understanding the past history of Class 2 is almost certainly a necessity for evaluating the present state of support for Class 2.0.

Because of the lack of a large number of 2.0 implementations, it simply isn't possible to tell at this stage how the standard will be implemented. Past experience tells us it is rare for standards to be implemented in exactly the same way by all vendors, even where the standard is officially approved. All fax modem developers and purchasers should be aware that there are no mechanisms for either official verification of compliance with the relevant standards, or for policing claims that devices comply fully with specific standards.

Verifying compliance with the standard

Just as the obvious test for Class 2.0 software is to see whether it works with a Class 2.0 modem, so also the only commonly available test of whether a modem is Class 2.0 compliant is to see whether it works with properly installed Class 2.0 fax modem software. If you think this is circular, you're quite correct. While researching this book, I came across one fax modem manufacturer that was looking for Class 2.0 fax software to test whether their modem was behaving correctly. While the danger is that the incestuous nature of this process will breed bugs, it is actually very difficult to know what either software houses or modem manufacturers can do except cross-test their products against.

Apart from the circularity of the testing procedures, it is unfortunately true that this type of compliance testing is less than rigorous. Since the command set for Class 2.0 (like that of Class 2) is so rich, it is unlikely that all available commands will be used. Adding this to differing levels of performance due to different computing platforms, we can see that compliance with Class 2.0 will be a matter of degree.

Most software for Class 2.0 has taken the safe option and is simply a straightforward conversion of existing Class 2 packages for a different set of commands.

Cross referencing between Classes 2 and 2.0

Tables 11.1–11.4 can be used to cross-reference between the two command sets. The notation is quite simple.

A ⊕ in the column before any feature indicates it is specified as optional. You cannot rely on a non-optional feature being present (bearing in mind the frequency of partial implementations), but on the other hand, certain optional features are universally implemented (such as the ability to receive).

A ✓ in the first column indicates that, apart from any differences in the name, the Class 2.0 version of a parameter is downwardly compatible with the Class 2 version.

Notice that all the unique Class 2.0 commands now consist of **AT+F** followed by two letters. Even where a command is functionally identical in both command sets, the name have changed to meet the new style. You should also be aware that certain actions previously handled by **AT+F** commands in Class 2 are now handled by a greatly extended set of phase C embedded commands. (This particularly applies to the old **AT+FET=** command to send a post page message.)

Table 11.1: Class 2/2.0 action commands

Same	CLASS 2	CLASS 2.0	Short Description
✓	ATA	ATA	Answer phone
✓	ATD<number>	ATD<numbers	Originate call (dial <number>)
	AT+FDT	AT+FDT	Continue with transmission
	AT+FDT=<codes>		Transmit with data conversion
	AT+FDR	AT+FDR	Continue with reception
	AT+FK	AT+FKS	Terminate session
	AT+FET=<code>		Send post-page message
		AT+FIP	Reset fax parameters

Table 11.2: Class 2/2.0 parameter setting commands

(unless otherwise stated these can be tested, read or set)

Same	Opt?	CLASS 2	Opt?	CLASS 2.0	Short Description
✓		AT+FCLASS		AT+FCLASS	Service class
✓		AT+FMFR?		AT+FMI?	Read only manufacturer ID
✓		AT+FMDL?		AT+FMM?	Read only modem ID
✓		AT+FREV?		AT+FMR?	Read only revision ID
✓		AT+FDCC=		AT+FCC=	Modem capabilities
✓		AT+FDIS=		AT+FIS=	Session capabilities
✓		AT+FDCS=		AT+FCS=	Read only current session parameters

Table 11.2:   (continued)

Same	Opt?	CLASS 2	Opt?	CLASS 2.0	Short Description
✓		AT+FLID=		AT+FLI=	Local ID
✓		AT+FCIG=		AT+FPI=	Polling ID
✓		AT+FSPL=		AT+FSP=	Request to poll
✓		AT+FLPL=		AT+FLP=	Document available for polling
✓		AT+FPTS=		AT+FPS=	Received page transfer status
✓		AT+FBUG=		AT+FBU=	HLDC frame content reporting
				AT+FNR=	Enable negotiation reporting
				AT+FIE=	Enable procedure interrupts
				AT+FLO=	Select flow control
				AT+FPR=	Serial port bit rate
				AT+FPP=	Packet Protocol
			⊕	AT+FNS=	non-standard negotiations
✓	⊕	AT+FCR=	⊕	AT+FCR=	Ability to receive
✓	⊕	AT+FAA	⊕	AT+FAA=	Adaptive answering
	⊕	AT+FBUF?	⊕	AT+FBS?	Read only buffer status
	⊕	AT+FTBC			Transmit block size
	⊕	AT+FRBC			Receive block size
	⊕	AT+FCQ		AT+FCQ=	Received quality checking
			⊕	AT+FRQ=	Receive quality thresholds
	⊕	AT+FBADMUL			Acceptable ratio of good/bad lines
	⊕	AT+FBADLIN			Maximum consecutive bad lines
✓	⊕	AT+FCTCRTY	⊕	AT+FRY=	Number of CTCs allowed in ECM
	⊕	AT+FPHCTO	⊕	AT+FCT=	Phase C timeout
✓	⊕	AT+FAXERR	⊕	AT+FHS=	Last hangup status code
✓	⊕	AT+FMINSP	⊕	AT+FMS=	Minimum phase C data transfer speed
	⊕	AT+FECM			Ability to do error correction
✓	⊕	AT+FBOR		AT+FBO=	Bit ordering
✓	⊕	AT+FREL		AT+FEA=	EOL alignment
			⊕	AT+FFC=	Mismatch checking/conversion
	⊕	AT+FDFFC			Check/convert compression formats
	⊕	AT+FWDFC			Check/convert width mismatches
	⊕	AT+FLNFC			Check/convert length mismatches
	⊕	AT+FVRFC			Check/convert resolution mismatches
				AT+FFD=	Set file transfer diagnostic message
				AT+FAP	Accept SUB, SEP and PWD
				AT+FSA=	Subaddress
				AT+FPA=	Selective polling address
				AT+FPW=	Password

Table 11.3:   Class 2/2.0 responses and reports

Same	CLASS 2	CLASS 2.0	Short Description
✓	+FDCS:	+FCS:	Report fax parameters from DCS frame
✓	+FDIS:	+FIS:	Report fax parameters from DIS frame
✓	+FDTC:	+FTC:	Report fax parameters from DTC frame
✓	+FPOLL	+FPO	Answering fax is pollable
	+FCFR		CFR sent, prepare to receive data
✓	+FTSI:	+FTI:	Transmitter's ID from TSI frame
✓	+FCSI:	+FCI:	Answerer's ID from CSI fram
✓	+FCIG:	+FPI:	Poller's ID from CIG fram
✓	+FNSF:	+FNF:	Contents of NSF fram
✓	+FNSS:	+FNS:	Contents of NSS fram
✓	+FNSC:	+FNC:	Contents of NSC fram
✓	+FHT:	+FHT:	Debug report of HDLC frames sen
✓	+FHR:	+FHR:	Debug report of HDLC frame received
✓	+FCON	+FCO	Fax connection established
✓	+FVOICE	+FVO	PRI-Q (transition to voice)
		+FDM	Data call detected : reset +FCLASS=0
	+FET:	+FET:	Report contents of post-page message
✓	+FPTS:	+FPS:	Received page transfer status
	+FPTS:		Transmitted page transfer status
	+FHNG:	+FHS:	Hangup status report
		+FFD:	Report FDM file diagnostic message
		+FSA:	Report SUB subaddress frame
		+FPA:	Report SEP selective polling address
		+FPW:	Report PWD password frame

Table 11.4:   Class 2/2.0 embedded phase C commands

Data transparency

`<dle><dle>`	one dle (16 decimal, 10h)
`<dle><1ah>`	two dle

Transmit commands

`<dle><2ch>`	end of page, return to phase C for another (MPS)
`<dle><3bh>`	end of page, return to phase B and renegotiate (EOM)
`<dle><2eh>`	end of page and session (EOP)
`<dle><21h>`	procedure interrupt (PRI-Q)
`<dle><3fh>`	request free space in transmit buffer
`<dle><etx>`	acknowledge modem abort request `<can>`

Receive commands

`<dle><etx>`	end of page
`<dle><41h>`	embedded `<soh>` in packet
`<dle><57h>`	embedded `<etb>` in packet
`<dle><4fh>`	marker for data loss due to overrun

Table 11.4: (continued)

Transmit format conversion	
<dle><61h>	vertical resolution normal
<dle><62h>	vertical resolution fine
<dle><63h>	A4 length
<dle><64h>	B4 length
<dle><65h>	unlimited length
<dle><66h>	215 mm width
<dle><67h>	255 mm width
<dle><68h>	313 mm width
<dle><69h>	151 mm width
<dle><6ah>	107 mm width
<dle><6bh>	1-D modified Huffman
<dle><6ch>	2-D modified Read
<dle><6dh>	2-D uncompressed mode
<dle><6eh>	2-D MMR (T.6)
Receive buffer status	
<dle><3fh>	request free space in transmit buffer
<dle><12h>	buffer empty
<dle><30h>	buffer less than 10% full
<dle><31h>	buffer 10-20% full
<dle><32h>	buffer 20-30% full
<dle><33h>	buffer 30-40% full
<dle><34h>	buffer 40-50% full
<dle><35h>	buffer 50-60% full
<dle><36h>	buffer 60-70% full
<dle><37h>	buffer 70-80% full
<dle><38h>	buffer 80-90% full
<dle><39h>	buffer more than 90% full

The following notes to the Class 2.0 versions of the above commands and responses are primarily concerned with cases where there are functional differences to the versions described either in the last chapter or in the NOTES directory on the accompanying disk. If a Class 2.0 command in the list above is not covered further, you can assume it is identical to the corresponding Class 2 command; for instance, there is no difference between the **+FLID** and **+FLI** local ID commands.

Transmitting a fax: AT+FDT

The basic differences between Class 2 and 2.0 when transmitting are as follows:

- **AT+FDT** takes no parameters. Where mismatch checking and data conversion is required, the modem is informed of the type of phase C data it is being supplied with by means of a series of embedded phase C commands shielded with `<dle>` characters. Together these comprise a superset of information previously contained in the Class 2 `AT+FDT=DF,VR,WD,LN` command. Mismatch checking and conversion is described in more detail later; see the `AT+FFC` command.

- The initial Class 2 XON prompt prior to transmission of phase C data is not used in Class 2.0. The only prompt given is `CONNECT <cr><lf>` in verbose mode or `0<cr>` if numeric result codes are being used. The phase C data can then be sent on by the controlling application software immediately.

- Class 2 block mode (see the disk notes) is not available.

- Phase C data is formatted in a similar fashion to Class 1 and 2 phase C data, with a `<dle>` character (decimal 16 or hexadecimal 10) being sent as `<dle><dle>`. There is an additional option to send two successive `<dle>` octets as `<dle><sub>`, where `<sub>` is decimal 26 or hexadecimal 1ah. A large number of additional control sequences shielded by `<dle>` octets can also be embedded in the data stream, notably for format conversion and to obtain buffer status; see the `AT+FFC` and `AT+FBS` commands. The application sending phase C data must also ensure that the bit ordering used is that expected by the modem; see the `AT+FBO` command.

- The fax modem can ask the transmitter to stop sending data with a `<can>`, where `<can>` is decimal 24 or hexadecimal 18. This could be due to unrecoverable formatting errors in the transmission, or a loss of the connection to the receiver. If a `<can>` is sent by a fax modem to a computer sending phase C data, the application should stop sending and acknowledge termination with a `<dle><etx>` sequence, where `<etx>` is decimal or hexadecimal 3. The modem responds with an `OK`. The Class 2.0 standard states that this terminates phase C. If it is unclear whether the call is terminated or what the reason for the cancellation was, an application should presumably use the `AT+FHS` command to find the hangup status, but this is not something that the standard specifies.

- The phase C data is terminated with an RTC sequence (six consecutive EOL codes) included in the data stream. Embedding EOL codes in the data is also done in Class 1; but all draft Class 2 standard fax modems inserted their own RTC codes and did not accept any in phase C data. The application sending the RTC must ensure that the bit ordering used for the EOL codes is expected by the modem; see the `AT+FBO` command.

- Neither the `<dle><etx>` sequence used to terminate data in Class 2 nor the `AT+FET` command to instruct the modem which post-message command to send are used in Class 2.0. Instead, the embedded RTC is followed by a special `<dle>` sequence that determines the post-page message to be sent immediately afterwards as part of phase D. The codes used are as follows:

1. If the page is to be followed by another, the terminating sequence is **<dle><mps>**, where **<mps>** is the ASCII character 44 decimal or 2c hexadecimal. The post page message sent is an MPS, and the modem issues an **OK** prompt if the page is acknowledged. The next step is to issue another **AT+FDT** command, and the session then carries on from the start of phase C again with a **CONNECT** message followed by more data.

2. If the page is the last of the session, the terminating sequence is **<dle><eop>**, where **<eop>** is the ASCII character 46 decimal or 2e hexadecimal. The post-page message sent is an EOP, and the modem issues an **OK** prompt if the page is acknowledged. The next step is to issue another **AT+FDT** command, when modem proceeds to phase E and sends a DCN frame before reporting back with the hangup status code (**FHS:0**) and a final **OK**.

3. If the page is to be followed by a return to the start of phase B to renegotiate parameters, the terminating sequence is **<dle><eom>**, where **<eom>** is the ASCII character 59 decimal or 3b hexadecimal. The post page message sent is an EOM, and the modem issues an **OK** prompt if the page is acknowledged. The receiver sends its CSI and DIS negotiating frames again, and the session resumes at the start of phase B.

4. Any of the above codes can be sent as PRI-Q versions by sending the embedded **<dle><pri>** command, where **<pri>** is the ASCII character 33 decimal or 21 hexadecimal. So **<dle><pri>** followed by a **<dle><eop>** sends a PRI-EOP frame requesting that the receiver switch to voice operation and acknowledge with a PIN or a PIP. The **<dle><pri>** embedded command need not necessarily immediately precede the terminating post-page message command. It could occur anywhere in the phase C data. The negotiation of the interrupt is left to the modem, using the procedure outlined in Figure 6.9. A successful procedure interrupt negotiation results in the **+FVO** message, indicating that voice communications can be resumed, but the modem at this stage remains off hook, with the **+FCLASS** parameter unchanged.

- The modem prompts for the next command with an **OK** if the post-page response is MCF, RTP or PIP; the page has been accepted. If the response is RTN or PIN, then the page has been rejected and the modem prompts with an **ERROR** message instead. The Class 2 **+FPTS:** page transfer status report is not given. RTP and RTN frames requiring renegotiation are handled transparently by the modem when the next **AT+FDT** command is issued. PIN or PIP interrupt requests are only acknowledged if procedure interrupts are enabled; see the **AT+FIE** command.

Table 11.5 replicates the simple two-page fax sending procedure shown in the last chapter, with the difference that it uses the Class 2.0 command set.

Table 11.5: Sample fax transmission using Class 2.0 commands

```
Pre-session modem setup
AT+FCLASS=2.0
                    OK
AT+FLI="ID string"                    set local ID
                    OK
AT+FCC=VR,BR,WD,LN,DF,EC,BF,ST        set any modem parameters
                    OK
AT+FNR=1,1,1,1                        enable phase B reporting
                    OK
Phase A : Call establishment
ATD<number>
                    +FCO              Fax preamble detected
                    +FNF:<nsf>        NSF detected
                    +FCI:<remote id>  CSI detected
                    +FIS:<codes>      DIS detected
                    OK                phase A complete
Phase B : Pre-message negotiations
AT+FDT
                    +FCS:<codes>      TSI and DCS sent
                                      training OK, CFR received
                                      phase B complete
Phase C : Message transmission
                    CONNECT           high-speed carrier, ready to start
                                      sending
<T.4 data>                            page 1 sent with final RTC
<dle><mps>                            end of page 1
Phase D : Post-page 1 exchanges

                                      another page to come, send MPS
                                      MCF detected
                    OK                phase C and phase D complete
AT+FDT                                now back to phase C for page 2
Phase C : Message transmission
                    CONNECT           high-speed carrier, ready to start
                                      sending
<T.4 data>                            page 2 is sent with final RTC sent
<dle><eop>                            end of page 2
Phase D : Post-page 2 exchanges

                                      no more pages, send EOP
                                      MCF detected
                    OK                phase C and phase D complete
Phase E : Call hangup
                    +FHS:00           DCN sent
                    OK                back on hook
                                      phase E complete
```

Underrun handling in Class 2.0

Unlike a Class 1 or 2 device, a Class 2.0 fax modem must be able to handle buffer underruns when fax data is being transmitted. The usual causes of buffer underrun are either badly written software or (more usually) a multi-tasking computer that permits another program to either hog the processor or disable the interrupt system. The specification states that delays of up to 5 s must be allowed for, during which the computer may be unable to transmit data; and that the modem can implement this feature in any way the manufacturer chooses, provided the T.series recommendations are adhered to. While a good case could be made out for the view that a computer that starves a fax program of resources for 5 s is one that has no business attempting to handle real-time communications, there are a number of ways of handling this:

- For T.4 data the easiest course of action is to ensure that all scan lines are buffered. The ITU state (T.4/3.2) that the maximum length of a transmitted scan line should be less than 5 s, and if this is exceeded a receiver must disconnect. This means that delays of up to 5 s can be accommodated provided transmission of the line with the underrun in it has not yet started. The simplest method is to delay transmission of the EOL for each line until the modem buffer has received the next line to be sent, including its EOL. The delay can be filled by sending bits with a value of 0 in exactly the same way as the T.4 recommendation says that fill can be used to meet the criteria for minimum scan line times.

- An alternative method, which can be used if a line has already begun, relies on the transmission software decoding data on the fly. Using this scheme, delays of up to 5 s can be accommodated provided transmission of any particular code word has not yet started. Under this method, a modem should not start to send the code for any run-length until a terminating code has been received into the modem buffer. If this means a delay in the middle of a line (between code words), the delay can be filled by sending alternate white and black (or black and white) run lengths of zero. Fill bits can't be used in the middle of a line, but fill run lengths are not prohibited.

- Although similar methods can be used for error-corrected T.4 or T.6 faxes (as all Class 2.0 modems compose HDLC frames 'on the fly'), a more difficult situation arises with non-fax transmissions where the data inside the frames cannot be padded out in this way. An example of this type of transmission is T.434 BFT binary file transfer. In these cases, frames must be buffered rather than lines, with no frame transmissions beginning until the data for the whole frame has been received in the buffer.

However, the ITU standards state that only 50 ms of flags can be used as fill between frames. If the 50 ms is about to elapse with no replacement frame becoming available, the only recourse is for the fax modem to terminate the partial page with three RCP frames and negotiate a new partial page in phase D by sending a PPS-NULL. Apart from the fact that a return to phase D introduces delay, there is some a scope for extending the delay, as use can be made of the 6 s T2 timeout between responses and commands in fax negotiations. This method can be used with T.4 and T.6 fax files, as well as with data file transmissions.

The complementary problem to buffer underrun (caused by not sending data to the modem quickly enough) is that of buffer overrun (caused by not retrieving data from the modem quickly enough). While this can affect fax transmissions where modem responses are lost, it is more usually seen to affect phase C data reception. The cure for overrun is more complicated than that for underrun, and requires a special receive packet protocol. See the section on the **AT+FPP** command.

Receiving a fax: AT+FDR

Compared to sending, there are few differences between fax reception sessions using Class 2 and 2.0 modems. Once the setup stage is passed, even the commands remain the same. Apart from the name change of some responses, the only differences are:

- The Class 2 **+FCFR** report indicating that a CFR frame has been sent to end a successful phase B negotiation has been scrapped. There is no Class 2.0 equivalent; the fact of a CFR being sent is implicit in the **CONNECT** response to an **AT+FDR** command issued in phase B.

- The RTC sequence sent by a transmitting fax is included in the phase C data the modem presents to the computer. Including the six EOL codes comprising the RTC in the data stream is also a feature of Class 1; but all draft Class 2 standard fax modems delete any received RTC codes and do not include them in the phase C data.

- The Class 2.0 **+FPS:** and Class 2 **+FPTS:** reports are slightly different. While the five parameters **<ppr>,<lc>,<blc>,<cblc>,<lbc>** are the same for both, all are compulsory in the Class 2.0 **+FPS:** report. Though no parameters are omitted, parameters which are either disabled or inapplicable are set to 0. This is clearly a better specification than that of the Class 2 **+FPTS:** report, which on at least one interpretation could take either 2, 3, 4 or 5 parameters.

The modem can spontaneously generate the values for **<ppr>** listed in Table 11.6, based on its own quality checking. If quality checking is disabled, **<ppr>** will always take the value 1. The copy quality checking commands in Class 2.0 are different to those in Class 2. See **AT+FCQ** and **AT+FRQ** for more details.

Table 11.6: Possible Class 2.0 FPS parameters

FPS:1	MCF	Page received OK
FPS:2	RTN	Page bad : renegotiate/retrain
FPS:3	RTP	Page good : renegotiate/retrain

The ability to modify the post-page response by writing a modified code back to **+FPS** is much the same as writing a code back to **+FPTS** with Class 2 modems; see the **AT+FPS** command.

- While the standard post-page message (ppm) parameters for the **+FET:** report of 0 for the MPS frame (return to phase C), 1 for the EOM frame (return to phase B) and 2 for the EOP frame (go to phase E) are identical to the Class 2 versions, the remaining non-standard ppm codes are different. The complete Class 2.0 ppm codes are shown in Table 11.7.

Table 11.7: Possible Class 2.0 FET parameters

	T.30 frame	Meaning
FET:0	MPS	another page follows
FET:1	EOM	another document follows
FET:2	EOP	no more pages
FET:3	PRI-MPS	procedure interrupt - another page follows
FET:4	PRI-EOM	procedure interrupt - return to phase B
FET:5	PRI-EOP	procedure interrupt - no more pages

The differences in **FET:** are caused by the scrapping of all post-block ECM reports. In Class 2, **FET:3** was used for the PPS-NUL frame, but only the codes for MPS, EOM and EOP are normally encountered. The PRI-Q codes are usually suppressed and replaced with the equivalent non-PRI codes 0, 1 and 2. See the **AT+FIE** command for more details.

Table 11.8 replicates the simple two-page fax reception procedure shown in the last chapter, but for the Class 2.0 command set.

Table 11.8:   Sample fax reception using Class 2.0 commands

Pre-session modem setup		
AT+FCLASS=2.0		
	OK	
AT+FLI=îID stringî		set local ID
	OK	
AT+FCC=VR,BR,WD,LN,DF,EC,BF,ST		set any modem parameters
	OK	
AT+FNR=1,1,1,1		enable phase B reporting
	OK	
AT+FCR=1		enable reception
	OK	
		now wait for phone to ring
Phase A : Call establishment		
	RING	modem reports a ring
ATA		answer the phone
		send V21 flags preamble
		send CSI and DIS
	+FCO	detect V21 flags preamble
Phase B : Pre-message negotiations		
	+FNS:<nsf>	NSS detected
	+FTI:<remote id>	TSI detected
	+FCS:<codes>	DCS detected
	TCF	
		training OK
	OK	phase B complete
AT+FDR		
	+FCS:<codes>	final DCS report
Phase C : Message reception		
	CONNECT	high-speed carrier detected
<dc2>		ready to start receiving data
	<T.4 data>	page data ending with the RTC
	<dle><etx>	end of page
		phase C complete
Phase D : Post-page 1 exchanges		
	+FPS:1,<lines>, 0,0,0	page status report
	+FET:0	MPS detected, another page to come
	OK	
AT+FDR		send MCF confirmation
Phase C : Message transmission		
	CONNECT	high-speed carrier detected
<dc2>		ready to start receiving data
	<T.4 data>	page data ending with the RT
	<dle><etx>	end of page
		phase C complete

Table 11.8: (continued)

Phase D : Post-page 2 exchanges		
	`+FPS:1,<lines>,`	page status report
	`0,0,0`	
	`+FET:2`	EOP detected, last page
	`OK`	
`AT+FDR`		
		send MCF confirmation
Phase E : Call hangup		
	`+FHS:00`	DCN sent
	`OK`	back on hook
		phase E complete

Cancelling a session: AT+FKS

The only difference between the Class 2 **AT+FK** and Class 2.0 **AT+FKS** command is that the latter also sets **AT+FCLASS=0** immediately after generating the **+FHS:** hang up status report. Note that issuing a **<can>** character (decimal 24, hexadecimal 18) when receiving data in phase C has the same effect, causing an implicit **AT+FK** in class 2 and **AT+FKS** in Class 2.0.

Reset fax parameters: AT+FIP

The **AT+FIP** command has no Class 2 equivalent. It initializes all the Class 2.0 parameters to their default settings. This command can optionally take a single numeric parameter indicating a default profile to use. The profiles are set by the manufacturer, and there is no Class 2.0 command enabling modems to set their own profiles. The **AT+FIP** can be issued at any time, with the new settings being used on the next suitable occasion.

Class 2.0 fax parameters: AT+FCC, AT+FIS, AT+FCS

While the various forms of the Class 2.0 **+FCC**, **+FIS** and **+FCS** parameters are identical to the corresponding Class 2 **+FDCC**, **+FDIS** and **+FDCS** versions, there are differences in the Class 2.0 fax parameter table.

Most of the changes are extensions designed to incorporate the latest enhancements to ITU recommendations. Note that the new table is downwardly compatible with the old one, and any parameters valid for Class 2 modems have the same effects when used with Class 2.0 modems. The only exceptions to this are the non-zero values for the EC error correction field, which hardly matter as ECM wasn't supported widely on Class 2 modems. The most recent version is shown in Table 11.9.

Table 11.9: Class 2.0 fax session parameters

Label	Function	DIS/DCS bits	Values		Description
VR	Resolution	bit 15	00		Normal, 3.85 1/mm
	(bitmapped)	"	01		Fine, 7.7 1/mm
		bit 41	# 02	*	Superfine 15.4 1/mm
		bit 43	# 04	*	16 d/mm x 15.4 1/mm
		bits 15,44	# 08	*	200 x 100 dpi
		"	# 10	*	200 x 200 dpi
		bits 41,44	# 20	*	200 x 400 dpi
		bit 42	# 40	*	300 x 300 dpi
BR	Bit Rate	bits 11-14	0		2400 bps V.27 ter
		"	1		4800 bps V.27 ter
		"	2	*	7200 bps V.29 or V.17
		"	3	*	9600 bps V.29 or V.17
		"	4	*	12000 bps V.17
		"	5	*	14400 bps V.17
WD	Page Width	bits 17-18	0		215 mm
		"	1	*	255 mm
		"	2	*	303 mm
		bit 34	3	*	151 mm
		bit 35	4	*	107 mm
LN	Page Length	bits 19-20	0		A4, 297 mm
		"	1	*	B4, 364 mm
		"	2	*	unlimited length
DF	Data format		0		1-D modified Huffman
		bit 16	1	*	2-D modified Read
		bit 26	2	*	2-D uncompressed mode
		bit 31	3	*	2-D MMR (T.6)
EC	Error checking	bit 27	0		Disable ECM
		bits 27-28	1	*	Enable ECM
		bit 67	# 2	*	T.30 half duplex
		"	# 3	*	T.30 full duplex
BF	File	bit 51	00		Disable file transfer
	Transfer	bits 51,53	01	*	Select BFT T.434
	(Bitmapped)	bits 51,54	# 02	*	DTM
		bits 51,55	# 04	*	Edifact
		bits 51,57	# 08	*	Basic Transfer Mode
		bits 51,60	# 10	*	Character Mode
		bits 51,62	# 20	*	Mixed Mode
		bits 51,65	# 40	*	Processable Mode
ST	Scan Time/line	bits 21-23			VR=normal VR=fine
		"	0		0 ms 0 ms
		"	1		5 ms 5 ms
		"	2		10 ms 5 ms
		"	3		10 ms 10 ms
		"	4		20 ms 10 ms
		"	5		20 ms 20 ms
		"	6		40 ms 20 ms
		"	7		40 ms 40 ms

Those items marked * are optional, while those marked # are included in the latest 2.0 drafts, reflecting the 1993 changes to T.30, but are not yet finally approved. It is unlikely that the next version of Class 2.0 will contain a table identical to that given above, since there is as yet no provision for faxing at rates of up to 28800 bps, which is the fastest speed defined in the latest ITU V.34 recommendation. It is possible that the next version of the Class 2.0 standard will be delayed so that it can incorporate the expected adjustments to T.30 that the new speeds will require.

The differences from the Class 2 fax parameter table are mostly due to adjustments made to keep the standard in line with changes in the ITU T.4 and T.30 recommendations. Refer to Chapters 2 and 4 for more details on the latest T.4 and T.30 options.

The VR and BF fields controlling the vertical resolution and file transfer capabilities, respectively, no longer consist of the simple Boolean values 0 or 1, but are now bitmapped. When the parameter table is used for DIS or DTC capability frames, multiple bits in these two fields can be set to indicate which options are supported. However, for reporting the DCS frame, indicating which of the possible options will actually be used, the VR and BF fields in +FCS: can only have one bit set.

Note that the values in the WD field no longer include the number of dots in the line: these should be taken from the horizontal resolution now included in the VR field. As well as the T.30 horizontal resolutions of R8 and R16 (approx. 8 and 16 dots/mm) the inch-based resolutions of 100–400 dpi can now be specified. Two areas of inconsistency that remain from the Class 2 specification are that uncompressed mode cannot be disabled for T.6 coding, and T.6 is not specifically stated to depend on error correction.

Error correction in Class 2.0

The only real incompatibility with Class 2 is in the EC field, which controls error correction options. The default of 0, meaning error correction is disabled, is the same both for Classes 2 and 2.0. However, the Class 2 command AT+FECM, which controlled both the ability to do error correction and how it was handled, has been scrapped. Error correction control is now set solely by the EC parameter. When error correction is enabled, it corresponds to the old Class 2 +FECM=2 setting, and is handled transparently by the modem. The +FECM=1 option no longer exists.

EC settings of 2 and 3 set half and full duplex ISDN capability, respectively, and have presumably been added for use in Class 2.0 ISDN adapters. ITU Recommendation T.30 (Annex C) defines the use of the group 3 fax session protocol on ISDN lines, and is outside the scope of this book. In the UK, ISDN is prohibitively expensive for non-corporate use as none of the telecommunications vendors encourage use by private individuals. T.30 states that the use of this protocol with full duplex modems is for further study. The problems with full duplex mode as defined in Annex C were outlined in on page 167.

There are no longer options for setting preferred ECM frame lengths. The examples given in the EIA 592 recommendation assume a 256 octet frame, but this doesn't seem to be stated clearly anywhere (possibly left to the manufacturer to decide).

Remember that all non-zero settings for the BF file transfer options require that error correction is enabled. This is not handled automatically.

Polling AT+FPI, AT+FLP, AT+FSP

Polling with a Class 2.0 modem can be handled in exactly the same way as with a Class 2 modem; only the names of the commands have altered. The **AT+FPI** command to set a polling ID is identical to the Class 2 **AT+FCIG** command; the **AT+FLP=1** command used to show that an answering fax is pollable sets bit 9 in the initial DIS frame in exactly the same way as the Class 2 **AT+FLPL=1** did; and the Class 2.0 **AT+FSP=1** command used to poll a remote fax behaves in almost exactly the same way as the Class 2 **AT+FSPL=1** command. The only difference between the two versions is that the Class 2.0 **+FPO** report (given to a calling fax to indicate that the answerer has a document to poll) is only given if **+FSP=1**. The equivalent Class 2 report +FPOLL has no such restriction listed in the SP-2388 Class 2 draft.

Selective polling and subaddressing: AT+FAP, AT+FSA, AT+FPA, AT+FPW

These commands concern with here are the additional security and addressing features introduced in the 1993 version of T.30. They have not yet been officially incorporated in EIA 592, but they are included in the latest draft and have also been submitted to the ITU as part of the T.Class2 proposal.

A global parameter that can be tested, read and set controls the additional feature. The **AT+FAP** command takes three parameters, **<sub>,<sep>,<pwd>**, each of which can be set to either 0 or 1. In terms of T.30, setting **<sub>** to 1 sets bit 49 of the negotiating frames and enables subaddressing; setting **<sep>** to 1 sets bit 47 and enables selective polling; while setting **<pwd>** to 1 sets bit 50 and enables security passwords. So the single command **AT+FAP=1,1,1** controls all three features. The **+FAP** parameters can be tested using **AT+FAP?** and read using **AT+FAP?** as well as being set.

The contents of these three frames are controlled by three new features. Subaddressing is controlled by **+FSA**, selective polling by **+FPA**, with passwords controlled by **+FPW**. There are in fact two separate security password frames, one of which is used when polling, with the other used when transmitting. They are both controlled by **+FPW**, in the same way as the **+FIS** ID controls both the TSI calling and CSI answering frames. The IDs **+FIS** and **+FTC** are really very similar to the three new features. All use 20-character strings padded with spaces, with null strings indicating that the corresponding frame will not be sent. All occur in four separate contexts (three commands and one report) with the new features shown in Table 11.10.

Table 11.10:   Class 2.0 support for new T.30 negotiating frames

	Subaddressing	Selective Poll	Passwords
Set contents of transmitted frames	AT+FSA="<sub>"	AT+FPA="<sep>"	AT+FPW="<pwd>"
Read contents of frames to be sent	AT+FSA?	AT+FPA?	AT+FPW?
Read range of possible values	AT+FSA=?	AT+FPA=?	AT+FPW=?
Report contents of received frames	+FSA:<sub>	+FPA:<sep>	+FPW:<pwd>

Sample sessions

The sample sessions in Tables 11.11 and 11.12 show sending and receiving polled faxes, together with use of the **+FAP** enabled polling features. In both cases, we only show the details up to the start of phase C, as from then on they progress identically to non-polled session.

Quality checking: AT+FCQ

We have seen many instances of command pairs that perform the same function for the Class 2 and 2.0 command sets, but have different names. An example of the reverse case is **+FCQ**. The Class 2.0 **AT+FCQ** command is not the same as the

Table 11.11: Polled reception

```
Pre-session modem setup
AT+FCLASS=2
                        OK
AT+FCC=VR,BR,WD,LN,DF,EC,BF,ST              set any modem parameters
                        OK
AT+FLI="ID string"                         set local ID
                        OK
AT+FPI="Polling ID"                        set polling ID
                        OK
AT+FNR=1,1,1,1                             enable phase B reporting
                        OK
AT+FCR=1                                   enable reception
                        OK
AT+FSP=1                                   enable polling
                        OK
AT+FAP=0,1,1                               enable selective polling/
                                           password
                        OK
AT+FPA="Document"                          select document
                        OK
AT+FPW="Password"                          set password
                        OK
Phase A : Call establishment
ATD<number>
                        +FCO               Fax preamble detected
Phase B : Pre-message negotiations
                        +FNF:<nsf>         NSF detected
                        +FCI:<remote id>   CSI detected
                        +FIS:<codes>       DIS detected
                        +FPO               the answerer is pollable
                        OK
AT+FDR                                     send PWD if DIS bit 50 set
                                           send SEP if DIS bit 47 set
                                           send CIG and DTC frames
                        +FTI:<remote id>   report TSI detected
                        +FCS:<codes>       report DCS detected
                                           TCF
                                           training OK
                        +FCS:<codes>       final DCS report
Phase C : Message reception
                        CONNECT            high-speed carrier detected
<dc2>                                      ready to start receiving data
                                           continue as normal
```

Table 11.12: Polled transmission

```
Pre-session modem setup
AT+FCLASS=2
                         OK
AT+FDCC=VR,BR,WD,LN,DF,EC,BF,ST          set any modem parameters
                         OK
AT+FLI="ID string"                       set local ID
                         OK
AT+FAP=0,1,1                             enable selective polling and
                                         password
                         OK
AT+FLP=1                                 enable polled transmission
                         OK
AT+FNR=1,1,1,1                           enable phase B reporting
                         OK              now wait for phone to ring
Phase A : Call establishment
                         RING            modem reports a ring
ATA                                      answer the phone
                                         send V21 flags preamble
                                         send CSI and DIS
                                            (with bits 47 and 50 set)
                         +FCO            detect V21 flags preamble
Phase B : Pre-message negotiations
                         +FPA:<Document>    SEP detected
                         +FPW:<Password>    PWD detected
                         +FPI:<Polling ID>  CIG detected
                         +FTC:<codes>       DTC detected
                         OK                 check password match
                                            check document available
AT+FDT                                      only if both ok
                         +FCS:<codes>       TSI and DCS sent
                                            training OK
                                            phase B complete
Phase C : Message transmission
                         CONNECT            high-speed carrier sent
                                            continue as normal with
                                            <Document>
```

Class 2 **AT+FCQ**. While it still controls copy quality checking for received documents, it does so in a different way. It also takes on the extra function of checking the correctness of transmitted data. The format of the command is

AT+FCQ=<rq>,<tq>

where the values of **<rq>** and **<tq>** are shown in Table 11.13.

Table 11.13: Class 2.0 quality checking and correction

rq=0	Receive copy quality checking is disabled
tq=0	Transmit copy quality checking is disabled
rq=1	Receive copy quality checking enabled, +FPS set accordingly
tq=1	Transmit data checked, <can> returned if data violates T.4 or T.6 formats
rq=2	Receive copy quality checking enabled, any errors corrected
tq=2	Transmit data checked, any errors corrected

The default for **<rq>** is 1 (received copy quality is enabled). This is an exception to the usual rule that parameter defaults are either 0 or are set by the manufacturer. The kind of received copy quality checking is controlled by the **+FRQ** command, which we will come to shortly. The default for **<tq>** is 0.

The other possible **+FCQ** values, **<tq1>**, **<rq2>** and **<tq2>**, are all optional. The most interesting of these additions are the options to correct errors which would be set using **AT+FCQ=2,2**. The method for correcting received data would normally take the form of either deleting bad lines or replacing them with previous good lines, but the procedure for correcting transmitted data (which the standard says is 'manufacturer specific') is the more intriguing.

It is not immediately apparent what sort of transmission error this could be aimed at. It is possible to run through a check list of specific items and correct things like wrongly formatted RTC sequences, using the wrong bit order and forgetting to do **<dle>** shielding, but it is hardly the job of a standard to compensate for bugs in software. Equally, defective hardware should be fixed rather than compensated for. In any event, I would be surprised if correcting badly formatted T.4 data proved to be a practical proposition. As for the correction of T.6 data, I don't see how it is practical at all.

Receive quality thresholds: AT+FRQ

The Class 2.0 AT+FRQ command comes into effect when the **<rq>** parameter is non-zero. It replaces both of the Class 2 commands **AT+FBADMUL** and **AT+FBADLIN** and takes two parameters. The first is the maximum number of consecutive bad lines, known as **<cbl>**. Unlike the old **+FBADLIN** parameter, this does not vary with the resolution. The second parameter is the minimum percentage of good lines, known as **<pgl>**. Note that this is not the same as the Class 2 **+FBADMUL** parameter, which was the maximum ratio of bad lines to good lines. While the defaults are manufacturer specific, the default of **+FCQ=0,0** means that these are normally turned off.

You should note that if **+FCQ=2,2** and autocorrection is enabled, it is not clear whether the **+FRQ** thresholds for page rejection are turned off. Although if **<rq>=2**, the fax modem will attempt to correct errors itself, the effect of any **+FRQ** settings is not specified.

Writing post-page message responses: AT+FPS

We described the `+FPS:<ppr>,<lc>,<blc>,<cblc>,<lbc>` report earlier in this chapter. The key difference from the Class 2 `+FPTS:` report lies in the fact that all the parameters are compulsory, with any deactivated ones reported as 0. The command `AT+FPS<ppr>` can be used to write a post page response value back to `+FPS` after evaluating the line count `<lc>` as compared to the total bad line count `<blc>`, consecutive bad line count `<cblc>` and lost byte count `<lbc>`. The acceptable values for doing this are shown in Table 11.14.

Table 11.14: Possible write-back values for the Class 2.0 page status

ppr value	T.30 frame	Meaning
AT+FPS=1	MCF	Page received OK
AT+FPS=2	RTN	Page bad : renegotiate/retrain
AT+FPS=3	RTP	Page good : renegotiate/retrain
AT+FPS=4	PIN	Page bad : interrupt requested
AT+FPS=5	PIP	Page good : interrupt requested

Procedure interrupts: AT+FIE

The Class 2.0 `AT+FIE` command has no Class 2 equivalent. Its purpose is to control how the modem responds to procedure interrupts. With the default of `AT+FIE=0`, procedure interrupts requested by a remote station are not merely disabled, they are not even reported. PRI-Q commands are replaced by non-PRI-Q equivalents, so if a PRI-EOM is received with `AT+FIE=0`, it is reported as a normal EOM by `FET:1`. Though the standard does not say so, PIN and PIP are presumably treated as RTN and RTP under these conditions.

When `AT+FIE=1`, procedure interrupts are negotiated according to the T.30 protocol, with a successful interrupt reported by the `+FVO` response. The modem remains off-hook, and the session can be resumed manually at the start of phase B with `ATD` and `ATA` commands, respectively. Received PRI-Q commands are included in the FET: report. PRI-EOM is reported as `FET:4`. PRI-Q responses are not reported, but the requests can be inferred from the fact that `+FVO` was negotiated.

The Class 2.0 standard does not specify how a fax modem should behave if the commands `AT+FPS=4` to send a PIN or `AT+FPS=5` to send a PIP are used when receiving a fax if `AT+FIE=0`. Neither does it state what happens if the embedded command `<dle><21h>` to request a procedure interrupt is used under the same circumstances. My guess is that the session would resume at the start of phase B, but I wouldn't be surprised if the modem disconnected.

Negotiation reporting: AT+FNR

This command is new to Class 2.0. It takes four parameters `<rpr>`, `<tpr>`, `<idr>`, `<nsr>` which control the reporting of receive parameters, transmit parameters, ID frames and non-standard negotiations. The default is for all reports to be off. The full meanings of the four parameters are given in Table 11.15.

Table 11.15: Class 2.0 negotiation reporting parameters

`<rpr>=0`	Parameter reports from DIS/DTC frames are suppressed
`<rpr>=1`	Received DIS/DTC frames reported via +FIS: and +FTC:
`<tpr>=0`	Parameter reports of the DCS frame are suppressed
`<tpr>=1`	Transmitted DCS frame reported via +FCS:
`<idr>=0`	ID strings are not reported.
`<idr>=1`	ID strings reported via +FTI: +FCI: and +FPI:
`<nsr>=0`	Non-standard frames are not reported.
`<nsr>=1`	Non-standard frames reported via +FNF: +FNS: and +FNC

I find the default of all reports off to be quite surprising, and anyone converting old Class 2 software for the newer standard will certainly need to switch some of them on again. The Class 2.0 specification does points out that the default of `<tpr>=0` means there is no way of detecting a mismatch between transmitted data and receiver's capabilities. There are two ways of avoiding any problems that arise as a consequence: the first is to ensure that all transmissions are normal resolution A4 length pages, coded in 1-D with 1728 dots on a 215 mm line; the second requires that format conversion must be present and enabled, and the required embedded format identifiers be placed at the start of the data stream. See the **AT+FFC** command for more on format conversion.

On a more general note, we can distinguish three types of optional Class 2.0 reports:

- The `<rpr>` and `<tpr>` **+FIS: +FTC:** and **+FCS:** reports are given in the form of standard Class 2.0 parameter strings. They are governed by the **+FNR** parameter.

- Reports of the ID strings **+FTI: +FCI:** and **+FPI** are given in standard ASCII, and are also governed by the **+FNR** parameter. The same ASCII format is also used for the **+FSA:** subaddressing frame, the **+FPA:** selective polling frame and the **+FPW:** password frame. Reporting of these is governed by the **+FAP** parameter.

- A third type of report is given as a string of hexadecimal digits. The reports of the non-standard frames **+FNF:**, **+FNS:** and **+FNC:** are governed by the **+FNR** parameter. Other reports given as hexadecimal strings include the **+FHT:** and **+FHR:** HDLC debugging reports which are governed by the **+FBU,** but these are unlike all others in that their content is affected by the **+FBO:** bit ordering parameter. Other hexadecimal reports include the almost non-standard **+FFD:** optionally used in binary file transfer, which is identical in form to the non-standard frames governed by **+FNR.** A related report which consists of a single hexadecimal digit is the **+FHS:** hangup status code.

Flow control: AT+FLO

Class 2.0 modems have a standardized way of specifying the flow control method to be used between the fax modem and computer. The command **AT+FLO** takes one parameter only:

- **AT+FLO=0** turns off all flow control. Buffer overruns must be prevented using a different method; see the **+FBS** and **+FPP** commands.

- **AT+FLO=1** turns on XON/XOFF flow control. This is the default. The Class 2.0 specification doesn't make it clear that this should be active in only one direction at a time; flow control is needed to control data output to the modem when transmitting faxes, but is used to control data input from the modem when receiving faxes. In both cases, the bit stream may contain apparent XON and XOFF characters which in reality are part of an image and need to be left alone.

- **AT+FLO=2** sets CTS/RTS flow control. The standard actually states that this means we 'use circuit 133 for flow control of the DCE by the DTE' and that we 'use circuit 106 for flow control of the DTE by the DCE'. Hardware flow control is optional and does not have to be implemented.

There is no **AT+FLO=3** command to turn on both XON/XOFF and CTS/RTS flow control as the parameter is not a bitmapped one. Unless **+FLO=2**, CTS from the modem is left on and the state of RTS is ignored.

The existence of the **AT+FLO** command does relieve a fax application programmer of the need to find out what modems use which commands for setting flow control, or of matching the selected type of flow control with that in the controlling computer.

Serial port bit rate: AT+FPR

This command controls the speed at which the computer and fax modem communicate. It takes a single parameter, which is the desired bit rate divided by 2400. There are two possible defaults: the first is the special case **AT+FPR=0**, which causes the modem to autobaud. It detects the speed from the AT prefix of each command it is sent, and uses the last speed it received as a command at for all its reports; the second default is **AT+FPR=8**. Since $2400 \times 8 = 19200$, this causes the modem to communicate with the modem at a fixed rate of 19200 bps while **AT+FCLASS=0**. Modems running at a fixed speed of 19200 when in fax mode pre-date the Class 2.0 specification by some years.

While autobauding is optional, values of 4 and 8 (9600 and 19200 bps) are mandatory. As usual, the **AT+FPR=?** form of the command will enable an application to determine the possible values. This is quite important in view of the fact that the proposed Class 1 equivalent of **+FPR** seems to take a hexadecimal parameter rather than a decimal one. The speed specified by **+FPR** takes effect immediately after the modem has finished issuing the **OK** result code.

Packet protocol: AT+FPP

The following quotation from the 2.0 specification describes what the packet protocol is designed to accomplish:

> 'For 19200 bit/s, the time between asynchronous characters is $1/1920 = 521\mu s$ (μs = microseconds). There are many processes in common DTE (e.g. personal computers) that cause serial input channels to be neglected for longer than $500\mu s$; data loss is a constant hazard. If a character is lost in the received data, the image will be impaired; if a character is lost in a DCE final result code, the connection may fail. This protocol permits recovery from such data loss so that images remain intact and the facsimile transfer will succeed.'

The problem being described is basically that of data overrun, whereby a character arrives at a computer serial port before the previous character has been removed. Since a decently written interrupt handler will easily be able to cope with data speeds well in excess of 19200 bps, data overrun (like the data underrun problem affecting transmission discussed earlier) is generally associated with software that doesn't merely neglect the serial input channel, but actually sabotages it by disabling the interrupt system. Processes that disable interrupts for longer that 500 ms are actually quite few. This is a very long time in computing terms. Modern CPUs can execute millions of instructions per second, and there should be no need for anyone to disable interrupts for a long time. However, there are programs that do this. The main culprits in common use run on IBM compatible PCs, and the one most often seen is Microsoft Windows, with disk caching programs coming a close second. Given that both these are widely used, it is hardly surprising that the problem needs addressing.

However, the need for a packet protocol on PCs was probably greater a few years ago. There are a number of reasons for this. Better hardware has played its part. The performance of entry-level PCs has risen dramatically over recent years, and continues to rise. It is hardly surprising that using a 16 Mhz 386 processor to handle tasks that were previously the province of 8 Mhz 286 systems has resulted in better real-time performance. Additionally, the increasing realization on the part of both users and software houses of the benefits of installing and supporting buffered UARTs like the National Semiconductor 16550AFN has enabled an instant hardware fix for lost data.

Better communications handlers in the operating system is another factor; early versions of Windows could not handle speeds of 9600 bps reliably, and didn't want to know about 16550 UARTs. Later versions offered some limited support for the received FIFO buffers on the 16550, while speeds of 19200 bps are officially supported by Microsoft in versions of Windows which include their own Microsoft at Work fax package. Windows 95 is said to include full 16550 configurability and should also be reliable at speeds of 38400 bps with even the most basic UART.

Perhaps the most important factor has been the fact the system software has been making better use of PC hardware. Instead of 386 PCs being used simply for running 8086 based software as fast as possible, programs now increasingly rely on pure 32-bit operation using more efficient memory architectures.

Nevertheless, the link between the computer and fax modem remains one of the weakest points in computer faxing. It is possible that we will see faster fax transmission speeds, either using the proposed standard for 28800 bps V.34 half duplex or using inexpensive ISDN adapters delivering 64K bps. This may make developers look again at methods of providing fax communications with some much-needed extra security against data being lost at the point of delivery.

The packet protocol is enabled by setting **AT+FPP=1**, and can be disabled with **AT+FPP=0**. While the default is 0, implementation of the protocol is not optional. Despite this, it would be wise to check if the modem supports it with the **AT+FPP=?** command.

Packet protocol affects the data flow from the fax modem to computer, and applies to everything transmitted by the modem, including responses and phase C data. The only exceptions are the single control characters XON and XOFF used for flow control, and the **<can>** used to abort a transmission. Like most blocked protocols, the Class 2.0 packet protocol is quite simple:

1. The maximum length of a packet is 254 data bytes plus one trailing control character.

2. A packet transfer begins with a single **<len>** byte containing the packet length.

3. The receiver acknowledges this with an **<ack>** character, which has a decimal and hexadecimal value of 6.

4. The transmitter then sends the number of bytes stated in the **<len>** followed by an **<etb>** (decimal 23 or hexadecimal 17) marking the end of the packet. This is not included in the count of characters for **<len>**.

5. When the receiver detects the **<etb>** it checks to see if the number of characters it received was the same as **<len>**. If it was, it responds with **<ack>** and

awaits the next packet; if the count of characters does not match `<len>`, it sends a `<nak>` (decimal 21 or hexadecimal 15) and waits for the packet to be re-sent.

6. If a continuous stream of data is to be sent, a packet can optionally end with the `<soh>` character (decimal and hexadecimal 1). This tells the receiver that the next packet has the same length as the last packet, and the exchange of `<len>` and `<ack>` bytes is then missed out.

7. The `<etb>` and `<soh>` characters are not allowed to occur inside packets. They are ORed with hexadecimal 40 and shielded with the usual `<dle>` character, so `<dle><41h>` stands for `<soh>` in the packet data and `<dle><57h>` stands for `<etb>`. As usual, `<dle>` characters are sent as `<dle><dle>` and two successive `<dle>` characters as `<dle><sub>`, where `<sub>` is decimal 26 or hexadecimal 1a.

8. Shielding is done before packetization; this means `<dle>` shielding characters are included in the count for the `<len>` byte. They can also straddle packet boundaries. If the last character in a packet was a `<dle>`, the first character of the next packet is the second half of a shielded pair.

9. If the receiver detects an error during the packet (for instance, the UART status register might indicate an overrun error), the packet can be aborted by sending an `<enq>` (decimal and hexadecimal 5) to the sender, who should terminate the packet with `<soh>` or `<etb>`, wait for the `<nak>` and then re-send. A timer should be implemented in case of a catastrophic error (like someone unplugging the modem in mid-session, or a network crashing when using a shared modem).

10. Ideally, the modem should buffer two packets worth of data and only send a packets when the length of the next one is known, so that the correct `<soh>` or `<etb>` can be used; but in the case of the last packet of phase C data, it is permitted to pad the packet out with 0 bytes after the `<dle><etx>` marker.

The standard states that the only problem the packet protocol is designed to solve is that of data loss. The computer to modem link is assumed error-free, as are all communications. It is not an error correcting protocol. The hardware assumption it depends upon is that where a data overrun occurs, the new character always overwrites older ones, i.e. the last character of a packet is guaranteed never to be lost.

As an example of how the packet protocol works for both responses and for fax data, a phase C data reception of a small fax containing 234 lines looks like as shown in Table 11.16; length bytes are shown as hexadecimal numbers and the arrows indicate direction of the data flow. We pick up the session immediately after any final FCS: report, and just before the **CONNECT** prompt, and continue until the post-message reports have been delivered in phase D.

Table 11.16: Class 2.0 packet protocol

computer	modem
Phase C	
	←<0bh>
<ack>→	
	←<cr><lf>CONNECT<cr><lf><etb>
<ack>→	
<dc2>→	
	←<feh>
<ack>→	
	←254 octets of data<soh>
<ack>→	
	←254 octets of data<soh>
<ack>→	
	←254 octets of data<soh>
<ack>→	
	←254 octets of data<soh>
<ack>→	
	←254 octets of data<soh>
<ack>→	
	←254 octets of data<soh>
<ack>→	
	←254 octets of data<etb>
<ack>→	
	←<66h>
<ack>→	
	←100 octets of data<dle><etx><etb>
<ack>→	
Phase D	
	←<14h>
<ack>→	
	←<cr><lf>+FPS:1,234,0,0,0<cr><lf><etb>
<ack>→	
	←<0ah>
<ack>→	
	←<cr><lf>+FET:0<cr><lf><etb>
<ack>→	
	←<06h>
<ack>→	
	←<cr><lf>OK<cr><lf><etb>
<ack>	

Both EIA 592 and EIA 605 include comprehensive state tables and example sessions for what is a very straightforward protocol.

The packet protocol is more than just a quick fix needed for when real-time communications applications (such as fax reception) become impossible on systems that don't have the capability to service input ports at the necessary speed. As we've explained, a better way of solving this problem may be to make the operating system itself more reliable; Microsoft have been doing this with Windows, and the latest versions can service serial ports every 250 ms if necessary. But no matter how good an operating system might be, a PC by definition has sole control over its resources, and on a multi-tasking system it is difficult to guard against applications that are hostile to real-time needs. Consequently, the packet protocol is something that multi-tasking systems could find useful.

Non-standard negotiations: AT+FNS

The command **AT+FNS=<id>** sets the content of a non-standard frame. The type of frame to be sent depends upon the context. The content of **+FNS** is sent either as an NSF frame on answering, as an NSC frame on polling or as an NSS frame on transmitting (if the answerer sent an NSF of its own). The **+FNS** is entered as a series of up to 90 hexadecimal characters inside quotes. Multiple uses of **+FNS** append successive strings to the pre-existing contents of +FNS. The command

AT+FNS=""

initializes this parameter to a null string. Once this is done, a sequence such as

AT+FNS="48 45 4C 4C 4F"

sends the string **"HELLO"** as a non-standard frame.

A non-standard negotiating frame, like other frames, can take a maximum of 3 sec ±15% to transmit. Since negotiations take place at the fixed speed of 300 bps, the maximum length of a frame must be regarded as 765 bits, equivalent to 95 octets and 5 bits. Since address, control and FCF facsimile control field octets take up three octets, with the FCS frame check sequence and the closing flag taking another three octets, the room for the FIF facsimile information is reduced to 89 octets and 5 bits. The maximum of 90 octets allowed by the **+FNS** command may be a little optimistic, especially when one considers that bit stuffing requirements mean data is not always sent transparently.

The exact number of bytes a fax modem can accept as part of the **+FNS** parameter is variable. Unusually, the command **AT+FNS?** returns the number of bytes the modem will accept, rather than the range. It is understood that the range can be any hexadecimal value from 00h to ffh.

Implementation of **+FNS** is optional. Programmers should also note that use of **+FNS** to set up non-standard facilities can contravene the provisions of the ITU-T series recommendations.

Embedded buffer handling and buffer status: AT+FBS?

The command **AT+FBS?** returns the transmit followed by the receive buffer sizes. No other information is given, and the figures returned by **AT+FBS?** remain constant.

The procedures for handling the modem's buffers in Class 2.0 are not the same as those in Class 2, and in some respects, they are inferior. There is no way of finding out what the flow control thresholds might be set to on any particular implementation of Class 2.0. As already mentioned, flow control thresholds can be just as important to the successful functioning of a fax modem as the buffer size. The benefits of a generous buffer can be nullified if the thresholds are set to stupid levels. While information about threshold levels is not always required by an application program, the levels set are valuable items of information for both purchasing decisions and systems design.

Receive buffer

The Class 2.0 recommendation states that a modem must have a receive buffer size sufficient to hold 3 s worth of data at maximum reception speed. In other words, if XON/XOFF flow control between the computer and modem is used to control the flow of data when a fax is received, the maximum time that can elapse between an XOFF and a subsequent XON is 3 s. The Class 2.0 recommendation states that at 9600 bps this is 3600 bytes.

As the modem receive buffer will not be empty after an XON/XOFF sequence, it is clear that the computer cannot rely on having the full 3 s latitude. The modem reports when its buffers are empty again by embedding a **<dc2>** code in the received data using the sequence **<dle><12h>**. Until this code arrives it may be considered to be unsafe to send another XOFF.

While the buffer may have filled up in 3 s, it takes longer to empty, as more data is coming in all the time. Assuming a fax is being received at 9600 bps and that data is transferred to the computer (with starts and stop bits added) at 19200 bps, the buffer will take almost exactly 8 s to empty. An application might need to issue another XON during this time. If this happens, the ? character (decimal 63, hexadecimal 3f) issued during phase C reception will enable a program to find out how full the receive buffers are, and discover how much latitude it has. The response from the modem once again comes embedded in the received data using **<dle>** shielding as shown in Table 11.17.

Whether it is safe to send an XOFF depends upon why it needs to be sent and how full the receive buffer is. The first XOFF in each session, and all XOFFs after the first where a **<dle><dc2>** arrived in the meantime, are guaranteed to have a full 3 sec worth of buffer space available.

After any flow control event, the letter O embedded in the data with **<dle><4fh>** can appear spontaneously in the data stream to mark a receive

Table 11.17: Class 2.0 embedded buffer status reports

`<dle><30h>`	0	`= buffer less than 10% full`
`<dle><31h>`	1	`= buffer 10-20% full`
`<dle><32h>`	2	`= buffer 20-30% full`
`<dle><33h>`	3	`= buffer 30-40% full`
`<dle><34h>`	4	`= buffer 40-50% full`
`<dle><35h>`	5	`= buffer 50-60% full`
`<dle><36h>`	6	`= buffer 60-70% full`
`<dle><37h>`	7	`= buffer 70-80% full`
`<dle><38h>`	8	`= buffer 80-90% full`
`<dle><39h>`	9	`= buffer 90%-100% full`

buffer overrun. The most common cause of this would be if an XOFF was issued and an XON didn't arrive before the buffer filled up. Alternatively, if hardware handshaking is used, RTS would have gone low for too long. In either case, the most recent data will have been discarded. The **`<dle><4fh>`** marks each spot where data is known to have been lost.

Transmit buffer

There is no recommended size for the transmit buffer. More seriously, there is no direct method of finding out the restart threshold used for the transmit buffer. This means there is no way of calculating whether any particular modem might cause problems given the latency of the response to an XON on any particular system. This must be considered a serious omission in the Class 2.0 specification.

Finding out the XOFF threshold is easier. When an application is sending fax data, it can embed the characters **`<dle><3fh>`** in the data stream to request the amount of space in the transmit buffer before a flow control event is triggered. Note that the character <3fh> is an ASCII **?** , and is referred to in the Class 2.0 specification as a **`<bc?>`** code. The application should then wait for the modem to respond. The response from the modem is a text string terminated by a **`<cr>`** showing the amount of free space in the buffer as a decimal number. One would hope that if flow control is turned off with **`AT+FLO=0`**, the figure returned is the total amount of free space left in the buffer, but it is not entirely clear from the documentation that this is so.

If the packet protocol is enabled, the text string may be padded out with spaces to avoid it having a length of 17 or 19 characters, as these are the decimal values for XON and XOFF, respectively.

To send data without provoking any flow control events, the technique we described on page 192 using the Class 2 **`+FBUF`** command, needs to be slightly adjusted for a Class 2.0 modem:

1. Issue a **AT+FBS?** command as part of the modem setup phase and store the resulting figure for transmit buffer size **<tbs>**. Issue the **AT+FLO=0** command to turn off flow control as it will not be used.

2. When the first phase C **CONNECT** message arrives, send **<tbs>** bytes of data followed by **<dle><3fh>**.

3. Wait for the response and convert the decimal string returned into a number.

4. If there is enough space available to hold all the remaining data, send it, ending with a **<dle><etx>**.

5. Else send that amount of data if it is available, ending with another **<dle> <3f4h>**, and then loop back to step 3.

Number of CTCs allowed while in ECM: AT+FRY

The Class 2.0 **+FRY** parameter has exactly the same function as the Class 2 **+FCTCRTY** parameter described in the disk notes, and determines the number of continue to correct (CTC) frames a transmitter will send after four consecutive partial page errors when in error correcting mode. However, the implementation is slightly different, and is certainly improved over the old version. The difference is that **AT+FRY** sets the total number of CTC frames that can be sent using a given speed. If it possible, a Class 2.0 fax modem automatically steps down in speed when the CTC count reaches zero.

As described in Chapter 7, all CTCs are accompanied by the same set of bits the negotiating frames use to determine the speed and modulation, and a transmitting fax has three options if four partial page errors occur: the first is to signal the end of retransmission (using the EOR frame); the second is to continue to correct at the same speed (using a CTC frame accompanied by the same speed as was set by the DCS frame); and the third option is to continue to correct, but to drop down in speed. The Class 2 **+FCTCRTY** parameter did not allow this last option. The Class 2.0 **+FRY** parameter makes it the automatic default.

As a consequence, an EOR frame is only sent when the transmitter can no longer drop down in speed This can be because the lowest speed of 2400 bps has been reached, or because the modem has been told to use a higher minimum speed. If the minimum speed is set to the same value as the connection, the ability to drop speed is lost and **+FRY** parameter behaves just like the old **+FCTCRTY** parameter. See the **AT+FMS** command for more on setting minimum speeds.

With the minimum speed at the default of 2400 bps and **+FRY** at the default of 0, four partial page errors at 9600 bps would result in another four attempts being made at 7200 bps. If all these were unsuccessful, the modem would try again at 4800 bps and finally at 2400 bps. The EOR frame would only be sent after a failure at 2400 bps. This speed would be kept for the remainder of the session.

The Class 2.0 recommendation points out that resending 16K once at 9600 bps would take about 20 s, and with **+FRY** set to 255, over an hour could be spent attempting a partial page with a bad line 'if the remote station was patient enough'. It's a matter of simple arithmetic that if every frame of a 64K block was always bad, if **+FRY=255**, and if the initial connection speed was 14400, the retransmission times alone would mean that well over 36 hours could elapse between the initial attempt and an EOR.

The conclusion I would draw is that if you can't send a partial page correctly after four attempts at all speeds down to 2400 bps, you are probably better off disconnecting and redialling in the hope of getting a better line.

Phase C timeout: AT+FCT

This command sets a timeout for the maximum permissible delay during the sending of phase C data from the computer to the fax modem. It is an enhancement of the old Class 2 **+FPHCTO** parameter. Some of the criticisms levelled against this in the disk notes have been addressed, but the command remains very unsatisfactory.

AT+FCT takes a hexadecimal parameter from 0 to ffh, which is interpreted as a timeout value in seconds (not in 100 ms units as with **+FPHCTO**). The default is 1eh, or 30 s. Whether a fax modem is transmitting or receiving phase C data, it will abort a transfer and send a DCN frame if this timeout expires after all the available data has been processed and no command has been received from an application.

The purpose of this command is unclear. The only use I can think of is that it enables a modem to disconnect the line if the computer or application controlling it crashes, hangs, has its plug pulled out or otherwise becomes unavailable. Even in this context, it is quite unnecessary. It is likely that any remote fax will have disconnected long before the 30 s is up. If it had been receiving, either the 5 s limit on line length or the 6 s timeout on a phase D command would cause a remote fax to disconnect; if the remote has finished sending, the three attempts to get a response from a post-message command within 3 s would have resulted in a disconnection within 10 s.

Hangup status code: AT+FHS:

The hangup status code in the Class 2.0 **+FHS:** report is almost identical to the Class 2 **+FHNG:** code. The difference is quite subtle: the codes given in the 1990 version of SP-2388, on which most Class 2 modems are based, were returned as decimal numbers, while codes in the approved PN-2388-B specification are returned as two-digit hexadecimal numbers.

While a Class 2.0 report of **FHS:50** looks the same as a Class 2 report of **FHNG:50**, the codes are actually different. The difference only really concerns the Receive phase D hangup codes, which are now numbered from A0 to BF instead of 100 to 119. The complete table looks much the same as the Class 2 version, but presumably the change to hexadecimal was made to accommodate more possible error codes in the future.

Codes C0–DF are reserved for future versions of the standards, while codes E0 to FF are reserved for manufacturers to implement as they wish. All unused codes are also reserved. Codes 0, 10, 20, 40, 50, 70, 90 and A0 are mandatory, and should be supported by all modems, while all the other codes are optional.

The class 2 **AT+FAXERR?** command which returned the last hangup status code has (quite sensibly) been replaced with the new **AT+FHS?** command, which otherwise behaves in the same way.

Minimum phase C data transfer speed: AT+FMS

This parameter is almost identical to the Class 2 **+FMINSP**. It sets the minimum speed the modem will negotiate for a session. However, unlike **+FMINSP**, the Class 2.0 version can explicitly take any valid value for BR. **AT+FMS=5** will thus force a fax modem to disconnect if it can't establish a 14400 bps connection.

The **+FMS** parameter assumes a greater role in error correction mode; it constrains the manner in which the modem will continue to correct partial pages.

Table 11.18: Class 2.0 +FHS status codes

Codes 00 - 0F are reserved for call placement and termination errors. Six codes are defined. 04 and 05 are new in Class 2.0.

00	Normal end of session
01	Call answered without successful fax handshake
02	Call aborted by user (via either +FK or CAN)
03	No Loop Current
04	Ringback detected, dial attempt timed out with no answer
05	Ringback detected, call answered but without a CED tone

Codes 10 - 1F are reserved for Transmit Phase A errors. Only two of these are defined.

10	Unspecified Phase A error
11	No answer or no DIS detected (T.30 T1 timeout)

Codes 20 - 3F are reserved for Transmit Phase B errors. Nine of these are defined.

20	Unspecified error
21	Remote cannot receive or send (bits 9 and 10 in DIS both 0)
22	Command frame error
23	Invalid command received

Table 11.18: (continued)

24	Response frame error
25	DCS sent three times without response
26	DIS/DTC received 3 times; DCS not recognised
27	Failure to train at either 2400 bps or +FMS value
28	Invalid response received

Codes 40 - 4F are reserved for Transmit Phase C errors. Eight codes are defined.

40	Unspecified error
41	Unspecified image format
42	Image conversion error
43	DTE to DCE data underflow (data not sent fast enough)
44	Unrecognised <dle> command
45	Wrong line length
46	Wrong page length
47	Wrong compression code

Codes 50 - 6F are reserved for Transmit Phase D errors. Nine of these are defined.

50	Unspecified error
51	Response frame error
52	No response to MPS repeated 3 times
53	Invalid response to MPS
54	No response to EOP repeated 3 times
55	Invalid response to EOP
56	No response to EOM repeated 3 times
57	Invalid response to EOM
58	Unable to continue after PIN or PIP

Codes 70 - 8F are reserved for Receive Phase B errors. Only five errors are defined.

70	Unspecified error
71	Response frame error
72	Command frame error
73	Expected frame not received (T.30 T2 timeout)
74	Failure to resume phase B after EOM (T.30 T1 timeout)

Codes 90 - 9F are reserved for Receive Phase C errors. Four of these are defined.

90	Unspecified error
91	Missing EOL after 5 seconds (line too long)
92	CRC or frame error in error corrected mode
93	Modem receive buffer overflowed

Codes A0 - BF are reserved for Receive Phase D errors. Four codes are defined.

A0	Unspecified error
A1	Response frame error
A2	Command frame error
A3	Unable to continue after PIN or PIP

Bit ordering: AT+FBO

The **+FBO** parameter takes exactly the same parameters and has exactly the same effect as the Class 2 **+FBOR**.

Note that the **+FBO** parameter effectively bit-maps reversed bit ordering, with bit 0 affecting phase C data and bit 1 affecting the **+FBU** reports of phase B and D frames.

There are five points to note in relation to bit ordering with Class 2.0 fax modems:

1. The bit ordering for hexadecimal reports in the non-standard frame reports is always reversed; the LSB of all octets is the first bit sent or received. This also applies to setting up non-standard frames with AT+FNS command.

2. The **+FBO** parameter does not affect control characters sent by the modem. Flow control and embedded **<dle>** sequences are always sent in the usual order.

3. The **+FBO** parameter does not apply to shielding or control characters sent to the modem. In other words, if an application is about to send a **<dle>** character occurring in a fax image to the modem, then that character is always to be shielded. An example of follows shortly, when we look at how RTC codes are sent.

4. The Class 2.0 command **AT+FEA=1** is used to instruct the modem to byte-align received EOL characters. The default of **+FEA=0** leaves EOL codes untouched. The **+FEA** parameter is identical to the Class 2 **+FREL** parameter.

5. The **+FBO** parameter must apply all to outgoing RTC sequences.

Note that the Class 2.0 recommendation is unclear whether the **+FEA** parameter is supposed to apply to incoming RTC sequences. This is only truly important if an incoming fax is going to be sent out again exactly as received. As 4.1.3/T.4 explicitly states that fill is permitted only between a line of data and an EOL, I think the safest course would be either to check that the final RTC sequence has no fill bits included or to ensure that **+FEA=0** when the data is received.

Table 11.19: Class 2.0 +FBO parameters

+FBO=0	selects direct bit order for all data (this is the default)
+FBO=1	selects reversed bit order only for phase C data
+FBO=1	selects reversed bit order only for phase B and phase D data
+FBO=3	selects reversed bit order for all data

Mismatch checking and data conversion: AT+FFC

This single command controls the capabilities of the modem for handling both mismatch checking and data conversion. It take four parameters,

```
AT+FFC=<vrc>,<dfc>,<lnc>,wdc>
```

where the `<vrc>` parameter controls mismatch checking and data conversion for vertical resolution, `<dfc>` handles data format, `<lnc>` handles page length and `<wdc>` handles page width. These parameters correspond almost exactly to Class 2 parameters `+FVRFC`, `+FDFFC`, `+FLNFC` and `+FWDFC`, which control the same mismatch and conversion functions:

- The default for all of them is `0`, indicating that no mismatch checking or format conversion is possible and that the application software must check the appropriate `+FCC` parameter and decide if there is a mismatch, and what to do if one occurs. Implementation of any `+FFC` functions is optional, and you should check with the `AT+FFC?` form of the command before using any data conversion functions on an unknown Class 2.0 modem.

- If any of the four `+FFC` parameters is set to `1`, any corresponding data mismatch will be automatically detected, and the session will be terminated.

- A value of `2` set for `<dfc>` or `<wdc>` and a value of `3` for `<vcr>` and `<lnc>` enables both mismatch checking and fully automatic data conversion if a mismatch is detected. As with Class 2 conversions, this is handled transparently, and intermediate phase D/C transitions at the end of any page breaks inserted by the fax modem are not reported.

- A value of `2` set for `<vcr>`, `<lnc>` enable mismatch checking with data conversion for 1-D data, but with session termination for a mismatch using 2-D data.

Table 11.20 summarizes the available codes and actions.

Whereas Class 2 modems were notified of the format they were being fed with by parameters included with a modified **AT+FDT** command, Class 2.0 use the more flexible approach of embedding commands in the transmitted phase C data stream, as shown in Table 11.21.

Note that although codes for both 2-D uncompressed mode and T.6 are provided, the specification is silent as to what the modem should do with this type of data. The specification has also not brought the vertical resolution conversion options into line with the latest T.30 enhancements included in the VR field for the Class 2.0 fax parameters. Notably, neither of the options for 15.4 l/mm superfine vertical resolution have embedded codes assigned to them, nor are the inch-based resolutions supported.

Table 11.20: Possible values for the Class 2.0 **AT+FFC** parameters

	0	1	2	3
vrc	no checking conversion ignored	checking enabled	1-D conversion enabled	2-D conversion enabled
dfc	no checking conversion ignored	checking enabled	conversion enabled	
lnc	no checking conversion ignored	checking enabled	1-D conversion enabled	2-D conversion nabled
wdc	no checking conversion ignored	checking enabled	conversion enabled	

Set file transfer diagnostic message: AT+FFD

File transfer diagnostic messages are an optional feature of the T.434 BFT protocol. This was discussed in Chapter 7, with more detail provided in the NOTES subdirectory on the accompanying disk. It aims to provide a standard method of transferring computer files between systems equipped with fax modems.

Table 11.21: Embedded command for including format information in Class 2.0 phase C data

`<dle><61h>`	< a > vertical resolution normal
`<dle><62h>`	< b > vertical resolution fine
`<dle><63h>`	< c > A4 length
`<dle><64h>`	< d > B4 length
`<dle><65h>`	< e > unlimited length
`<dle><66h>`	< f > 215 mm width
`<dle><67h>`	< g > 255 mm width
`<dle><68h>`	< h > 313 mm width
`<dle><69h>`	< i > 151 mm width
`<dle><6ah>`	< j > 107 mm width
`<dle><6bh>`	< k > 1-D modified Huffman
`<dle><6ch>`	< l > 2-D modified Read
`<dle><6dh>`	< m > 2-D uncompressed mode
`<dle><6eh>`	< n > 2-D MMR (T.6)

Contents of Disk Notes

Further Reading and Sources of Information

Books

The following short list of books doesn't claim to be either authoritative or exhaustive. It simply contains titles which either I've found useful or which contain the type of information which would be needed by someone wanting to develop the code on the disk.

1. *Advanced Windows Programming*, Martin Heller, Wiley 1992.
   A Windows programming textbook with a very useful graphics bias.

2. *Bitmapped Graphics Programming in C++*, Marv Luse, Addision Wesley 1993.
   Excellent detail on many different graphics formats.

3. *C Programmer's Guide to Serial Communication 2nd Edition*, Joe Campbell, Sams 1994.
   The mammoth second edition of this classic now has an extremely useful fax section. A book I wish I'd written.

4. *C++ Communications Utilities*, Michael Holmes and Bob Flanders, Ziff-Davis 1993.
   Includes code of a C++ communications program with fax capability.

5. *Complete Modem Reference*, Gilbert Held, Wiley 1991.
   A useful introduction to modem technology.

6. *Data Communications, Computer Networks and Open Systems 3rd Edition*, Fred Halsall, Addison-Wesley 1992.
   This is the best general-purpose communications textbook I've come across.

7. *Fax: Digital Facsimile Technology and Applications 2nd Edition*, Dennis Bodson, Kenneth McConnell, Richard Schaphorst, Artech House 1992.
   Excellent technical book on fax history, standards, machines and technology.

8. *Programmer's Technical Reference: Data and Fax Communications*, Robert Hummel, Ziff-Davis 1993.
   A comprehensive reference book on modem technology and command sets, including fax.

9. *Serial Communications: A C++ Developer's Guide*, Mark Nelson, M&T 1992.

10. *Windows Programmer's Guide to Serial Communications*, Timothy Monk, Sams 1992.

If anyone wants to recode the low-level communications functions in the code on disk, these last two books would be good places to start.

Standards

1. Up to date information can be obtained direct from the ITU. You can contact them at :

   International Telecommunication Union
   General Secretariat - Sales Section
   Place des Nations, CH-1211 Geneva 20
   Switzerland

   Phone: +41 22 730 5285
   Fax: +41 22 730 5194
   E-mail: itudoc@itu.ch

   The ITU either charge by the page for hard-copy versions of their standards, or alternatively provide access to them on-line by subscription only. The list of T-Series Recommendations is available free of charge from the ITU gopher server located at info.itu.ch, and includes the latest specifications not covered in this book, such as T.42, which covers the new Group 3 colour and greyscale fax standard.

2. The EIA/TIA can be contacted at :

   Telecommunications Industry Association
   1722 Eye St. NW
   Suite 440
   Washington DC 20006
   U.S.A

   Phone: +1 202 457 4942

Callers are usually referred to the main distributor for the EIA/TIA standards, who are :

Global Engineering Documents
2805 McGaw Ave
Irvin, CA 92713
Phone: +1 800 854 7179
Fax: +1 202 331 0960

The EIA/TIA standards are also chargable.

3. Proprietary and unofficial standards are usually available free of charge. The CAS and TIFF specifications are widely available electronically, and are included on the disk. Specifications and programming manuals for the Rockwell implementations of the EIA fax standards and command sets for their modems are often available at no cost through Rockwell distributors, or on various WWW sites such as http://www.tokyo.rockwell.com/. Many fax modem manufacturers, including Rockwell, Cirrus and Zyxel, make information available on-line either through direct dial BBS systems or else through the Internet.

Cyberspace

Much of the information we could give here becomes dated very quickly. There are a number of well-established Usenet groups, notably

comp.dcom.fax
comp.dcom.modems
alt.fax

which are reasonably reliable source of informed news, advice, information and gossip. More and more fax information is becoming available via the World Wide Web. Any Web navigator will be able to get you to a suitable starting point where you can follow links, but Web sites worth visiting (as of September 1995) include:

http://www.faximum.com/	Faximum Software (for an fax FAQ)
http://hydra.phy.cam.ac.uk/alan/faxfaq.htm	(for a fax programming FAQ)
http://www.gammalink.com/	Gammafax
http://www.gentech.com	Genoa Technolgy, Inc.
http://www.grayfax.com/	Gray Associates
http://www.itu.ch/	ITU
http://www.zyxel.com/	Zyxel
http://www.Delrina.com/	Delrina
http://www.vix.com/hylafax/	Sam Leffler's well-regarded Hylafax software (also via ftp://sgi.com/sgi/fax/)

Index

Index compiled by Sheila Shephard